INDUSTRIAL WATER RESOURCE MANAGEMENT

Challenges in Water Management Series

Editor:

Justin Taberham
Independent Consultant and Environmental Advisor, London, UK

Other titles in the series:

Water Resources: A New Management Architecture
Michael Norton, Sandra Ryan and Alexander Lane
2017
ISBN: 978-1-118-79390-9

Urban Water security
Robert C. Brears
2017
ISBN: 978-1-119-13172-4

INDUSTRIAL WATER RESOURCE MANAGEMENT: CHALLENGES AND OPPORTUNITIES FOR CORPORATE WATER STEWARDSHIP

PRADIP K. SENGUPTA

WILEY

Registered Offices
John Wiley & Sons Ltd, The Atrium, Southern Gate, Chichester, West Sussex, PO19 8SQ, UK

Editorial Office
111 River Street, Hoboken, NJ 07030, USA
9600 Garsington Road, Oxford, OX4 2DQ, UK
The Atrium, Southern Gate, Chichester, West Sussex, PO19 8SQ, UK

For details of our global editorial offices, customer services, and more information about Wiley products visit us at www.wiley.com.

Wiley also publishes its books in a variety of electronic formats and by print-on-demand. Some content that appears in standard print versions of this book may not be available in other formats.

Library of Congress Cataloging-in-Publication Data

Names: Sengupta, Pradip K., 1946– author.
Title: Industrial water resource management : challenges and opportunities for corporate
    water stewardship / Pradip K Sengupta.
Description: Hoboken, NJ : John Wiley & Sons, 2017. | Series: Challenges in water
    management series | Includes bibliographical references and index.
Identifiers: LCCN 2017012527 (print) | LCCN 2017027061 (ebook) |
    ISBN 9781119272472 (pdf) | ISBN 9781119272465 (epub) |
    ISBN 9781119272502 (cloth)
Subjects: LCSH: Industrial water supply–Management.
Classification: LCC TD353 (ebook) | LCC TD353 .S375 2017 (print) |
    DDC 363.6/1068–dc23
LC record available at https://lccn.loc.gov/2017012527

Cover Design: Wiley
Cover Image: West Point Water Treatment Plant in Discovery Park, Seattle, USA,
© 2016 King County.

Set in 10/12pt Melior by SPi Global, Pondicherry, India
Printed in Singapore by C.O.S. Printers Pte Ltd

10 9 8 7 6 5 4 3 2 1

To my children

Contents

Series Editor Foreword – Challenges in Water Management

The World Bank in 2014 noted:

> Water is one of the most basic human needs. With impacts on agriculture, education, energy, health, gender equity, and livelihood, water management underlies the most basic development challenges. Water is under unprecedented pressures as growing populations and economies demand more of it. Practically every development challenge of the 21st century – food security, managing rapid urbanization, energy security, environmental protection, adapting to climate change – requires urgent attention to water resources management.
>
> Yet already, groundwater is being depleted faster than it is being replenished, and worsening water quality degrades the environment and adds to costs. The pressures on water resources are expected to worsen because of climate change. There is ample evidence that climate change will increase hydrologic variability, resulting in extreme weather events such as droughts floods, and major storms. It will continue to have a profound impact on economies, health, lives, and livelihoods. The poorest people will suffer most.

It is clear there are numerous challenges in water management in the 21st century. In the 20th century, most elements of water management had their own distinct set of organisations, skill sets, preferred approaches and professionals. The overlying issue of industrial pollution of water resources was managed from a point source perspective.

However, it has become accepted that water management has to be seen from a holistic viewpoint and managed in an integrated manner. Our current key challenges include:

- The impact of climate change on water management, its many facets and challenges – extreme weather, developing resilience, storm-water management, future development, and risks to infrastructure
- Implementing river basin/watershed/catchment management in a way that is effective and deliverable
- Water management and food and energy security
- The policy, legislation, and regulatory framework that is required to rise to these challenges

- Social aspects of water management – equitable use and allocation of water resources, the potential for water wars, stakeholder engagement, valuing water and the ecosystems that depend upon it

This series highlights cutting-edge material in the global water management sector from a practitioner as well as an academic viewpoint. The issues covered in this series are of critical interest to advanced-level undergraduates and Masters students as well as industry, investors, and the media.

Justin Taberham, CEnv
Series Editor
www.justintaberham.com

Foreword

For all the technological marvels available today, the defining element of the Third Millennium as of date is, perhaps, crisis management. Nothing can be taken for granted anymore; all raw material extractions are carried out now under careful management plans to avoid future crises. Like the growing list of endangered animals, there is a growing list of dwindling resources; the latest entry into that endangered list is fresh water.

In January 2016, the UN Secretary General, Ban Ki-moon, convened an emergency meeting of nation states to focus on water crisis. He said, "*Water is a precious resource, crucial to realising the sustainable development goals, which at their heart aim to eradicate poverty.*" Freshwater resources of the planet Earth are put to enormous pressure by the mad rush for economic growth which, as is well known, started with the Industrial Revolution and had a free run for more than two hundred years. It is only in the last couple of decades that the word 'development' has undergone a metonymic transformation into 'sustainable development'; as a consequence of which science and technology is now fully prepared to mitigate water crisis; but tragically, the preparedness is a typical instance in which the soul is willing but the flesh is weak: Science has identified parameters of water crisis and its reversal methods; technology has come up with innovative solutions; the only reason why the freshwater crisis continues to be a crisis is because it now rests upon what appears to be a lackadaisical response of industry and agriculture.

The term *corporate water stewardship* (CWS) has entered into the vocabulary of the corporate world. This is a clear demonstration that the corporate sector is aware of the water problem, but it invests only nominally in CWS initiatives. At a functional level, the corporate sector invests so much more in business expansion that the far from successful implementation of CWS it is aggravating water crisis. The lackadaisical response of the corporate world is enigmatic because CWS is specifically formulated to cater to *all* stakeholders: from a corporate bigwig contemplating water related business risks to a marginal farmer praying for rain. The obvious question, then, is to ask why despite being aware, and indeed a victim, of water crisis the corporate bigwigs do not invest meaningfully in CWS? In a world riven by money power, the obvious answer is that the corporate world is much too preoccupied with the profit curve to have any time or energy left for CWS. The obvious answer, however, may not be the correct one, no matter how fervently it is endorsed by suffering millions. A far more insightful answer would be that the corporate world is neither lackadaisical nor tight-fisted about CWS. The most likely reason why the corporate world is shooting itself in the foot is because what it knows about CWS is declarative knowledge; the corporate world has no procedural knowledge of how to go about solving the water crisis. With all its good intentions, the corporate cannot solve the water crisis until it has a firsthand understanding of the problem. And, that firsthand understanding or procedural knowledge can only begin with reading up on what CWS aims to achieve and how.

This book, by my friend Pradip, will lead the reader step by step through the entire narrative of water crisis and its solutions. It is a comprehensive treatment that pans from basic elements of hydrology to various tools available for water resource assessment and delivery of water stewardship; clearly outlining at each step the long-term benefits of CWS. The chapters are so designed that they can both be read sequentially or at random because each chapter is complete in itself. The reader is at liberty to choose which chapter to begin with, which is a degree of freedom few books will allow. To sum up the book in just one line, I would say this book makes Corporate Water Stewardship an achievable utopia.

Samaresh Sanyal
Kolkata, India

Preface

The online edition of Merriam-Webster dictionary defines the word 'stewardship' as 'the activity or job of protecting and being responsible for something'. This goes a long way towards explaining the subject matter of this book, that is, the responsibility of the corporate world to society by way of water stewardship; which, in recent times, has become an essential part of corporate management termed *corporate water stewardship* (CSW).

In 1776, Adam Smith introduced the concept of the 'invisible hand' in his book *The Wealth of Nations*, which instantly became the driving force of industrialisation and capitalism. The philosophy propounds that every individual strives to make as much money as he can; he neither intends to promote public interest, nor knows how much he is promoting it "he intends only his own gain, and he is in this, as in many other cases, led by an invisible hand to promote an end which was not a part of his intention. The fact that it was not a part of his intention does not always worsen the situation for the society. By pursuing his own interest he frequently promotes that of the society more effectually than when he really intends to promote it. I have never known much good have been done by those who affected to trade for the *public good*" (italics mine). What Adam Smith could not have foreseen two and a half centuries ago is the ambivalence of rapid industrialisation, which is plainly visible now: Aside from accelerating modernization (see Chaplin's *Modern Times*), the invisible hand is surreptitiously leading the world to unliveable conditions (which is hardly surprising, seeing that the maximum impetus to technological advancement came from the two World Wars and the Cold War that followed).

It is high time for the world to become aware of the water problem before the problem becomes irreversible. This book aims to promote such awareness, and I hope that my Muse will grant me such simplicity so that my readers can identify with the subject in hand and the book may acquire another status – not that of a polemic, nor of a mere source of topical reference but of turning it into a well-described text with an attempt to present it to, and make it understood by large numbers of the public; that, then, is my priority in which simplicity of presentation is just as important as the idea of water stewardship vis-à-vis industrialization.

When I started my career as a hydro-geologist in a state government department of India in the mid-1970s, the concept that water is a 'resource', like gold, for instance, was incomprehensible to users; in my home state, West Bengal, water is quite abundant, by virtue of an average rainfall of more than 1500 mm per year and groundwater level close enough to the surface to make dug wells and tube wells affordable even to low-income groups. The West Bengal landscape is dotted by huge waterbodies used for irrigation and fisheries. Most of the rivers are perennial and navigable in the rainy season. Availability of usable water was taken for granted, except in the relatively thinly populated western districts, which are chronically draught-prone. The Gangetic alluvial plain is one of the richest aquifers in the world. During the 1970s, the government of West Bengal was implementing the 'Minor Irrigation Policy' by installing thousands of deep and shallow tube wells across the state. Groundwater was exploited as an apparently inexhaustible resource and was made available to small and marginal farmers at subsidised rates. It was a boon to the farmers. Needless to say, however, such good times do not last forever. By the end of the 1980s, the water problem was first noticed when arsenic in toxic concentrations was detected in the groundwater of eastern districts. It was also observed that groundwater levels had gone down in many areas, so opportunistically, many privately owned irrigation tube wells came up and farmers had to purchase water at a premium. This was the time when water started becoming an unorganised business sector not just in West Bengal but in India as a whole.

It was during these times of rising water problems that I visited some of the other states of India, some of which differed from West Bengal in terms of hydrogeological conditions and, therefore, water culture in general. The groundwater level in some states was so deep that heavy-duty submersible pumps were required to lift water from low-yielding aquifers. Those states are still suffering from a chronic shortage of water due to over-extraction, erratic rainfall and poor aquifer storage. In my home state too, the hydrological scenario has changed rapidly under the pressure of the minor irrigation sector; aquifer levels went far below the centrifugal pumping limits pushing extraction costs beyond the affordability of small and marginal farmers, who constitute the vast majority of the agricultural sector. Heavy industry was not a big player in the water market at that time because their number in West Bengal was lower compared to other states. Heavy industries started exploring West Bengal prospects in the 1990s, and it was land, not water, which was the main hurdle. Resistance to industrialisation came mainly from small and marginal farmers who, as mentioned earlier, constitute the majority and depended almost solely on agriculture for livelihood; farmer resistance, therefore, had the support of the civil society. Consequently, industries were allotted land in non-agricultural belts, that is in hard-rock or laterite-covered areas deficient in water. Though the land was made available, water scarcity of this land compelled industries to take appropriate measures to secure their water supply, which meant industries had to draw water from distal aquifers and rivers, which in turn, formed the lifeline of the agrarian population in the source region. It soon became clear that industries could take water for granted. Perforce, therefore, industrialists had to assure rural people that industries would not snatch their glass of water and, for a substantive assurance campaign, industrialists had to undertake a quantified assessment of water like any other valuable resource. Industry, agriculture and society thus became entwined in issues relating to water; the issues being over-withdrawal, quality deterioration by contamination, and monopolisation of water sources.

In agriculture-based economies of South and Southeast Asia, industrialisation collided head on with farming communities and domestic water consumption. What was once considered to be a boon now became a polarising force separating two fundamentally important economic sectors: agriculture and manufacturing. The only way to surmount, or at least mitigate the divisive force of water is informed management of water distribution, that is, water stewardship. With hindsight, it can be appreciated that the modified avatar of water was an inevitable consequence of the steam-powered Industrial Revolution set in motion by colonising countries. Moreover, there is also the climate change set in motion by industrialisation, compounding the job of water stewardship.

The British Geological Survey document, *Groundwater Information Sheet* on *The Impact of Industrial Activity* (Water Aid, 2008) termed industries as an environmental pressure. It claims that industries draw much more water than they consume and are fraught with polluted effluent discharge to the environment. Untreated wastewater has the potential to cause irreversible damage to the ecosystem. As an instrument of economic development, industries need to act as facilitators of social wellbeing; this is a tough task, but it is crucial to sustainability. There are laws and enforcement machinery, however, to compel industries to move on the right track. In developed countries and also in some developing countries, there are stringent laws to compel industries to ensure pollution-free discharge. But safeguarding the remaining water reserve is not enough. Usable water like any other geological resource is finite, but population growth is unbounded. Water stewardship is mandated to keep economic growth running apace with growing population.

The introductory chapter (Chapter 1) of this book provides an overview of the status of global water, how it is shared by different sectors of the society like domestic, agriculture and industry, and the various effects forced upon the ecosystem due to such sharing processes. The introduction also deals with other topics like why water is a critical business issue for companies and why water crisis is a risk to their operations and brands. The chapter also discusses how industries should manage water from an ethical standpoint and how water should be priced and evaluated. Pre-war industries, and especially the war industries during World War II, operated without restraint; but now industries are facing social and political pressure to operate under regulating norms. Since manufacturing industries are major consumers of fresh water, it is incumbent upon them to foster green growth through water stewardship.

In Chapter 2 global trends of business are put into perspective against the backdrop of water crisis; that is, how business relates to global water crisis and how increased consumption leads to conflicts as a consequence of population growth; the impact of the ever-increasing demand of water for industrial growth is also discussed. Also offered for discussion are examples of water-related conflicts in different parts of the world. The unsustainability of the 'business as usual' model is compared with emerging paradigms that are attempting to redefine sustainable business practice. The United Nations agenda on sustainable development has also been discussed. Finally, the concept Integrated Water Resource Management (IWRM) is introduced.

In Chapter 3 a simplified and very brief introduction to hydrology is provided for decision makers on water management. Fundamentals of the hydrological cycle and redistribution of water to various storages and flows are discussed in light of such hydrological terms as watershed, river basin, aquifer and so on. Focus has been given on activities at the interface of water with industry, society, ecology and water cycle. A brief discussion on soil-water interaction is incorporated to help readers understand the concept of 'green water'. This chapter also gives special emphasis to river basin–watershed interaction so that readers can appreciate the delicate balance between river basin dynamics, surface water resource and water withdrawal. Ground water being a major source of industrial water, methods of measurement of groundwater resource are discussed in detail.

In Chapter 4 water stewardship is explained. The methods of developing a corporate water strategy and how stakeholders can be involved in industrial water management are thoroughly discussed citing examples of various engagement policies. The role of global organisations that deal with corporate water stewardship and extend technical guidance to sustainable business has been emphasised. The issues, challenges and opportunities in water stewardship are highlighted, and global standard of corporate water stewardship is incorporated.

In Chapter 5 water framework laws or directives required for formulating policies of national water governance are discussed. While introducing the concepts of water framework and water policy, the social, environmental and political drivers of the concept are highlighted. Water framework and water governance of nine countries, both from the developed and developing world are discussed as case studies.

In Chapter 6 water quality and water pollution are discussed, with water quality standards of different countries in the first part of this chapter. Chemical and physical parameters of water quality and their major sources are listed and how those chemicals are potential hazards to human health and environment are appended in different tables. Water quality standards for human consumption and food-beverage industries are listed. National guidelines on the drinking-water quality of some countries are compared. Quality guidelines on intake water and effluent water that are followed in different countries are discussed. In the second part of this chapter, different aspects of water pollution caused by industries are discussed. Effects of industrial pollution on the social, environmental and economic environment that cause conflicts, loss of reputation and brand value of the industries are also discussed.

In Chapter 7 the methods of withdrawal of water from surface water and groundwater sources along with processes of drilling and developing tube wells are discussed. This is followed by a discussion on water conveyance systems and methods of water purification. Industrial methods of storing and distributing water are also discussed with special reference to EPANET 2.0 software.

In Chapter 8, different methods of source water assessment are dealt with. The reader will find here the different methods of quantitative assessment of both surface and groundwater resources of a drainage basin.

In Chapter 9, available tools for water accounting and disclosure are introduced. Methods of water profiling by industries are discussed, along with methods of calculating a 'business water footprint' and how it is useful in estimating corporate water sustainability. The chapter also discusses corporate water disclosure documents and sustainability assessments.

In Chapter 10 the focus is on water conservation which is discussed in terms of (a) water overuse from the point of view of benchmarking, (b) identification of water loss through water audit, (c) prevention of loss through leakage management by retrofitting, (d) water saving in agriculture-based industries and (e) augmentation of water resources through rain-water harvesting (RWH) and artificial groundwater recharge. The chapter explains the use of water auditing software developed by the American Water Works Association (AWWA).

In Chapter 11 corporate social responsibility in the field of water management is discussed with illustration from case studies of several industrial organisations.

Water stewardship is an emerging subject crying out for global attention. And the most effective way of drawing attention to the need for water stewardship is through citing examples from countries already affected or in the firing line of a water crisis. I have mostly used the internet (and some publications) as is the global practice, for fishing out illustrative examples, and I have also drawn from my own field experience.

Water stewardship, as a discipline, has not been well established in most of the countries where industrial development has made deep environmental footprints. Alarmists project industries as a scourge of freshwater resources (which may be partly true), but the larger truth is that corporate bodies are becoming increasingly more conscientious about the environmental sustainability of big business. More importantly, the national economy is buttressed by industries. And, industries generate employment in massive numbers, which makes industries a de facto agency for government's employment-generation plans. Reinforced by water stewardship, industries can be instrumental in achieving the green growth targets of a nation. I hope this book will guide industries and policy makers in that direction.

Pradip K. Sengupta
Kolkata, India

Acknowledgements

The formulation of this book proceeded with the delightful experience of a remarkable event: the birth of my first grandchild, Myra in the household of my daughter, Rai. The composition of a book can be correlated to the prolonged course of gestation that ultimately leads to the birth of a new entity.

A book cannot take the required shape of a 'magic lantern' without an immense incentive and stimulus from the outside. While I entered into an agreement with John Wiley & Sons, my friend Sibabrata Basu Roy presented me with a book on Industrial Water Resource Management that has provided inspiring guidance in fashioning the various sections of this book. A lot of inspiration and invaluable constructive criticism was offered by my son, Shinjan who, from the United States, extended all possible logistic help. This book would have never seen the light of day without the constant support of my friend Samaresh Sanyal who not only gave me friendly advice but also religiously revised the content of all the chapters with care and patience. I'm equally indebted to Ananya Sasaru, my young friend for editing the language of the text that greatly assisted in my project. I am also grateful to my friends Arunabha Das, Sandeep Gupta, Debasish Guha and Arijit Das, who provided me with various photographs, which were essential to frame the content of this book. I profited immensely due to the permission granted from individuals and organisations for using their original texts and diagrams in my book. In this context, I would like to humbly acknowledge Dr. K.V.S.G.Murali Krishna, Dr. V.V. Rao, Dr. Colin Mayfield, Dr. A.Y. Hoekstra, Eddy Leks, Joachim Baumgaertner and M. Owasis Khattak. The organisations that granted me permissions to use their material in my book are SD Knowledge Platform, CEO Water Mandate, *Policy Quarterly*, the American Water Works Association, World Meteorological Organisation, The United Nations, Food and Agriculture Organisation of the UN, UNDP, TIDE, Leks & Co., Alliance for Water Stewardship, King County and Water Footprint Network. Their generous support has enormously facilitated the act of organizing this book. I must also mention in this connection that I reprinted a good amount of public domain materials from Wikimedia, the United States Geological Survey, the United

States Department of Agriculture, the United States Environment Protection Agency, the Environmental Protection Agency of Ireland, the Government of Australia, India WRIS and HM Government of the United Kingdom. While writing this book I consulted many publications in both printed and digital media, which I have listed at the end of each chapter. I am also thankful to Justin Taberham, independent editor of this series who offered me the opportunity to write a book for John Wiley & Sons. I sincerely thank Ramya Raghavan, the editor at John Wiley & Sons who managed to spare her valuable time in commenting upon the drafts and earlier versions of the manuscript. In this adventuring process upon which I had been engaged for quite some time, I would like to take the opportunity to express my gratitude to everyone who supported me throughout.

At this stage of life, when children are away from us, my wife deserved my attention which I could not pay while writing this book. I wish to express my sincere gratitude to Shyamali, my wife for her overall support.

Pradip K. Sengupta
Kolkata, India

1 Introduction

1.1 The context

In the closing years of the 20th century, big business had a rude awakening to the fact that its obsession with the profit curve was causing environmental crises, which was affecting the curve itself. Commenting on the results of CDP Global Water Report 2014 (CDP, 2014), CEO Paul Simpson says 'The availability of water poses a strategic risk to a large and growing number of companies.' Industries are facing competition in regions of water scarcity, and as a consequence public grievances are beginning to force cancellation of licences to operate. It has taken the shape of a global phenomenon because the underprivileged are becoming more and more aware of their right to water. Additionally, global warming (even though the underprivileged are not well informed on climate change) looms larger with time. In short, the profiteering tendency of big business is being directly challenged by (1) resource depletion, (2) non-delivery of the social responsibilities, (3) management crises in the value chain (4) rising costs of water, (5) climate uncertainties, and (6) political and social pressure.

In the past, industries were considered in isolation, that is, as entities separated from the ecosphere although, in fact, industrial dynamics is dependent on information flow between the industry and its environs (Odongo, 2014) because they share common resources with other user sectors (see Figure 1.1). The underlying concept is simple: Industries depend on services provided by the ecosystem; consequently, an industrial system should be viewed as a subsystem that impacts the biosphere, hydrosphere, and lithosphere.

Industries often act upon the idea that they are free to operate merely by complying with Environmental Protection Agency guidelines and the water laws of the country in which they operate. But simply complying with guidelines is not enough to sustain a business, especially in areas known for water

Figure 1.1 Industry and agriculture, sharing the same resource.

crisis (Aghababyan, 2006). Industry's own way of fulfilling social responsibilities often fails to convince the people when both industry and people confront severe resource depletion. Given below is a case in India that has given rise to newer questions on water stewardship delivered by industries.

1.1.1   The story of Coca-Cola in India

Mehdiganj near Varanasi in Uttar Pradesh (India) is mainly an agricultural belt. In 2003, with temporary permission to operate with a production target of 20,000 crates per day, Coca-Cola started its bottling plant, which increased its production to 36,000 crates per day by 2009 (India Resource Centre, 2014). In 1999, Mehdiganj was declared a safe category block[1] by the Central Groundwater Authority. By 2004, after only a few years, it turned into a critical block in terms of the groundwater resource. A report published on 19 June 2014 states that (The Ecologist, 2014) in pursuance of a Supreme Court ruling the 'local authority has passed an order to evict Coca-Cola.' The Coca-Cola Company, however, has published documents in defence against the charges drawn against them (Coca-Cola India, no date) claiming that the charges are nothing but myths and that they have abided by the laws and statutes.

The court verdict against Coca-Cola is a clear demonstration that big corporate houses have to look beyond profiteering and proactively participate in broader catchment-level resource management. Formulating or practicing their own policies and simply abiding by the terms of their legal licence will not suffice. Industries will lose the social licence (Lassonde, 2014) if resources begin to get endangered by their activities. The Mehdigunj case underscores the responsibility of multi-national companies to respect societal needs alongside protecting shareholder's interest (Sengupta, 2014). Goods produced to cater to the desires of the privileged class may have to confront questions such as how ethical it is to make use of scarce resources that are supremely essential for the larger underprivileged population for the sake of producing something, rendered non-essential by another group. The difference between the concepts of 'need' and 'want' must be emphasised to the society. The derivative of this question is also equally important to manufacturers because consumers, of such products as are deemed non-essential in the world at large, do not like to see their brand manufacturers falling foul of the law. The bottom line is a fundamental question: Can the corporate sector be allowed to secure their production requirements only on the basis of complying with local or national statutes? Clearly not, as exemplified in the aforesaid court verdict. Big business has to factor in safeguarding the rights of the underprivileged to remain in business.

Manufacturers are thus accordingly awakened to their responsibility to the two most powerful factors that have the power to affect business sustainability: (1) accountability to shareholders and consumers, who feel affronted if the manufacturer is punished by the court of law; and (2) preventing people from protesting in the area of their operation by following the law both in letter and spirit. In the context of water stewardship, for example in the beverage industry, manufacturers have to keep a keen eye on the water they are exporting from catchment vis-à-vis effectiveness of their replenishment plans so that the total water reserve is not irreversibly damaged or severely depleted. Manufacturers are trying to compensate withdrawal by providing rainwater harvesting installations to the community. The question is how these two are balanced. Herein rests the importance of source-water assessment, which is vital to balancing resource extraction and input—a balancing act vital to avoid possible confrontation between business and society.

Coca-Cola is merely one example; it is by no means unique. The corporate sector is facing popular resistance throughout the world. A few more examples are given in Box 1.1

Corporate-society conflict over water

- In July 2012, residents of Shifang in the southwestern province of Sichuan, China, protested against a proposed USD 1.64 billion molybdenum copper plant, fearing negative impact on the environment and their health. The protest turned violent, and 17 people were injured. Ultimately, local officials announced that the project would be halted (Economy, 2013).
- In 2004, a Peruvian gold mine had to suspend its production due to people's protest about their water concerns (Walton, 2016).
- In 1999, the people of Cochabamba in Bolivia protested against privatisation of water in favour of industries, and the conflict led to violence (Otto and Böhm, 2006).
- In Jakarta, popular protest against privatisation and corporate control over water won a legal battle on the grounds that the Public-Private Partnerships (PPP) were negligent in fulfilling the people's right to water (The Transnational Institute, 2015).

BOX 1.1

The essence of modern-day water management requires balancing sustainability and economic growth. Coca-Cola is one of the leading companies among those who profess to propagate and apply the concepts of water stewardship. Despite resources and experiences that Coca-Cola invested in the water problem, they failed the Mehdinagar test. But *why*? A fuller understanding needs a deeper look into the problem.

We shall seek the answer to the italicised '*why*' in this book and try to identify the drivers, risks and opportunities of water stewardship within the framework of industrial water management.

1.2 Water goals in the 21st Century

The 'twentieth century has witnessed an unprecedented rise in human population, from 2.8 billion in 1955 to 5.3 billion in 1990 and is expected (Mary, 2016) to reach between 7.9 and 9.1 billion by 2025' (Gardner-Outlaw and Engelman, no date). This phenomenal population growth demands an increase in production rates from mines, power plants, beverage industry, and so on. Paradoxically, the industrial growth rate is higher in developing countries in comparison to developed countries, the simple reason being the industrialists' preference for such countries that easily provide them with cheap labour and weak/flexible environmental laws.

The United Nations World Development Report 2015 under the United Nation World Water Assessment Programme (WWAP, 2015) states that during the 19th and 20th centuries, development of water resources was largely driven by rapidly increasing demands for food, fibre and energy. A growing middle class, with improved living standards and increased purchasing capacity, boosted industrial growth that led to sharp increases in water use, which inevitably culminated in worldwide alarm cries against unsustainable growth rates. The crisis is more pronounced in places where water use is wasteful, distribution is inequitable and management is poor.

Increased meat consumption, building larger homes, using more motor vehicles, gadgets and appliances caused an increase in water consumption. While this may sound unconnected, it holds true, because, every product and service has its own virtual water content, and increased consumption increases the water footprint.

Savenije, Hoekstra and Zaag (2014) are of the opinion that the natural distribution of the freshwater resource on the face of the earth 'has changed as a result of human intervention and efforts to manage water'. Though changes are viewed in terms of utilising water for industrial growth and production of energy, the direct hydrological impact is seen as a diversion of natural water flow to facilitate supply to urban areas, industries and for agriculture. River basins have suffered a loss of perennial flow, depletion of resources and contamination from point or diffused sources. Such changes have also affected the global climate, biodiversity, land cover change and serious depletion of quality freshwater.

The science of water, or hydrology, is relatively new because it was developed in response to the recent population boom and the consequent rapid industrialisation. Diversion of river courses for harnessing hydropower was made possible by 'increasing knowledge of water engineering, large-scale water supply, flood mitigation and irrigation' (Savenije, Hoekstra and Zaag, 2014). But there is no such thing as a free lunch; the question arises, who is paying for the unregulated industrial use and how? 'Unregulated industrial use' is a euphemism for depriving marginal people of their meagre needs of fresh water. When pushed to the wall, the invisible and silent majority of marginal people join hands in large enough numbers to show up on the radar of the media, and the government is eventually forced to introduce regulatory acts. Interdisciplinary water management, where scientists, engineers, ecologists, decision-making managers, politicians, sociologists and the media work in tandem has emerged as a post-colonial paradigm; it is termed Integrated Water Resource Management (IWRM).

IWRM recommendations actually gathered momentum in 2002 from a far more serious problem that remained unnoticed in the 1970s, namely environmental degradation. This hitherto unrecognised problem came into focus in the Global Environmental Summit (United Nations, 1992b) where the Millennium Development Goal was proposed. In the proposal, unprecedented scarcity of water and other resources, skyrocketing food prices (KPMG, 2012), and escalating energy security issues were emphasised in light of projected population growth. Some key indicators of development were proposed in Agenda 21 of UN 1992, and Section 30.3 (strengthening the role of business and industry) stated that the highest priority should be given to 'environmental management' by society in general and by the corporate sector in particular. The role that big business must play in sustainable development is the crux of the matter discussed in Section 30.

In the context of water, the aforementioned UN document highlights 'responsible care' as the key element in water stewardship. Participating governments drew from IWRM recommendations in their approach document, which was presented in the 2002 World Summit on Sustainable Development (WSSD) held in Johannesburg. One hundred and ninety-three countries agreed to the Johannesburg Plan of Implementation (JPOI), in which Article 25 calls for 'the development and implementation of IWRM and water efficiency strategies, plans and programmes at national and regional levels with national-level IWRM plans to be developed by 2005' The salient features of Article 26 of JPOI are (United Nations, 1992a):

a Develop and implement national/regional strategies, plans and programmes with regard to integrated river basin, watershed and groundwater management and introduce measures to improve the efficiency of water infrastructure to reduce losses and increase recycling of water.
b Employ the full range of policy instruments, including regulation, monitoring, voluntary measures, market and information-based tools, land use management and cost recovery of water services, without cost recovery objectives becoming a barrier to access to safe water by poor people, and adopt an integrated water basin approach.

    c Improve the efficient use of water resources and promote their allocation among competing uses in a way that gives priority to the satisfaction of basic human needs and balances the requirement of preserving or restoring ecosystems and their functions, in particular in fragile environments, with human domestic, industrial and agriculture needs, including safeguarding drinking water quality.

    d Develop programmes for mitigating the effects of extreme water-related events.

    e Support the diffusion of technology and capacity-building for non-conventional water resources and conservation technologies, to developing countries and regions facing water scarcity conditions or subject to drought and desertification, through technical and financial support and capacity-building.

    f Support, where appropriate, efforts and programmes for energy-efficient, sustainable and cost-effective desalination of seawater, water recycling and water harvesting from coastal fogs in developing countries, through such measures as technological, technical and financial assistance and other modalities.

    g Facilitate the establishment of public-private partnerships and other forms of partnership that give priority to the needs of the poor, within stable and transparent national regulatory frameworks provided by Governments, while respecting local conditions, involving all concerned stakeholders, and monitoring the performance and improving accountability of public institutions and private companies (Source: Sustainable Development Knowledge Platform, United Nations. Printed with permission from the United Nations.)

IWRM was accepted as a paradigm and 'particularly recommended in the final statement of the ministers at the International Conference on Water and the Environment in 1992 (called the Dublin principles).' The fundamental objectives under IWRM are: (1) Social equity, (2) Economic efficiency, (3) Ecological sustainability (*Integrated water resources management*, 2015).

These two world summits on environmental degradation emphasised water management plans especially with regard to industrialisation. The points of emphasis were:

    1 Development of a national strategy or vision on water.

    2 Implementation of regulations by letter and spirit on water withdrawal, use and discharge.

    3 Pro-poor prioritisation of industrial water needs.

    4 Promoting awareness among stakeholders and civil society towards industrial water management. This concept is crucial to areas of chronic water shortage to avoid conflict between people and industry.

How industries should look upon the water, therefore, becomes a big question. The United Nations in its document *WWDR 2014* has specifically outlined a visionary recommendation for water-intensive industries (see Box 1.2). Byers *et al.* (2002) observed that industry today is constantly striving to operate more efficiently. Most successful plants claim to be sincere in their search for safe, socially responsible, economically profitable and environmentally sustainable methods. While this is true in some isolated

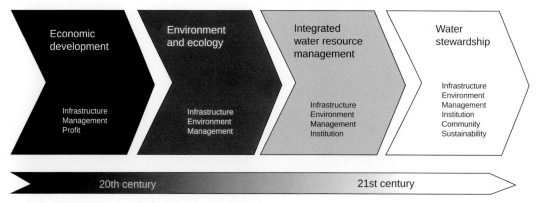

Figure 1.2 Paradigm shifts in water management.

instances, by and large, industries are focused on profits and merely pay lip service as a smoke screen to avoid the additional expenditure necessary for the implementation of IWRM guidelines.

Industries, as a rule, are more focused on short-term profit as opposed to the long-term assessment of economic growth vis-à-vis the environmental cost, a focus which is suicidal in a way because environmental degradation never fails to claim its pound of flesh. Small-scale industries in developing nations, which generally face step-motherly treatment from the government, therefore, feel justified in turning a blind eye to the pollutants they release into the environment.

However, there is also a bright side. Big businesses are shifting their focus from mere profit-making activities towards more sustainable business models that can be beneficial for both economic development and social well-being. The 20th-century economic development model is being metamorphosed into a more sustainable water stewardship model that the 21st century will witness. See Figure 1.2.

1.3 Water ethics

'Water is not a commercial product like any other but, rather, a heritage that must be protected, defended and treated as such.' EU Water Framework Directive (Bloech, 2010).

The water crisis of a country is assessed by various indicators, which have found that water and poverty are interlinked (Sullivan, Meigh, and Fediew, 2002). The indicators used for examining the extent of water crisis are water availability (surface and groundwater), demand (domestic, agriculture or industrial), access (distance to source and legal rights), institutional capacity (human and financial capacity to manage the system), price and affordability of water and environment (hydrological functions of ecosystems), vulnerability to water-related disasters and water quality improvement.

Water infrastructure and provisioning of water are indicators of both economic and institutional capacity of a country.

Big business generally considers water as a raw material while the ethical value of water is often ignored. The question of water ethics comes to mind only when faced with development projects dependent on water, such as hydroelectric projects, river linking, trans-boundary water transfer and the like, all of which are planned for the benefit of stakeholders in keeping with the interests of the silent majority. People who are dependent on the river basin are not factored into the planning of business development projects. This deliberate oversight borders on criminal negligence, because even a single deep tube well installed for industrial use adversely affects the drinking water security of local people. That being said, the scale of impact of large-scale businesses, comprising factories, offices and residential quarters, is left to the imagination of the reader. Admittedly, big projects recruit some local people for menial jobs, but their numbers constitute a minute percentage of people whose water security is compromised.

Ethical conceptualisation of water evolved alongside social changes brought about by technology. Hassan (2011) identified eight modern day paradigms: (1) spiritual and religious, (2) aesthetic-recreational, (3) ethical, (4) scientific, (5) ecological, (6) technological, notably hydraulic engineering, (7) financial and economic, and (8) management policies. The paradigm shifts that have taken place from ancient to modern times follow the path of modernisation of society. In the course of further evolution of water ethics, many modern societies have discarded some of the values listed above for reasons of practicality and have focused on the more scientific, technological, economic and managerial considerations.

The sub-commission COMEST, a committee appointed by UNESCO to examine the question of water ethics, has identified some fundamental principles. It states that water is a fundamental pillar that 'human dignity' rests on. The COMEST (2005) document clearly states that all individuals including the poorest are entitled to 'participate in the water planning and management'. The other principles of entitlement such as 'solidarity', 'human equality', 'common good' and 'stewardship', 'transparency' (UNESCO, 1999) and the right to information are also important in that data-sharing process. The disclosure of water-related information empowers the poor by making them aware of their rights to water. The document also advocates educating stakeholders who can influence water management (Liu *et al.*, 2011).

Consider the case of river-linking projects in India (Bansal, 2014). The defining character of India is not just its territorial vastness and the pressure of a vast population. India cannot be understood if the varied hydro-ecological zones and the consequent ethnic diversity are not included in its description. Rainfall is unevenly distributed, so that there is a chronic drought in some parts while another part of the country enjoys the world's highest rainfall. Both surface and groundwater resources are thus unevenly distributed. The government of India has taken a policy decision to link the river basins through canals and transfer water from one basin to another to feed the water-deficient regions of India.

Let us examine the ethical base of the proposal. Controversies arise when human enterprise causes irreversible destruction of natural habitats and attendant extinction of species. The use of the plant and animal kingdoms as a substrate for economic growth at the cost of life forms becoming extinct is challenged by conservationists fighting to protect animal rights, but they are vastly outnumbered by the champions of industry who continue with widespread habitat destruction behind a facade of sustainable growth. Animal-rights controversies make headlines when the media deem them newsworthy. The trouble is that there are also people sharing water and other ecosystem services of river basins targeted for linkage (interestingly tribal cultures have shared that space with local flora and fauna for millennia without disturbing ecological balance; this remarkable coexistence is not an example of sustainability for the policy makers of sustainable development).

According to Groenfeldt (2013), it is not wise to add further stress on an 'already depleted water ecosystem' by constructing new infrastructures without factoring in the environmental cost into the projected profits from development. One would have expected the internet-savvy third-millennium urban population to be alert to the environmental cost of their lifestyle. I, personally, am aghast at the amount of campaigning required to sensitise big business and its beneficiaries to the enormous losses caused by environmental degradation. To shorten a long story, decision makers need to balance environmental cost, ethical value and economic gain. Hopefully time, the great healer will restore the balance someday.

There are several tools and methods to support the knowledge base of decision makers (notably, historical water-related databases, consumptions pattern and practice, geographical information system (GIS) software, flownet analysis tools, and water-assessment tools). These tools are mostly technical in nature and require quantification (of such parameters as groundwater resource estimation, minor irrigation census, industrial water demand assessment tools, etc.). Decision makers must necessarily be conversant with these tools to reinforce and justify their decisions. The only drawback of such tools and methods is that they do not factor in local people and their water culture, a drawback subsequently addressed in the newly formulated policy of 'water stewardship'. In the absence of water stewardship guidelines, decision makers have traditionally applied their uninformed ideas (read self-serving interests), which were often found woefully short in ethical content leading to conflicts.

Since water stewardship is still in its infancy, it faces various unresolved ethical issues at various stages of big project implementation. For effective water stewardship, managers should focus on ethical values instead of lopsided arguments highlighting the economic value of big projects. Not all ethical issues merit prioritisation. The rural population is not intellectually incapable of appreciating development. They are flexible enough to readjust or even replace antediluvian water-intensive rituals to accommodate development plans in their ancestral lands, provided of course that the drive for development does not literally drive away the poor from their rural habitat to cities to make a living as domestic helpers

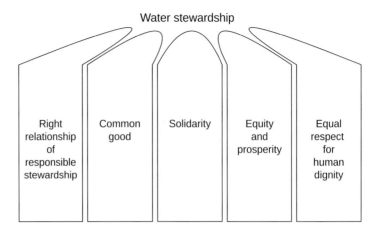

Figure 1.3 Ethical framework of water stewardship. Adapted from Jennings, Heltne and Kintzele (2009).

or exploited and underpaid daily wage labourers or even as beggars. Prevention of exodus from rural India to cities is arguably the greatest challenge to water stewardship.

With these ethical orientations and practical considerations in mind Jennings, Heltne, and Kintzele (2009) have suggested five principles of water ethics, which have the potential to elevate water stewardship from a mere strategy to a socially redeeming force. See Figure 1.3.

1.4   Value of water

The value of water is closely related to water ethics, that is, how a person is using water and how water is related to his identity. Water has different values to different persons or entities. In the context of water management for industries, water is seen in terms of economic value; whereas for a fisherman or a sailor water is as valuable as his own life.

Indian philosophy considers water as an economic, environmental, social and spiritual good, but water is not evaluated separately in each of these valuation concepts. Indian philosophy, like similar philosophies throughout the world, considers it as the most valuable natural resource to be shared and cared for. In the Indian way of thinking, both standing and flowing water have always been considered as the most vital common property to be shared by all that inhabit the region around a particular water source. Water has often been valued on par with livestock in ancient India. It has been stated, 'Water is equivalent to the cattle wealth' (Rig Veda, 10/19). The value of water evolves along with the changing economic system in an evolving society. Table 1.1 shows the value of water for different entities.

Table 1.1 Value of water.

System	Value	How valued
Ecosystem	The most essential fluid necessary for sustenance.	Valued as a natural asset and invaluable resource, hence it needs protection. The cost of conservation.
Human society	Water has a great social and entertainment value. It is also essential for religious activities. Ownership of a source of water is also an asset.	The value in terms of price or the cost of making it available for use, such as revenue paid to the municipal water supply or the cost of installing a personal tube well, pump and distribution system and cost of maintenance. The price of bottled water.
Farmers	Water for irrigation is essential for farming. Farmers value water in terms of crop production.	A commodity, irrigation has a cost. Valued as the cost of lifting or purchasing water from the irrigation authority or from the private irrigation service provider.
Business	Considers water as an economic good and natural capital.	Water has a cost; it can earn revenue and enhance brand value. The cost assessment at every stage of the business value chain is essential.

1.4.1   Water valuation

The Dublin Statement on Water and Sanitation Development pays attention to the economic value of water (United Nations, 1992a). Principle 4 states:

> Water has an economic value in all its competing uses and should be recognised as an economic good. Within this principle, it is vital to recognise first the basic right of all human beings to have access to clean water and sanitation at an affordable price. Past failure to recognise the economic value of water has led to wasteful and environmentally damaging uses of the resource. Managing water as an economic good is an important way of achieving efficient and equitable use, and of encouraging conservation and protection of water resources.

Water value is not a simple generic term, but a term that has different meanings for different people. Water value changes not only with time and space but also with social and political changes. Industries are more concerned with the economic, rather than the social and ethical values of water, a fact which might turn into a costly liability to business. If such values, apparently abstract, are neglected or treated with little seriousness, society itself becomes a threat to the project. For example, a site selected for water withdrawal from a river may be a place of worship, which can interfere with business plans.

Morgan and Orr (2015) have proposed a balance-sheet approach to water valuation. In their model they have included (a) on to the asset side, corporate water-related-built capital for both green and grey infrastructure, ownership of water reserves, community goodwill, brand value, social value and assets created through operation; and (b) on the liability side, water-related social and legal liabilities.

The move from 'water management' to 'water stewardship' includes the consideration of values in an upstream supply chain, industrial operations, and downstream product use (Sarni, 2015). The business world looks upon water as a form of a commodity that needs an appropriate valuation. At every stage of operation, capital investment and recurring costs are involved. The World Business Council for Sustainable Development (WBCSD) defines water valuation (WBCSD, 2013) as the amount an individual is willing to pay for it, regardless of whether it is for social or economic purposes. In industry, the value of water is understood to be the expenditure incurred to procure it, cost for conservation, and cost for implementing water-related programmes under corporate social responsibility, maintaining a trade-off between the corporate body and stakeholders. These costs are measurable in terms of money and the volume of water purchased.

Another method of valuation is based on the cost of water productivity in USD for each kilolitre of water consumed. This valuation varies from one country to another depending on geographical and climatic conditions (e.g. temperature and evaporation), the cost of raw materials, market conditions and willingness to pay by the market (which is, of course, sensitive to USD exchange value in the country concerned). In most countries, the true value of water is not tagged with the product, and its cost is not recovered from the market. This practice can be greatly improved by adopting water steward-ship and technological advancements in extraction and recycling of water.

Valuation of water, either as product or service is calculated according to the value it carries in its life cycle. The valuation is calculated with three variables; namely cost, value and price. Cost involves manufacturing or procurement expenditure. For example, the cost of production of one litre of bottled water sourced from a lake involves expenditure on withdrawal, purification and bottling, which amounts to USD 1.2 which is then sold at USD 2.5 to generate revenue and is price-tagged accordingly. In other words, the value of that one litre of bottled water is the purchase price. Such valuation based on production cost and selling price is not applicable to cultures that inhabit arid regions, where there is little difference between water and blood. Among the three aforementioned parameters, the value of water can only be appreciated from the lifestyle of people that inhabit climatically water-starved regions, such as the Kalahari.

1.4.2 Application of water valuation

One must note however that there is more to the world than the Kalahari. Globally, water valuation has several applications in industries as indicated by the WBCSD (2013) document on water valuation. Efficient water stewardship builds upon how an industry values its water resources in improving its business, delivering essential services to society and building a reputation for effectiveness in resolving problems. Water valuation helps industries to consolidate its relationship with both community and regulatory bodies. The main drivers in water valuation are:

a   The role of water in proposed investments.
b   Application of new technologies in water treatment, use, reducing loss and tracking unaccounted water consumption. This will generate more value per unit volume of water consumed.

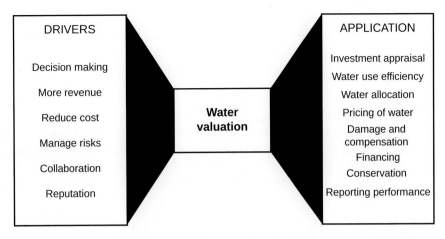

Figure 1.4 Drivers and applications of water valuation. Adapted from WBCSD (2013)

    c Quantified need-based water allocation to different sectors of business.

    d Water valuation is vital to quantifying the cost of domestic water consumption against revenue earned by water consuming industries (Porter and Kramer, 2011).

    e Water valuation determines the pricing of domestic water usage, water services and products.

    f Governments are known to charge users for polluting or damaging the environment (Wabunoha, no date). Water evaluation is essential for calculating damage compensation.

    g Help in sustainable financing to create enabling environments (United Nations, 2014b).

    h Water valuation is helpful in the assessment of impacts, where businesses must take conservation actions to offset degradation (Kreitler *et al.*, 2015).

    i Water accounting and disclosure requires water valuation.

The drivers and application of water valuation in business are shown in Figure 1.4.

1.5 Water and energy nexus

Water and energy are closely related and interdependent. Availability and predictability of water resources can directly affect energy systems. Energy is required to withdraw water from wells, bodies of water, rivers or any other sources or storage. Without energy, water can neither be drawn nor transported from the place of collection to the point of use. The energy input may be mechanical, electrical or manual. The household perception of water transportation is from ground-level storage to the overhead tank, for which a one horsepower pump will generally suffice. 'Each kilowatt-hour (kWh) of thermoelectric generation requires the withdrawal of approximately 25 gallons of water' (US-EPA, 2016a). Outside the cosy world of household needs, water transportation is enormously energy intensive; irrigating a wheat field, for instance, consumes a lot of power. It takes about 3.5 kW to lift 4000 gallons or 20 kilolitres of water from a low duty deep tube well in India. Torcellini, Long and Judkoff (2003) in a report of the US Power Department states, 'Eighty-nine per cent of electricity in the United States' is produced from thermal power plants. Thermal power plants use water for cooling. On an average, '1.8 litres of fresh water is evaporated per kWh of electricity production' and is considered fit for consumption. In hydroelectric power generation, water evaporates from the reservoir surface. 'In Arizona, for example, 7.85 gallons is lost to evaporation per kWh of power consumed' (US-EPA, 2016a). The power generation–water supply system may sound tautological because water is crucial to power generation and power is crucial to the water supply.

Needless to say that the tautology is broken by the fact that power generated by water usage is much more than the power required for water supply, but the tautology nevertheless highlights two very important points:

1 Water loss in either stage, i.e. during power generation or water supply, puts pressure on the water source, which in turn through a circuitous route puts pressure on the pocket. And most importantly,
2 Integration of agencies responsible for water-energy policy formulation with planning management agencies.

Resolving water and energy-related issues, and conservation of both in a complementary way has been put on the priority list by many industries in advanced countries. Big industries are trying to conserve water by investing in technologically advanced cooling systems, and by recycling and reusing treated effluence. A beginning has been made. But, the key question is whether corporate social responsibility (CSR) and water stewardship are functioning at par with the scale of the problem created by the corporate sector? Clearly not, as is plainly evident in the familiar picture of thousands of dead fish floating belly up in polluted rivers. Even more shocking is the post-modern phenomenon of farmer suicide in India in water-starved regions; strangely, the phenomenon started contemporaneously with economic reforms in 1990. World trade has brought together the affluent 'Global North' and its poor country cousin the 'Global South' to an interdependent platform. This interdependence has to expand beyond profit making and include sustainable use of water. That is the essence of water stewardship.

Water constraints lead to energy constraints. The vulnerabilities may arise from a depletion of quality and quantity of source water. Climate change can lead to water crises resulting in a reduced river flow or depressed groundwater levels. The water allocation policy of the government can do much to help in this direction. For example, the government of Chile is very strict regarding water allocation rights to the mining companies. A few years back, a copper mine, Xstrata's Collahuasi, was directed to reduce water extraction from 750 litres per second to 300 litres per second (McKinsey, 2009). The company was forced to invest in alternative technologies to adapt to the reduced supply.

Population growth, consequent rise in food production and especially, urbanisation, with its wasteful water consumption habits, put enormous stress on water and energy requirement. There are several scenarios conceived by scientists and planners to reduce the stress.

The International Energy Agency (IEA, 2014) has modelled the world water withdrawal and consumption in a country-wise manner from 2010 to 2035. Taking 2010 as the baseline, it is estimated that the water withdrawal may increase by 0.7% in 2035 whereas consumption of water will reduce by 1.2%. The reason for reduced consumption is attributed to the development of water conservation technologies. In European industries, water withdrawal and consumption are estimated to decrease by 0.9% and 1.4% respectively; whereas in the USA both withdrawal and consumption may increase by 0.1% and 2.5% respectively. Industrial growth in India will result in an increase in both withdrawal and consumption by 1.6% and 3.5% respectively. At present, industrial water consumption in China is almost 2.5 times higher than that of India; water resources in China are likely to be further stressed by a 3.5% increase in consumption. The world figure of water withdrawal and consumption is shown in Figure 1.5.

Clearly, water consumption in relation to global energy demand is increasing at an alarming rate and firmly heading towards the limit of sustainability. Some large industries are trying to conserve energy to minimise their water requirement. The US Department of Energy has identified six strategic pillars to provide the necessary support to water-energy nexus.

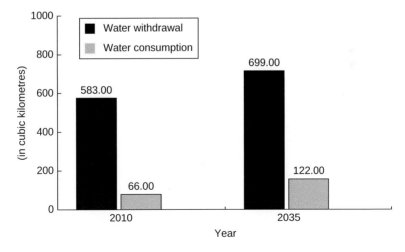

Figure 1.5 Predicted change in water withdrawal and consumption for industries from 2010 to 2035. Based on Shiklomanov (2000).

1.5.1 Impact of energy production on water resources

Hydroelectric power

Hydroelectric power plants utilise the potential energy of upstream water in reservoirs to drive turbines; the water then returns to the river downstream. Thus, hydroelectric power generation per se does not consume water (Davies *et al.*, 2009). But, by the construction of dams, reservoirs and diverting the natural flow, hydroelectric power plants do affect the natural order, notably by increased evaporation from reservoirs and submergence of vast areas in the upstream tract.

Thermal power

Thermal power plants, on the other hand, consume water in the cooling towers. Water consumption can be reduced by investing in improved cooling technology, but there is another factor involved. The higher temperature of water discharged from cooling towers increases ambient temperature causing thermal pollution of the environment.

Bio-fuel

Bio-fuel production sounds environment-friendly, but it needs large quantities of water.

Bio-fuel was initially championed as an environment-friendly alternative, but switching from fossil fuel to bio-fuel did not go down well with the automobile industry. Additionally, bio-fuel production is a big consumer of water. Consumption of water per litre of bio-fuel production is estimated by FAO (2008) as 2000 litres for sugarcane, 1357 litres for maize, 2364 litres for oil palm and 3333 litres for rapeseed; 30 to 50 litres of water is required per Giga Jule (GJ) energy yield. Also to be taken into consideration is water contamination caused by fertilisers and pesticides in bio-fuel crop cultivation. But, for all that, bio-fuel remains a formidable contender to replace fossil fuel, awaiting design improvements in internal combustion engines and improved agriculture.

Mining

Water use in energy and mineral mining is also an important factor in understanding the water-energy nexus since a good deal of water pollution also happens in mining. Water use in mines and attendant pollution are given below:

- **Oil and gas:** Water is used for drilling, hydraulic fracturing, oil-sand mining and in refineries. Secondary and enhanced oil recovery also consumes a substantial amount of water. Water pollution takes place from seepage and the produced water, contaminating both groundwater and surface water.
- **Coal:** Coal mining is a big user of water. Mechanical cutting of coal and related dust suppression need water. Water is also required in coal washing. Mining drainage and produced water are the main contaminants.

1.6   Global water stress

Annual water resource (AWR) and the need for improved technology are facing water-related stress trying to keep pace with population growth. The problem is global, but the magnitude of the problem is specific to each country. The human development report of the United Nations (UNDP, 2006) reveals that water overuse is the main cause of environmental degradation. As a result, major river basins are suffering from widespread and intensive agriculture development. Large river basins in China, India and North America are victims of rapid urbanisation and industrial growth. Most of the rivers in China are polluted by industrial waste; the scale of pollution can be appreciated from the fact that over 26 billion tonnes of wastewater was discharged into the Yangtze River in just one year (Aiyar, 2007). Hydrologists and social scientists define water stress in many ways and have tried to classify countries according to the stress on freshwater resources. Therein lies the communication gap. Evaluating freshwater depletion rate and contamination levels can only be measured in numbers, which do not make sense to the poor people from rural areas who are always positioned at the receiving end of the water problem. What the rural poor do understand is the fundamental truth, and, as opposed to numbers, they understand it from increasing physical hardship in everyday life as a consequence of the degraded environment. It is a world divided by scientific elitism. For effective water stewardship, it is incumbent upon social scientists to build a language bridge between the rural poor and scientific water conservation.

One of the popular indicators of water stress and scarcity is the Falkenmark indicator. It is perhaps the most widely used measure of water stress. Under this index, countries are categorised into no stress, stress, scarcity, and absolute scarcity areas (Table 1.2). The index thresholds 1,700m^3 and 1000m^3 per capita per year are used as lower limits of water-stressed and scarce areas, respectively (Falkenmark, Lundquist, and Widstrand, 1989).

The renewable freshwater resource of a country derives from the geographical, geological and geomorphological character of the territory, and as such the renewable freshwater resources of a country are more or less fixed. But the demand is ever increasing due to growth in different sectors.

Table 1.2 Water barrier differentiations.

Index (m^3 per capita)	Category/condition
>1,700	No Stress
1,000–1,700	Stress
500–1,000	Scarcity
<500	Absolute scarcity

Source: Falkenmark, Lundqvist and Widstrand (1989)

Shiklomanav (2000) estimated the renewable freshwater resources of various countries and calculated the per capita availability in each of the countries. His data was based on 1995 statistics.

Another scarcity index has been proposed, which compares annual water withdrawals relative to total available water resources (White, 2012). Using this approach a county is called water-scarce if annual withdrawals are between 20–40% of annual supply, and severely water scarce if they exceed 40%'.

A third approach to measuring water scarcity has been proposed by IWMI, which considers the water infrastructure, consumptive use and adaptive capacity of a country (White, 2012) as guiding factors. Using this approach, countries are classified into four groups (Seckler *et al.*, 1998) based on the growth percentage of water withdrawal (between 1990 and 2025). Group 1 consists of countries where water withdrawal is higher than 50% of the annual resource. Countries belonging to Groups 2 to 5 have water withdrawal less than 50%. Countries rated between 2 and 5 have been further classified on the basis of growth percentage. According to the development rate, countries that fall under Group 2 have more than 100% growth, Group 3 has 25 to 99% growth, Group 4 has growth below 25%, and Group 5 are countries having no, or negative, increase in projected water withdrawals.

Luo, Young and Reig (2015) measure water stress in terms of total annual water withdrawal (municipal, industrial and agricultural) expressed as a percentage of the total annual available blue water. Higher values indicate more competition among users. The classification is shown in Table 1.3.

Figure 1.6 shows the world map with countries classified according to the stress index.

'Aqueduct' is a water-related data portal that accumulates and disseminates water information by country. These data and information are also available as GIS-based interactive maps, which are very helpful to decision makers at the country level. Aqueduct utilised spatial aggregation methodology that projects sub-catchment level data onto the country level. The scope for modelling with Aqueduct enables the user to score 167 countries under the 'business-as-usual', 'pessimistic' and 'optimistic scenarios' for the years 2020, 2030 and 2040.

Table 1.3 Water stress index.

Score	Value
0–1	Low (<10%)
1–2	Low to medium (10–20%)
2–3	Medium to high (20–40%)
3–4	High (40–80%)
4–5	Very high (>80%)

Source: Luo, Young and Reig (2015) Printed under CC BY 3.0 licence.

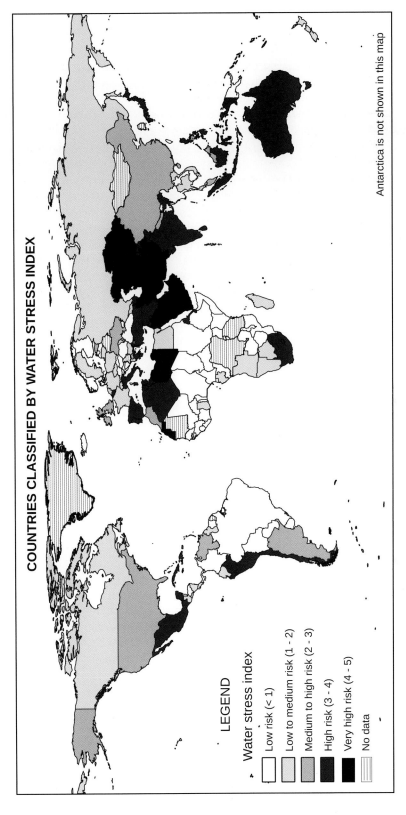

Figure 1.6 Map of countries classified according to the stress index. Data source: Luo, Young and Reig (2015).

1.7 Industrial impact on water resource

Industrial activities in a watershed have water-related impacts that a company needs to address in all stages of its value chain. These operations are as follows.

1.7.1 Impact on the quantity of the source water

One of the most challenging local water issues that arise from industrial modification of surface cover and high water withdrawal is the depletion of the quantity of water in natural ecosystems. For instance, high growth in the brown sector economy (tourism, hotel, power generation) in Kathmandu Valley has covered a large part of groundwater recharge zone of the valley by 'impervious roads and buildings for commercial and residential purposes' which has already posed a threat to the groundwater. Groundwater level, as a consequence, has dropped from 9 metres to as deep as 68 metres within a couple of decades as observed in 2001. The issue will develop social, ecological or regulatory challenges when companies will have an increasingly difficult time accessing water and may face reduced and/or unreliable water allocations.

1.7.2 Hydro-morphological impact

Hydro-morphological impact (Eleftheria *et al.*, 2004) is defined as a set of impacts on the land surface due to the creation of water infrastructures that change the direction, volume and hydrological activities of a flowing water resource. The morphological impact is also visible when modified river courses increase either erosion or deposition in its channel. A classic example of an adverse impact of constructing dykes on the embankment of a river can be given: The embankment on the Tista River has increased deposition on the river bed, thereby raising both the river bed height and the flood level. The river is now flowing at a height much above its original floodplain.

Over-exploitation of groundwater often triggers geomorphological changes like land subsidence. For example, 'in the United States, more than 17,000 square miles in 45 States, an area roughly the size of New Hampshire and Vermont combined, have been directly affected by subsidence' (USGS, 2016).

1.7.3 Quality impact

Industrial activities do change the water quality of the environment. In 2011, the Chinese government monitored water quality of rivers and found that 43% of rivers are so polluted that they are dangerous for human contact (Hays, 2013); needless to say, rivers so polluted have caused irreversible changes in the ecology in terms of species extinction. Contamination of surface water (and groundwater) by industrial effluents is a well-known killer as evident in the news of thousands of dead fish floating belly up in ponds, lakes and rivers.

1.7.4   Impact on the access to water by the stakeholders

Allocation of water resource as a whole is a point of serious concern for the community. The leasing out of rivers for industrial use deprives common people of their traditional water sources. In March 2007, the Chhattisgarh government was taken to court 'for allowing a private company to appropriate the waters of the Sheonath river' (Putul, 2012) because the people were not allowed access to the river for their daily water needs.

1.7.5   Affordability of water

Depletion of water resources has an adverse economic effect on the poor, especially on communities in water-stressed regions. The affordability issue develops into a socio-political conflict when the poor are compelled to search for alternative sources, which are either costlier or labour intensive. In the face of such hardship, the poor continue to grapple until their backs are against the wall, giving rise to conflicts.

1.8   Water sustainability

Water is an integral part of the environmental sustainability since it plays a vital role in the survival of nature, which includes not only our planet Earth but also the human enterprises that thrive on water sustainability. Whether the human enterprise is sustainable or not can be understood from certain indicators which are explained by US-EPA in *A Framework for Sustainability Indicators at EPA* (Fiksel, Eason and Fredrickson, 2014). Sustainable business enterprise rests on the harmonisation of three components: (1) The environment, as a source of water and other natural resources; (2) the society, which shares water and other natural resources with business enterprises, more often than not grudgingly because environmental degradation it causes; and finally, (3) business enterprise itself, which is profit-driven and draws equally from both environmental nature as well as human nature for its economic potential.

'Sustainability is based on a simple principle: Everything that we need for our survival and well-being depends, either directly or indirectly, on our natural environment. To pursue sustainability is to create and maintain the conditions under which humans and nature can exist in productive harmony to support present and future generations' (US-EPA, 2016b). US-EPA has proposed a taxonomic system for resource flow indicators, with

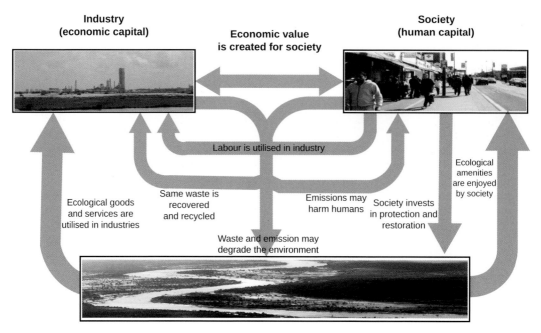

Figure 1.7 System taxonomy for resource flow indicators, with examples of specific metrics for material intensity, recovery and impact. Adapted from Fiksel, Eason and Frederickson (2010).

examples of specific metrics for material intensity, recovery and impact. See Figure 1.7.

The essential conditions for achieving a state of 'comfortable business' are continuous assessment of opportunities and risk, maintaining efficient water stewardship and compliance with all regulatory statutes. The risk-opportunity matrix in water sustainability is given in Table 1.4.

Sikdar (2003) has described sustainability as a function of (1) the economic aspect, (2) social aspect, and (3) environmental aspect. Where economic and social aspect are functioning then only socio-economic sustainability is achieved. See Figure 1.8.

Apart from economic, social and environmental aspects, two other aspects that are essential for consideration are the political and technological aspects. In several cases, especially in developing countries, political interventions account for multiple hindrances for maintaining smooth business operations. For example, a company has conducted a socio-economic survey in its area of operation and decided to extend a drinking water facility under a CSR scheme in selected villages. However, this move might not be politically approved, and political interventions may compel the corporation to modify its decision.

Technological advancement is often counted on for achieving sustainability through tools, data and instruments for resource estimation and environmental impact assessment. Remote-sensing data and advanced mapping tools can be of great help in finding suitable resources and estimating the capability of the resource to cater to the projected requirement. For example, (a) installation of meteorological observation systems on the company

Table 1.4 Risk and opportunity matrix in sustainable business.

Risk	Opportunities	Response
Deterioration of source-water quantity and quality	Re-creating business value of water	• Rainwater harvesting • Source-water protection measures
Rising cost of water	Investment in improved engineering and management solution in the value chain	• Water audit and water conservation measures • Reduction of water footprint
Change in water allocation status	Advocacy and lobbying with the authority	• Adapt to the new condition
More stringent government regulations	Improving water accounting and disclosure systems	• Engagement of experts • Training of staff
Stakeholders' resentment	Stakeholders engagement in water stewardship	• Promoting the wellbeing of the local workforce by providing better water and sanitation in the communities surrounding the factories (WSUP, 2015) • Engage employees to deliver on water goal

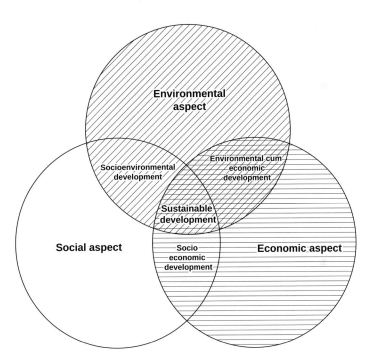

Figure 1.8 Three-dimensional pillars of sustainability. Adapted from Fiksel, Eason and Frederickson (2010).

premises will empower management with valuable weather data, and (b) an observation well in the company's area will help to monitor the fluctuation of groundwater level, which will further help them to decide groundwater withdrawal plans.

1.9   Impact of climate change

The US-EPA (2013) has defined climate change as 'substantial change in measures of climate (such as temperature or precipitation) lasting for an extended period (decades or longer).' Climate change is a naturally recurring phenomenon connected with variations in solar radiation. The earth has gone through several cycles of ice age followed by warm climate over the last couple of million years. This is known from shreds of geological evidence. The US-EPA definition of 'climate change' derives from alarm bells that scientists have been ringing since the 1980s; scientists believe that the climate change we are witnessing now has more to do with unregulated industrial activity than the periodicity of solar activity.

The present-day climate change is manifested in several phenomena, such as a rise in average global temperature, sea level rise, warming of oceans, retreating glaciers and shrinkage of ice caps. Extreme climatic events like drought and flood are on the rise because of global warming. The print and electronic media and, of course the internet, is rife with 'pop science' articles on global warming caused by greenhouse gases, notably CO_2. Scientists estimated that since the Industrial Revolution, atmospheric CO_2 has gone up by 40% from 280 ppm in the mid-18th century to 407 ppm as of 2016. The present concentration is estimated to be the highest in the last 800,000 years. The CO_2 increase is largely due to the burning of fossil fuels and deforestation. There are scientists who are critical of the alarmist approach to global warming, which they consider lead to hasty conclusions founded upon inadequate or misinterpreted data. But the majority believes, me among them, that it is better to err on the side of caution and take preventive steps. Blessed is he who sits on a pin for he shall rise again—a little moderation in programmed power consumption worldwide will go a long way in reversing the anthropogenic causes of global warming. Consider, for instance, the millions and millions of marginal people inhabiting coastal regions and low-lying countries (e.g. Bangladesh) who live under the threat of being rendered homeless by inundation because the sea level is rising at the rate of 3.38 mm per year (NASA, 2016). They may not drown after all, but what good can scientific integrity, critical of alarmists, do to relieve them of the anxiety of drowning?

The world climate is a chaotic phenomenon; that is, it is delicately balanced on the interdependence of multiple variables, all of which are so far beyond accurate measurement-capacity of present-day technology that the climate is extremely unlikely to become a deterministic phenomenon in the foreseeable future. What science does conclude at its present level of understanding is that even local problems can have ripple effects and cause catastrophic changes in the hydrological cycle, which in turn affect the monsoon cycle. And, adverse effects on the monsoon cycle can wreak havoc on water-dependent economies. Indeed, Mr Pranab Mukherji, the former Finance Minister (and now the President) of India, once said in his budget speech in Parliament that 'the monsoon is the real Finance Minister of India.' That statement sums up the scale of the problem water shortage can cause. Suffice it to say that any water crisis is creating problems within and between countries, and these problems will get deeper by the day if not brought under the umbrella of water stewardship.

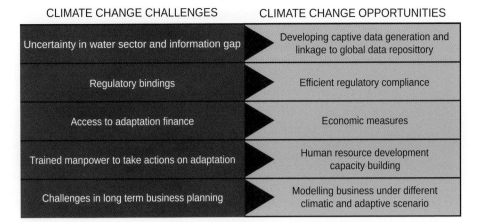

Figure 1.9 Climate adaptation framework for business. Adapted from United Nations (2015).

Industries are now aware that any water crisis cannot be mitigated without addressing climate change. 'Just as business operations are important, it is imperative to consider the impact of climate change on business operations, and potential impacts on sustainability', says ICIMOD (2016). Countries under the framework of the United Nations discussed water crisis in a convention on climate change in 1997. In that convention, an agreement named the Kyoto Protocol (United Nations, 2014a) was signed by the parties to agree on internationally binding emission-reduction targets. This protocol entered into force in 2005. This protocol is mostly binding on the industries.

If the climate change impact is not properly understood, business will face challenges as never before. The major challenges they will face are uncertainty in water availability and extreme climatic conditions like flood and drought. It is the right time for industries to address climate change in their business policy and business models.

The impact of climate change and the binding rules imposed by different regulatory authorities have a direct impact on business and have the potential for managing both challenges and opportunities. Corporate adaptation to climate change can make business more sustainable and socially responsible (Four Twenty Seven, 2012). Responsible corporate adaptation to climate change is a function of several parameters in the business environment. Climate adaptation framework for business is shown in Figure 1.9.

1.10   Dimensions in industrial water management

The famous remark of Mahatma Gandhi bears repeating in this context: 'The world has enough for everyone's need, but not enough for everyone's greed' because the quest for sustainable development rests on the nexus between government and big business promoting competitive consumerism, and the nexus rests fundamentally on the frenetic pitch of media advertisements. Manifestly, the combined wisdom of governments and big business believe in brinkmanship. The nexus works on the principle of blurring the difference between credit society and inclusive economy. Nothing illustrates the ills of advertisement-promoted consumerism like the phenomenal growth in the sale of, and, misuse of android phones among rural youth. What we are witnessing now is a paradigm shift from the invisible hand to water conservation, gradually tightening its grip on economic growth, a shift balanced on the introduction of water stewardship. Water stewardship is not about crippling economic growth by cutting off the water supply to industries; rather, it is about harmonising all end users of water across economic boundaries. As things stand now, it is more or less a rule of thumb that rich countries will reduce water stress by technological intervention without remedying underlying causes (Vörösmarty *et al.*, 2010), and, technologically deficient poor countries will continue to suffer. It is an extreme form of the false science of Social Darwinism because it justifies denying the poor (80% of the world population) of something as fundamental as water. Water stewardship is a multidisciplinary science that aims to correct the disparity in water availability.

The paradigm shift in water withdrawal from pro-growth to pro-ecology is taking only baby steps due to the constant efforts of pressure group researchers and activists. Pioneers of the pro-ecology movement fought for reform because they believed that the destruction of forests, fertile soil, wildlife and water resources would lead to the downfall of society (Chapman, 2010). During the 1970s, the face of environmentalism shifted to civil action with special-tactical groups forcing environmental issues onto the national agenda. After several global conferences on the environment and three major environmental summits, protection of water resources, maintaining the quality of runoff and protection of soil quality became major conservation issues.

Although the environmental movement started with the participation of focus groups from the educated middle class during the 1980s, local environmental groups draw their members from a broad cross section of class and occupational categories (Freudenberg and Steinsapir, 1991). Not only in America but also in the developing countries, movements against water pollution and environmental degradation were launched by heterogeneous organisations comprising activists, science clubs and, most importantly, local people.

Another major driver of the environmental movement was the blossoming of *Not in My Backyard* or the NIMBY campaign (Freudenberg and Steinsapir, 1991). Industries became the major target of environmental activism. The case against the Sundarban Fertiliser Company is an example of an environmental movement initiated by common people. The movement started in the mid-1980s before the establishment of industries in a thickly populated village in South 24 Parganas district of West Bengal (Dutta Banik, Basu and De, 2007). The movement was organised to protest against the construction of the proposed factory, stating that it was (1) close to thickly populated villages, (2) impinging on agricultural land, (3) close to an important waterway in the integrated network of Sundarban, and (4) within a mile of a deep tube well, which was the main source of drinking water for the entire area. The business community must recognise that resources must be conserved with utmost care and business expansion should be restricted to the extent of ecological limit (WBCSD, 2010).

Gleick (2000) has proposed 'two approaches' in water management, which include an increase in efficiency, both to meet the current needs as well as to develop the water allocation system. This may demand new sources of water supply, which can be met by innovative approaches and small projects in water sourcing. Best results will come from capacity-building in the corporate sector in terms of internal water management and source-water management. Both hydrological and engineering knowledge-bases are necessary to meet the requirements for efficient water management.

The major focus areas in corporate water management are summarised below.

1.10.1   Global perspective

There are companies which have factories and production centres distributed across many countries. Moreover, in their supply chain, they procure products from many countries. These industries must recognise that the impact of their activities on water resource is not a local issue anymore. It has to be addressed in a global perspective as well as with mitigation programmes at every point of the operations. Assessing local and global water situations along with political and regulatory issues is very important for multinational companies. This will help them to take informed decisions on how and where interventions are required to improve the machinery, management, relations with stakeholders and community, and vision towards corporate water stewardship.

1.10.2   Water accounting

This is one of the major tasks the industries should perform periodically. In developed countries, water accounting has been mandated through several business codes of conduct and methodologies. Water-footprint study will also help industries in assessing their impact on global and local water situations, both quantitatively and qualitatively. In developing countries, the concept is just emerging. For example, in India some of the major companies like Lafarge, Tata Motors, Tata Chemicals and extracting industries like Essar have started water audit and water footprint study. The minister of railways, India, has ordered water audits to be performed in all railway stations (*Deccan Chronicle*, 18 November 2014). Through water audit within the facilities and ancillary infrastructure, the risks of water loss can be assessed. An industry can benchmark its water use for processing, cooling and housekeeping. The water footprint is a tool to assess the amount of real and virtual water use in the supply chain, product life cycle and the environmental impact of discharge in terms of grey-water footprint.

1.10.3   Water stewardship

Corporate water stewardship is an approach of the companies to identify and manage water-related business risks especially, in the context of shared water. Protecting water resources in a sustainable way comes under the purview of corporate stewardship. In a water-intensive industry when public water governance cannot ensure water security, corporate stewardship can play a big role in achieving the water sustainability of the locality with the participation of stakeholders. Throughout the supply and production chain, water plays a pivotal role. Thus, conservative, judicious and time-specific actions are crucial to successful businesses, and so they have to set specific goals to eliminate, as far as possible, water wastage by modernisation of machinery, capacity building among staff and the setting of targets for water-related CSR activities, all of which are to be periodically checked by success indicators to determine the progress of water management.

1.10.4   Adaptive management

Adaptive management of water is a new paradigm that refers to a complete system of water management which includes all interconnected systems (Pahl-Wostl *et al.*, 2007) like water withdrawal, purification, allocation, licence, consumption, discharge and so on. It fulfils a complete set of social and economic functions. Adaptive management also includes social behaviour, legal requirements and institutional arrangements. It deals with both the organised sector and the unorganised sector. Adaptive management of water use is an inclusive management that covers three components related to water. These are (1) human components that actually pursue water-use solutions through engineered and social methods, (2) physical components that regulate water availability in time and space through the water cycle, and (3) biological and geochemical components. According to Pahl-Wostl (2006), the idea of adaptive water management refers to a systematic improvement of water management, policies and practices by learning from the experience gathered from the present system of management.

1.11   Green growth and green business

The term *green growth* was adopted in the *G20 Seoul Summit Leaders' Declaration* (2010). Green growth is a strategy of quality development, enabling countries to leapfrog old technologies. International cooperation in technology transfer and capacity building were also among the major issues, discussed in the summit.

The term *green growth* is used in the policy statement of a government or a business to describe a path of economic growth that uses natural resources in a sustainable manner. The term is used globally to provide an alternative to economic growth through ecologically destructive industries. Different global-level organisations have defined green growth in different ways, but all of them have advocated three basic points: (1) it is a policy, (2) it aims at the well-being and improvement of humankind coupled with steady economic growth, and (3) it fosters social equity.

The concept of green growth is applicable to every sphere of human activity starting from a single individual to the level of the national government, in which the role of the former can hardly be overemphasised given that the world population is in excess six billion people. Needless to mention, domestic water wastage is, in most cases, an act of the privileged population. For example, reused water draining out from the kitchen sink or an unplugged bathtub used for gardening is, in a microcosm, a green growth model. On a larger scale, green growth policy requires big businesses to adopt production methods that do not affect the needs of marginal people and protects ecologically sensitive locations. According to the usual scenario, green growth, at this stage of economic growth and business, has to face many challenges.

1.11.1   The challenges of green growth

The most impacting challenges of green growth are discussed below.

Fund

Retrofitting existing technology for improved water and energy conservation needs a good amount of capital investment. A successful application of policy must pay it back within a reasonable time.

Government policy

Government support to green growth should be based on *suo moto* policy framing (as opposed to being a consequence of public agitation). It may well require constitutional amendments in a country. For example, Ecuador has included in its constitution a clause establishing the 'Right of Nature' which is a radical improvement over regulatory laws and empowers green growth drives (Global Alliance for the Rights of Nature, 2016).

Business policy

Corporate sectors should also include green growth proactively in their business policies and involve their employees, supply chain and stakeholders to subscribe to those policies.

Natural capital

Water is one of the major natural capitals of business. Conservation of natural capital is a major challenge in business intended for green growth.

Climate

Climate change or climate impact may have serious effects on water resource and hence, on business.

Green growth is a concept that embraces stakeholders of almost every sphere of the society. The government will act as the maker of policy and law, and remain vigilant about the implementation of the laws. The potential risks in implementing green growth policy are inefficiency, corruption and poor accountability at government level. This may lead to excessive environmental degradation in many countries (Barbier, 2011).

1.11.2   Natural capital concept

According to Dasgupta (2007), the ecosystem, like road, bridge and other infrastructures, is to be viewed as a capital asset but it differs in three ways: (1) depreciation of natural capital is frequently irreversible (or at best the systems take a long time to recover); (2) except in a very limited sense, it is not possible to replace a depleted or degraded ecosystem by a new one; and (3) ecosystems can collapse abruptly, without much prior warning. Mining of minerals, oil and so on is an example of depreciation of natural capital.

Increasing costs associated with diminishing natural capital assets are not reflected in the market. The price of any product, whether it is coming from a managed or degraded ecosystem does not carry any indication of the environmental impact caused by its production. The idea is to give the brand's conscious consumer, a choice. If a green product (i.e. where the manufacturing process is eco-friendly) is labelled suitably to indicate its eco-sensitive manufacturing as brand value, consumer preference will directly incentivise green growth.

1.11.3   Green growth policy fundamentals

A green growth policy of a country must ensure:

1 Strong and efficient governance that considers the ecosystem as natural capital and remains vigilant that the natural capital is exploited in such a way that environmental recovery is amenable to natural processes.
2 Every developmental project of the corporate sector must recognise that the impact of development is not local; it affects the ecosystem as a whole. In addition to local effects, business planning must act as a crucial factor in effects on the river basin as a whole.
3 Economic factors should be considered in a manner that accommodates the full spectrum of interests (*Water and green growth*, 2012).
4 Active public participation in water stewardship and ecosystem restoration.
5 Ensure availability of potable water for every member, as a part of their fundamental rights. See Figure 1.10.

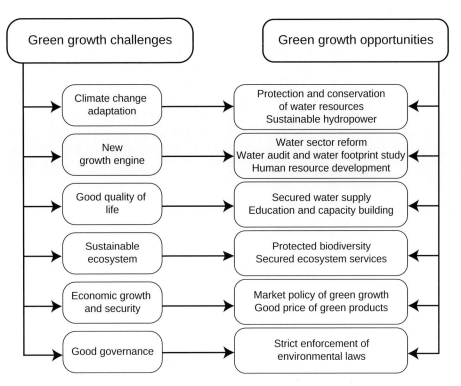

Figure 1.10 Water and green growth. Adapted from K-water Institute and World Water Council (2015).

1.11.4 Indicators of green growth

Researchers in Korea have developed a set of indicators—Environmental Sustainability Indicator (ESI), Environment Performance Indicator (EPI) and Environmental Vulnerability Index (EVI)—which they have termed the Water and Green Growth Index (WGGI). It includes different dimensions of green growth in water management (Kang *et al.*, 2012).

The OECD (2011) has identified four groups of indicators: (1) environmental and resource productivity indicators that monitor the use of natural products in the value chain, (2) natural asset-base indicators that monitor how the environment is functioning as environmental service provider and as a resource pool, (3) environmental quality of life indicators, and (4) economic opportunities and policy response indicators, which include policies, measures, opportunities and the function of corporations in investment, innovation, trade, training and so on.

1.12 Conclusion

Water is now a precious raw material for industries, especially in power generation, which demonstrates, with greater emphasis, the fact that water has a price and an ethical value. Industrial growth suffers if the environmental and societal values of water are ignored. Sustainable industrial development is now linked with IWRM and is crucial to green growth. Effective application of green growth, in turn, is crucially dependent on water literacy. It is, therefore, incumbent upon the corporate sector to adopt IWRM and spread water literacy among the management and stakeholders for its own wellbeing.

Challenges notwithstanding, green growth is an emerging and inescapable reality for the good of mankind. Powerful agencies of diverse kinds, such as reputable thinkers, scientists, engineers, elected governments and their opposition, are all aware of the economic returns of green growth, even though the idea of green growth is only in a nascent state of application today. These agencies are also well aware of and in agreement with those long-term economic returns of green growth—far in excess of beneficial effects visible today. The only spanner in the works for green growth is the corporate sector because of its obsession with short-term gains. The blindness of the corporate sector, compounded by clout, to the enormous growth opportunity embedded in green growth substantiates the two-and-a-half century old idea of the invisible hand which still remains valid. This is perhaps just as well because nothing will kill the idea of invisible hand more effectively than the idea of green growth.

Note

1. In India groundwater is assessed under groundwater-assessment units of land, which are equivalent to lower-level administrative units, called blocks. *Safe block* means groundwater development/withdrawal is not more than 70% of annual renewable groundwater resource.

Bibliography

Aghababyan, N. (2006) *The global water crisis*. Available at http://puredrinkablewater.weebly.com/the-global-water-crisis.html (accessed: 9 January 2016).

Aiyar, P. (2007) 'Water woes', *Frontline*, 24(12).

Bansal, S. (2014) National river linking project: Dream or disaster? Available at: http://www.indiawaterportal.org/articles/national-river-linking-project-dream-or-disaster (accessed: 9 January 2016).

Barbier, E. (2011) 'The policy challenges for green economy and sustainable economic development', *Natural resources forum*, 35(3): 233–245.doi: 10.1111/j.1477-8947.2011.01397.x.

Bloech, H. (2010) *The water framework directive: an overview*. Available at http://www.wecf.eu/download/2010/02/JournalEnvLawREVSep2009.pdf (accessed: 22 October 2016).

Byers, W., Lindgren, G., Noling, C. and Peters, D. (2002) *Industrial water management: a systems approach*. United States: American Institute of Chemical Engineers.

CDP (2014) *From water risk to value creation*. Available at https://www.cdp.net/CDPResults/CDP-Global-Water-Report-2014.pdf (accessed: 21 January 2016).

Chapman, R. (2010) *Culture wars : an encyclopedia of issues, viewpoints, and voices*. Armonk, NY: M.E. Sharpe.

Coca-Cola India (no date) *About Hindustan Coca-Cola's Varanasi plant.* Available at https://www.coca-colaindia.com/facts-hindustan-coca-colas-plant-operations-mehandiganj-varanasi-uttar-pradesh-india/(accessed: 19 March 2017).

COMEST (2005) *Best ethical practices in water use.* Available at http://unesdoc.unesco.org/images/0013/001344/134430e.pdf (accessed: 9 January 2016).

Dasgupta, P. (2007) 'Nature in economics', *Environmental and resource economics*, 39(1): 1–7. doi: 10.1007/s10640-007-9178-4.

Davies, E.G.R.G.R., Akhtar, M.K., Mcbean, G.A. and Simonovic, S.P. (2009) 'Integrated assessment, water resources, and science policy communication', 2nd Climate Change Technology Conference: ResearchGate. Available at https://www.researchgate.net/publication/241064598_INTEGRATED_ASSESSMENT_WATER_RESOURCES_AND_SCIENCE_POLICY_COMMUNICATION (accessed: 26 April 2016).

Dutta Banik, S., Basu, S.K. and De, A.K. (2007) *Environment concerns and perspectives.* APH Publishing Corporation.

The Ecologist (2014) *India: Coca cola bottling plant shut down.* Available at http://www.theecologist.org/News/news_round_up/2444501/india_coca_cola_bottling_plant_shut_down.html (accessed: 9 January 2016).

Economy, E. (2013) *China's environmental governance crisis.* Available at http://www.cecc.gov/sites/chinacommission.house.gov/files/documents/hearings/2013/CECC%20Hearing%20-%20Food%20Safety%20-%20Elizabeth%20Economy%20Written%20Statement.pdf (accessed: 17 June 2016).

Eleftheria, K., Wenke, H., Kampa, E. and Hansen, W. (2004) *Heavily modified water bodies: synthesis of 34 case studies in Europe.* Germany: Springer-Verlag GmbH & Co., Berlin and Heidelberg.

US-EPA (2016a) *Learn about sustainability.* Available at http://www.epa.gov/sustainability/learn-about-sustainability#what (accessed: 9 January 2016).

US-EPA (2016b) *Water-energy connection.* Available at https://www3.epa.gov/region9/waterinfrastructure/waterenergy.html (accessed: 16 June 2016).

US-EPA (2013) *Causes of climate change.* Available at https://www3.epa.gov/climatechange/science/causes.html (accessed: 27 April 2016).

Falkenmark, M., Lundquist, J. and Widstrand, C. (1989) 'Macro-scale Water Scarcity Requires Micro-scale Approaches: Aspects of Vulnerability in Semi-arid Development', *AGRIS.*, 13(4): 258–267.

FAO (2008) '5. Biofuels: Prospects, Risks and Opportunities', in *The state of food and agriculture, biofuels: prospects, risks and opportunities.* FAO, p. 64.

Fiksel, J., Eason, T. and Fredrickson, H. (2014) *A framework for sustainability indicators at EPA., U.S. Environmental Protection Agency, Washington, DC., EPA/600/R/12/687, 2013.* Available at http://cfpub.epa.gov/si/si_public_record_report.cfm?dirEntryId=254270 (accessed: 9 January 2016).

Four Twenty Seven (2012) *The business case for responsible corporate adaptation—four twenty seven.* Available at http://427mt.com/resources/responsible-corporate-adaptation (accessed: 9 January 2016).

Freudenberg, N. and Steinsapir, C. (1991) 'Not in our backyards: The grassroots environmental movement', *Society & Natural Resources*, 4(3): 235–245. doi: 10.1080/08941929109380757.

Gardner-Outlaw, T. and Engelman, R. (no date) *Sustaining water, easing scarcity: a second update revised data for the population action international report, sustaining water: Population and the future of renewable water supplies.* Available at http://pai.org/wp-content/uploads/2012/01/Sustaining_Water_Easing_Scarcity_-_Full_Report.pdf (accessed: 9 January 2016).

G20 Seoul Summit Leaders' Declaration (2010) Available at http://online.wsj.com/public/resources/documents/G20COMMUN1110.pdf (accessed: 9 January 2016).

Gleick, P.H. (2000) 'The Changing Water Paradigm: A Look at Twenty-first Century Water Resources Development', *Water international*, 25(1): 127–138.

Global Alliance for the Rights of Nature (2016) *Ecuador adopts rights of nature in constitution*. Available at http://therightsofnature.org/ecuador-rights/(accessed: 9 January 2016).

Groenfeldt, D. (2013) *Corporate water ethics archives: notes for a new water culture*. Available at http://blog.waterculture.org/?cat=9 (accessed: 9 January 2016).

Hassan, F. (2011) *Water history for our times*. Available at http://unesdoc.unesco.org/images/0021/002108/210879e.pdf (accessed: 9 January 2016).

Hays, J. (2013) *Water pollution in China*. Available at http://factsanddetails.com/china/cat10/sub66/item391.html (accessed: 6 May 2016).

ICIMOD (2016) *Climate change and business*. Available at http://www.icimod.org/?q=3488 (accessed: 12 October 2016).

IEA (2014) *World energy outlook 2014*. Available at http://www.iea.org/publications/freepublications/publication/WEO2014.pdf (accessed 21 March 2017)

India Resource Centre (2014) *IRC– coca-cola – Mehdiganj – the issues*. Available at http://www.indiaresource.org/campaigns/coke/2013/mehdiganjfact.html (accessed: 23 December 2015).

Integrated water resources management (2015) in *Wikipedia*. Available at https://en.wikipedia.org/wiki/Integrated_water_resources_management (accessed: 9 April 2016).

International Energy Agency (2014) *Water for energy is energy becoming a thirstier resource?* Available at http://www.worldenergyoutlook.org/media/weowebsite/2012/WEO_2012_Water_Excerpt.pdf (accessed: 9 January 2016).

Jennings, B., Heltne, P. and Kintzele, K. (2009) 'Principles of water ethics', *Minding Nature*, 2(2): 25–29.

K-water Institute and World Water Council (2015) *Water and green growth, beyond the theory for sustainable future*. Available at http://www.worldwatercouncil.org/fileadmin/world_water_council/documents/publications/forum_documents/Water_and_Green_Growth_vol_1.pdf (accessed: 10 January 2016).

Kang, S., Lee, S., Kang., B., Park, H., Kang, H., Kim, C., Choi, D., Park, S., Noh, H.J., Rho, P., Lee, D.-R., Lee, J.H., Chae, H., Cho, Y., Kim, D., Lee, G.-C., Lee J., S., Lim, J.L., Lim, J., Lim, K.-S., Park, J. hyeog, Park, J.-E., Ryu, M.H., Seo Y., I.S., Shin, e.-S., Clench, C. and Brewster, M.M. (2012) *Water and green growth*. Available at http://www.worldwatercouncil.org/fileadmin/wwc/Library/Publications_and_reports/2.Green_Growth_Report_Edition1.pdf (accessed: 9 January 2016).

KPMG (2012) *A new perspective on the changing business environment*. Available at http://www.kpmg.com/nz/en/issuesandinsights/articlespublications/pages/new-perspective-business.aspx (accessed: 26 April 2016).

Kreitler, J., Schloss, C.A., Soong, O., Hannah, L. and Davis, F.W. (2015) 'Conservation planning for offsetting the impacts of development: A case study of Biodiversity and renewable energy in the Mojave desert', *PLOS ONE.*, 10(11), p. e0140226. doi: 10.1371/journal.pone.0140226.

Lassonde, P. (2014) *The social licence to operate*. Available at http://socialicence.com/index.html (accessed: 9 April 2016).

Liu, J., Dorjderem, A., Fu, J., Lei, X. and Macer, D. (2011) *Water ethics and water resource management*. Available at http://philpapers.org/rec/LIUWEA (accessed: 9 January 2016).

Luo, T., Young, R.S. and Reig, P. (2016) *Aqueduct projected water stress country rankings*. Available at: http://www.wri.org/publication/aqueduct-projected-water-stress-country-rankings (accessed: 9 April 2016).

Mary, V. (2016) *A watershed approach for sustainable ecosystem management of Meenachil River basin with emphasis on remote sensing and GIS*. PhD Thesis. Available at http://shodhganga.inflibnet.ac.in:8080/jspui/bitstream/10603/73771/19/9_introduction.pdf (accessed: 26 April 2016).

McKinsey (2009) *The business opportunity in water conservation.* Available at http://www.mckinsey.com/business-functions/sustainability-and-resource-productivity/our-insights/the-business-opportunity-in-water-conservation (accessed: 27 April 2016).

Morgan, A.J. and Orr, S. (2015) *The value of water.* Available at http://d2ouvy59p0dg6k.cloudfront.net/downloads/the_value_of_water_discussion_draft_final_august_2015.pdf (accessed: 23 January 2016).

NASA (2016) *Global climate change.* Available at http://climate.nasa.gov/(accessed: 9 April 2016).

Odongo, B.J. (2014b) *Assessing water supply and demand management in industries and commercial enterprises in Athi River town, Machakos County.* Available at http://ir-library.ku.ac.ke/bitstream/handle/123456789/12805/Assessing%20water%20supply%20and%20demand%20management%20in%20industries.....pdf?sequence=1&isAllowed=y (accessed: 9 April 2016).

OECD (2011) 'The measurement framework', *towards green growth: monitoring progress,* 17–30. doi: 10.1787/9789264111356-5-en.

Otto, B. and Böhm, S. (2006) '"The people" and resistance against international business', *Critical perspectives on international business,* 2(4): 299–320.doi: 10.1108/17422040610706631.

Pahl-Wostl, C. (2006) 'Transitions towards adaptive management of water facing climate and global change', *Water resources management,* 21(1): 49–62. doi: 10.1007/s11269-006-9040-4.

Pahl-Wostl, C., Sendzimir, J., Jeffrey, P., Aerts, J., Berkamp, G. and Cross, K. (2007) 'Managing Change toward Adaptive Water Management through Social Learning', *Ecology and society,* 12(122).

Porter, M.E. and Kramer, M.R. (2011) *Creating shared value.* Available at https://hbr.org/2011/01/the-big-idea-creating-shared-value (accessed: 9 April 2016).

Putul, A.P. (2012) *Privatisation unlimited: rivers for sale in Chhattisgarh.* Available at http://infochangeindia.org/index.php?option=com_content&view=article&id=6943:privatisation-unlimited-rivers-for-sale-in-chhattisgarh-&catid=142:analysis&Itemid=49 (accessed: 17 March 2017).

Sarni, W. (2015) *Water: Turning a value-chain risk into an ecosystem opportunity.* Available at http://www.greenbiz.com/article/water-turning-value-chain-risk-ecosystem-opportunity (accessed: 9 January 2016).

Savenije, H.H.G., Hoekstra, A.Y. and van der Zaag, P. (2014) 'Evolving water science in the Anthropocene', *Hydrology and earth system sciences,* 18(1): 319–332.doi: 10.5194/hess-18-319-2014.

Seckler, D., Amarasinghe, U., Molden, D., De Silva, R. and Barker, R. (1990) 'World water demand and supply, 1990 to 2025: Scenarios and issues', *Research report 19.* Colombo, Sri Lanka: International Water Management Institute.

Sengupta, P. (2014) *Water crisis - closure of coca-cola factories in India.* Available at http://pradip-watercrisis.blogspot.in/2014/06/closure-of-cocacola-factories-in-india.html (accessed: 16 June 2016).

Shiklomanov, I.A. (2000) 'Appraisal and Assessment of World Water Resources', *Water int.* 25, 11–32. doi:10.1080/02508060008686794.

Sikdar, S.K. (2003) 'Sustainable development and sustainability metrics. AIChE J', *AIChE Journal,* 49(8): 1928–1932. doi: http//dx..org/10.1002/aic.690490802.

Sullivan, C.A., Meigh, J.R. and Fediew, T.S. (2002) *Derivation and testing of the water poverty index phase 1.* Available at http://nora.nerc.ac.uk/503246/1/WaterPovertyIndex_Phase1_2002_Final%20Report.pdf (accessed: 9 January 2016).

Torcellini, P., Long, N. and Judkoff, R. (2003) *Consumptive water use for U.S. Power production.* Available at http://www.nrel.gov/docs/fy04osti/33905.pdf (accessed: 9 January 2016).

The Transnational Institute (2015) *Jakarta court cancels world's biggest water privatisation after 18 year failure.* Available at https://www.tni.org/en/pressrelease/

jakarta-court-cancels-worlds-biggest-water-privatisation-after-18-year-failure (accessed: 17 June 2016).

UNESCO (1999) *UNESCO natural sciences portal*. Available at http://www.unesco.org/science/comest.shtml (accessed: 26 April 2016).

US Department of Energy (2014) *The water-energy nexus: challenges and opportunities overview and summary*. Available at http://www.energy.gov/sites/prod/files/2014/06/f16/Water%20Energy%20Nexus%20Report%20June%202014.pdf (accessed: 9 April 2016).

UNDP and Human Development Report Office (2006) *Human development report 2006*. Available at http://hdr.undp.org/sites/default/files/reports/267/hdr06-complete.pdf (accessed: 12 January 2016).

United Nations (1992b) *The Dublin statement on water and sustainable development*. Available at http://www.un-documents.net/h2o-dub.htm (accessed: 23 January 2016).

United Nations (1992a) *United Nations conference on environment & development*. Available at https://sustainabledevelopment.un.org/content/documents/Agenda21.pdf (accessed: 9 January 2016).

United Nations (2014a) *Kyoto protocol*. Available at http://unfccc.int/kyoto_protocol/items/2830.php (accessed: 9 January 2016).

United Nations (2014b) *Sustainable development financing*. Available at http://www.un.org/esa/ffd/wp-content/uploads/2014/10/ICESDF.pdf (accessed: 9 April 2016).

United Nations (2015) *The business case for responsible corporate adaptation: strengthening private sector and community resilience*. Available at http://427mt.com/wp-content/uploads/2015/12/Caring-For-Climate-Business-Case-Responsible-Corporate-Adaptation-2015.pdf (accessed: 9 January 2016).

USGS (2016) Land Subsidence in the United States. Available at https://water.usgs.gov/ogw/pubs/fs00165/ (accessed 31 October 2016)

Vörösmarty, C.J., McIntyre, P.B., Gessner, M.O., Dudgeon, D., Prusevich, A., Green, P., Glidden, S., Bunn, S.E., Sullivan, C.A., Liermann, C.R. and Davies, P.M. (2010) 'Global threats to human water security and river biodiversity', *Nature*, 467(7315): 555–561.doi: 10.1038/nature09440.

Wabunoha, R.A. (no date) *Legal aspects in calculating environmental damage and compensation*. Available at http://www.unep.org/delc/Portals/119/events/compesantion-environmental-damage.pdf (accessed: 9 April 2016).

Walton, B. (2016) *Conga mine in Peru halted by water concerns, civic opposition*. Available at http://www.circleofblue.org/2016/south-america/conga-mine-peru-halted-water-concerns-civic-opposition/(accessed: 17 June 2016).

Water and green growth (2012) Available at http://www.worldwatercouncil.org/fileadmin/wwc/Library/Publications_and_reports/2.Green_Growth_Report_Edition1.pdf (accessed: 9 January 2016).

WBCSD (2013) *Business guide to water valuation*. Geneva, Switzerland: WBCSD.

WBCSD (2010) *Water for business, initiatives guiding sustainable water management in the private sector*. Available at http://www.bcsd.org.tw/sites/default/files/node/domain_tool/678.file.2161.pdf (accessed: 9 January 2016).

WSUP (2015) *Creating business value and development impact in the WASH sector*. Available at http://www.wsup.com/2015/05/29/creating-business-value-and-development-impact-in-the-wash-sector/(accessed: 27 April 2016).

WWAP (2015) *The United Nations world water development report 2015: water for a sustainable world; 2015*. Available at http://unesdoc.unesco.org/images/0023/002318/231823E.pdf (accessed: 19 January 2016).

White, C. (2012) *Understanding water scarcity: definitions and measurements*. Available at http://www.iwmi.cgiar.org/News_Room/pdf/Understanding_water_scarcity.pdf (accessed: 9 April 2016).

2

Water Scenarios and Business Models of The Twenty-first Century

2.1 Water scenario

Industries all over the world are facing severe risks in the water sector because of over-exploitation of resources, climate changes, social resistance and water pollution. Industries as well as governments are trying to develop strategies to manage water efficiently. To develop an efficient water strategy, the emerging trends regarding water use, water risks, water impacts and water management must be reviewed. This will enable us to address the question as to how businesses will cope with the emerging water crisis, and what should be the business model in different scenarios from the growth perspective. We cite the research of Shiklomanov (2000), who has studied the changing demand for water through ages. According to him there was a remarkable change in the nature of demand during the end of the 20th century. The volume of freshwater withdrawal in the beginning of 20th century was 580 km^3/year and at the turn of the century it became as large as 3,973km^3/year.

According to Gleik (2000), the water management paradigm of the 20th century (which was mainly driven by the demand of economic development) was subsequently challenged by emerging social norms, environmental ethics and political interests. To cope with these changes, water resource assessment became an important activity.

Each country has its own water assessment policies and methods, and they generate their own data and reports regarding the status of available water and sectoral water withdrawal. There are international organisations like the Food and Agriculture Organisation (FAO), UNESCO and the World Resource Organisation (WRI) that act as data repositories and compile and

Industrial Water Resource Management: Challenges and Opportunities for Corporate Water Stewardship, First Edition. Pradip K. Sengupta.

assess the countrywide status of the water resource. The factors considered in such assessments are:

1 The internal water resources that are generated from the precipitation within the territory of the country.
2 The external inflow of water that comes from upstream neighbouring countries via trans-boundary river basins, and is added to the internal resource.
3 The water that flows into other downstream countries, i.e. the external outflow which is subtracted from the total resource of the country.

Water management of a country broadly depends on the estimation of natural renewable water resources, actual renewable water resources, exploitable water resources, and non-renewable water resources. See Figure 2.1.

- **Natural renewable water resources:** Natural renewable water resources of a country are the sum total of its internal renewable resources. These include both groundwater and surface water, and the external resources it receives through trans-boundary inflow.
- **Exploitable water resources:** These also called available water resources. The natural renewable water resources estimated for a 'water year' are not allowed to be exploited *completely* through human consumption like drinking, irrigation or industrial use. A part of the resources is kept reserved for the environmental flow and allocation to downstream countries. The quantity of exploitable water resources is always kept less than the natural renewable water resources.
- **Non-renewable water resources:** These constitute resources that are not renewable within human time-scale (FAO, 2003); for example, groundwater in the very deep aquifers of the Sundarbans, which is virtually disconnected from the water cycle and therefore is not recharged by present-day rainfall. These aquifers, if exploited, may go dry.

Water allocation for upstream and downstream countries is maintained and regulated by international agreements and treaties. For example, the

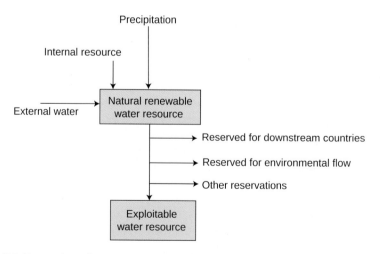

Figure 2.1 Dynamics of water resources of a country.

Orange River basin is shared by Botswana, Namibia, South Africa and Lesotho; the water allocation is governed by Orange Senque River Commission (ORASECOM, 2006).

2.1.1   Countrywise water scenario

The water scenario of some selected countries is reviewed and compared based on the long-term availability of water resources vis-à-vis per capita availability data in 1995 and 2014. The Central Intelligence Agency (CIA, no date) has published a list of countries and their annual renewable water resources based on the data provided by FAO (FAO.org). According to the CIA, 'this entry provides the long-term average water availability for a country in cubic kilometres of precipitation, recharged groundwater, and surface inflows from surrounding countries.' This assessment excludes water resource totals that have been reserved for upstream or downstream countries through international agreements. These values are averages, and they may vary from year to year due to short-term and long-term climatic and weather variations. A map showing the country wide availability of total renewable water resource is given in Figure 2.2.

A constant stress on the available water resources of different countries is increasing at a very high rate, as a consequence of population growth, which has led to a reduction in per capita water availability.

A recent study by the author with forty-eight selected countries based on countrywide available water resources and population growth reveals that the per capita water availability in 1995 and 2014 has been reduced in all countries except for Belarus, Russia, Albania, Estonia, Georgia, Lithuania, Latvia and Moldova. In African countries like Chad, Gambia, Nigeria and Sudan per capita water availability has reduced by almost 50% within twenty years. In Table 2.1 a list of countries which have suffered reduction of per capita water availability by more than 30% is given.

The most important part of water management is to analyse the trend of water withdrawal and its use in different sectors. Considering population growth rate, industrial advancement and growing agricultural demand, Shikomanav (2000) analysed sectorwise water withdrawal and consumption from 1990 to 2025. See Table 2.2

From the above table, the percentage of water consumption against withdrawal is computed for each sector. See Table 2.3.

Figure 2.3 shows that the gap between water withdrawal and consumption has a narrowing trend in agricultural and industrial sectors whereas the trend of municipal consumption of water against withdrawal shows that grey water footprint in municipal sector is increasing. In this graph, increased percentage of water use reflects water efficiency.

The water management scenario across various borders of the world has evolved over the years. While on the one hand, some developing countries like Ghana and Indonesia have managed to maintain a sustainable water resource system without much management intervention; other countries in Asia and Africa have not performed as well due to unequal distribution and incompetent management. Brazil and Russia are

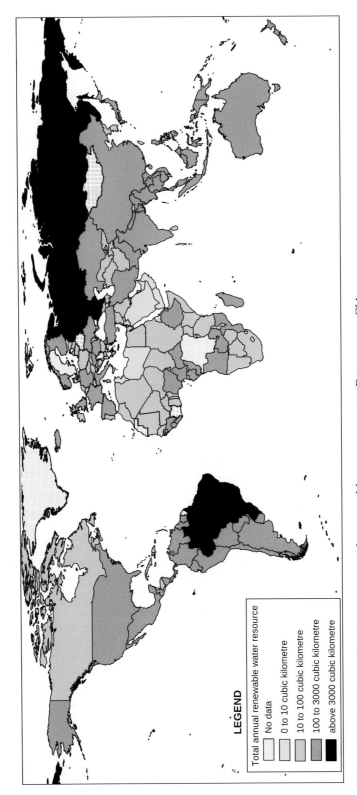

Figure 2.2 World map showing country wide renewable water resource. Data source: CIA.

Table 2.1 Country wisewater availability in 1995 and 2014 (total actual resource/population).

Country	* Area km²	** Population in 1995 (10⁶)	*** Population in 2014 (10⁶)	**** Total water resources (km³)	Per capita available water 1995 (m³/year)	Per capita available water 2014 (m³/year)
Chad	12,84,000	6.18	13.2	43	6958	3258
Nigeria	9,23,768	8.85	18.5	286.2	32339	15470
Senegal	1,96,190	8.1	14.5	39.4	4864	2717
Gambia	1,00,000	1.08	1.91	8	7407	4188
Bolivia	10,98,580	7.2	10.8	622	86389	57593
Honduras	1,12,090	5.49	8.1	95.5	17395	11790
Mali	12,40,000	10.5	15.3	100	9524	6536
Uzbekistan	4,47,400	20.3	29.3	50.4	2483	1720
Costa Rica	51,100	3.42	4.9	112.4	32865	22939
Ecuador	28,000	11.2	16	432	38571	27000

Source: *List of countries by area* (2016) **Shiklomanov (2000); ***Worldometers (2016) ****FAO (2003).

considered to be the world's highest water-resourced countries, with Brazil marginally ahead by a few points. However, lately a major freshwater lake in Russia (Aral Sea) has dried up and urban areas of Brazil have shown traces of drought, thereby affecting the long-preserved water management system. In terms of their internal and external renewable water resources, nine countries, so far, have been declared to be rich in water resource. See Figure 2.4.

India, being one of the countries rich in water, has a rather low (1576 m³/year) per capita water availability. Twelve countries are classified as water-deficient countries with less than 12 cubic km per year: namely, Israel, Jordan, Libyan Arab Jamahiriya, Mauritania, Cape Verde, Djibouti, United Arab Emirates, Qatar, Malta, Gaza Strip, Bahrain and Kuwait (FAO, 2003).

Every country has its own specific water issues. Basic data regarding water issues of some of the major countries are listed below:

- **India:** Basic scenario is low productivity of water in agricultural sector and high exploitation of groundwater, which has resulted in transforming 14% of assessment blocks into over-exploited and 15% into critical or semi-critical blocks. There is a projected requirement of 1180 billion cubic meters (BCM) against water availability of 1130 BCM (Government of India, Ministry of Water Resources, 2011).
- **China:** China is undergoing very rapid industrial and urban growth, thereby reducing the natural flow of rivers, resulting in an inevitable water crisis. China's water efficiency is poor in terms of water consumed per USD of GDP created, and as Lucy Carmody has famously pointed out, 'Water efficiency in China is poor compared to its G20 peers' (Carmody, 2010). What is even

Table 2.2 Dynamics of water use in the world by sectors of economic activity (km³/year).

Sector	Assessment									Forecast	
	1990	1940	1950	1960	1970	1980	1990	1995	2000	2010	2025
Population (million)			2,542	3,029	3,603	4,410	5,285	5,735	6,181	7,113	7,877
Irrigated Land Use (10⁶ ha)	47.3	75.9	101	142	169	198	243	253	264	288	329
Agriculture	513	895	1080	1481	1743	2112	2425	2504	2605	2817	3189
	321	586	722	1005	1186	1445	1691	1753	1834	1987	2552
Municipal	21.5	58.9	86.7	118	160	219	305	344	384	472	607
	4.61	12.5	16.7	20.6	28.5	38.3	45.0	49.8	52.8	60.8	74.1
Industrial	43.7	127	204	339	547	713	735	752	776	908	1170
	4.81	11.9	19.1	30.6	51.0	70.9	78.8	82.6	87.9	117	169
Reservoirs	0.30	7.00	11.1	30.2	76.1	131	167	188	208	235	269
Total (rounded)	579	1,088	1,382	1,968	2,526	3,175	3,633	3,788	3,973	4,431	5,235
	331	617	768	1,086	1,341	1,686	1,982	2,074	2,182	2,399	2,764

Remarks: First line—water withdrawal; second—water consumption.

Source: Shiklomanov (2000). © Taylor & Francis Ltd. Printed with permission from Taylor & Francis Ltd.

Table 2.3 Percentage water withdrawal consumed by different sectors.

Sector	1940	1950	1960	1970	1980	1990	1995	2000	2010	2025
Percentage of consumption (Agriculture)	65%	67%	68%	68%	68%	70%	70%	70%	71%	80%
Percentage of consumption (Municipal)	21%	19%	17%	18%	17%	15%	14%	14%	13%	12%
Percentage of consumption (Industrial)	9%	9%	9%	9%	10%	11%	11%	11%	13%	14%

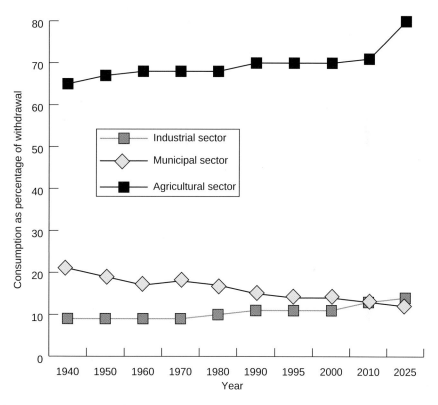

Figure 2.3 Percentage of water consumption against withdrawal from 1940 to 2025 (projected).

more alarming is that 30% of the river water in China is severely polluted and considered unsuitable for any use.

- **Brazil (São Paulo state):** The country has the world's highest renewable resource of 8.23 trillion cubic metres (Water Politics, 2007). Brazil is highly urbanised and industrialised with 80% of its population living in megacities

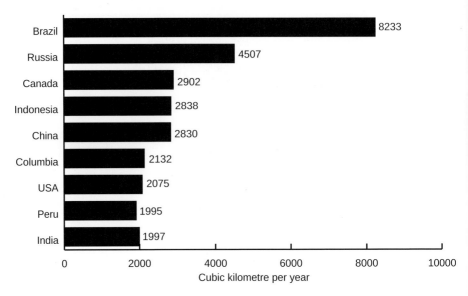

Figure 2.4 Total annual renewable water resources of water rich countries. Data source: FAO (2003).

on the east coast (Maddocks, Shiao and Mann, 2014). High concentration of population in localised areas of the east coast demands multi-sector activities generating quantity-quality issues in São Paolo.

- **Mexico:** On the one hand, the country is vulnerable to severe economic losses due to water-related disasters, but on the other, Mexico has increased its water productivity eightfold during 2002–2007 (UN-Water, 2013a).
- **South Africa:** Due to fast-growing urban requirements, it is estimated that by 2030 South Africa's water demand will reach 17.7 BCM while its current resources are only 15 BCM. South Africa purchases water from neighbouring countries, but locally available potable water is reduced by low rainfall, climate change and pollution, caused by mines. (Boccaletti, Stuchtey and Olst, 2010). Sharing river basins with other countries is yet another major concern of South Africa, which calls for delicate and diplomatic handling.
- **Indonesia:** Unlike South Africa, Indonesia has no trans-boundary issues because its requirement is fully internal and is supported by the 2700 mm of average annual rainfall. The industries use 18.5% of its total water withdrawal. However, the major issues in Indonesia are water pollution and lack of sanitation, which should be taken into consideration (*Indonesia water crisis assessment*, no date). See Figure 2.5.
- **Bangladesh:** Annual renewable resource of Bangladesh is 1227 km^3 which amounts to per capita availability of about 8051 m^3. The problem in the context of Bangladesh water scenario is unique to the country. Its industrial consumption of renewable water resource is 2% (UN-Water, 2013b), but industries pollute 20% of the combined resource of surface and groundwater.
- **Ghana:** Ghana is rich in its surface water resources. However, poor management policies and inadequate funds have severely affected the country, to the extent of handicapping it. The water demand for 2020 is projected to be 5 BCM out of total surface water resources of 41 BCM. Ghana is fully capable of meeting its future water demands (UNESCO, 2015), if the water policy and economy succeed in providing the required support.

Figure 2.5 Thickly populated riverbank, Yogyakarta, Indonesia. Domestic wastewater is directly drained into river.

2.2 Water indicators

Following the country-wise description given above, let us now consider the global water scenario in the light of water stress indicators. To begin with, global business faces severe water risk, as indicated by the water statistics of different countries published by World Resource Institute (Aqueduct). These indicators (Gassert *et al.*, 2013) include the following.

2.2.1 *Baseline water stress*

Baseline water stress is the ratio of total annual water withdrawal (municipal, industrial and agricultural) and available annual blue water; it is expressed as a percentage, in which a higher percentage indicates more competition among users. The global statistics of water stress (see Table 2.4) shows that about 46% countries are under medium to extremely high water stress (Reig, Maddocks and Gassert, 2013). Industries in eighty-five countries, mainly in South Asia, Middle East, Africa and Latin America, are under medium to extremely high stress.

Table 2.4 Countries under water stress.

Score	Value	Number of countries
0–1	Low (<10%)	64
1–2	Low to medium (10–20%)	32
2–3	Medium to high (20–40%)	15
3–4	High (40–80%)	32
4–5	Extremely high (>80%)	36
Total		179

Source: Gassert *et al.* (2013).

Table 2.5 Countries under inter-annual variability.

Score	Value	Number of countries
0–1	Low (<0.25)	30
1–2	Low to medium (0.25–0.5)	85
2–3	Medium to high (0.5–0.75)	36
3–4	High (0.75–1.0)	13
[4–5]	Extremely high (>1.0)	6
Total		170

Source: Gassert *et al.* (2013).

2.2.2   Inter-annual variability

This parameter measures the variation in water supply between years. The uncertainty in annual water availability adversely impacts business planning. Statistics show that 55 out of 170 countries are under medium to extremely high inter-annual variability. See Table 2.5

2.2.3   Water conflict

Since the advent of civilisation, one of the major reasons for political and social conflict has been water. The scales and issues of conflict have varied from country to country, largely because the definition of right to water and distribution prioritisation were not properly formulated. Some of the conflicts generated on development issues are listed in Box 2.1. In such situations, especially when it is operating in a 'high risk zone', business will be highly impaired (Pacific Institute, 2013). To combat this situation, companies need to pre-assess (a) the operating environment, (b) water situation, (c) gaps between supply and demand (Revision World, 2007) and (d) social impacts that may give rise to conflicts.

Examples of water conflict

India: In 2009, in Madhya Pradesh, India, three people were killed in violent clashes over access to water (Singh, 2009).

Yemen: 2013. The Marzouh and Qaradh tribal people of Yemen depend on locally available spring water. When these springs started drying up, the tribes lost six members to violent clashes (Friedman, 2016).

Canada: Demonstrators started a movement against hydraulic fracturing for shale gas in Canada. But the movement ran aground because the verdict from the New Brunswick court went against the demonstrators (Lucas, 2013).

Brazil: In the Brasilia region, the main economy is based on cattle breeding and dairy. Although Brazil is the top most water-rich country in the world, consecutive years of severe draught gave rise to violent conflicts which, in later years, came to be known as 'water wars'. An average of one person per day lost his or her life in the war (Source: Catholic Online, 2016).

Somalia: Water, pasture and politics have been at the centre of sporadic clashes between the Hodo-ogle and Bur-dinle clans, who lived in the central region of Somalia. In these clashes, at least ten people have been killed in 2015. (Abdirahman, 2015)

Iran: In March 2013, violent clashes erupted in the town of Varzaneh in Iran's Esfahan province over the issue of government's decision to divert water from Esfahan province to Yazd province. The people of Esfahan province considered it unfair because the move was politically motivated and a step to deprive the Esfahan people from their right to water. The clash led to bloodshed and the arrest of several people (Javedanfar, 2013).

Peru: During the past two decades, the Ica Valley, bordered by sand dunes has been transformed into an agricultural province, which is the result of the efforts of agricultural families. The aquifer of the valley is the sole source of water, and that aquifer is under threat. Conflict in this region developed when an agro-based company, Agrícola La Venta, tried to drill water wells in this region and laid pipelines to supply water to their own fields. The farmers have united to guard their aquifer against the intervention by the company (James, 2016).

China: In 2004, a Chinese man was executed because he protested against the construction of a hydro-electric dam in the southwestern province of Sichuan, which would flood several villages in that province. This is one of the thousand protests and clashes in China on the issues of land grabbing, illegal pollution and related corruption (Coonan, 2006).

BOX 2.1

2.2.4   River basins and aquifers under threat and conflict

From 1999 to 2008, there were approximately 2000 conflicts over the sharing of water in international river basins. Despite such crisis, there are incidents of collateral and mutual collaboration and agreements over water which resolved many of the disputes (Peek, 2014). According to Wolf *et al.*, 'there were 1831 water conflicts over trans-boundary basins from 1950 to 2000' (*Water conflict*, 2016). In water conflict, water acts as an irritant when parties sharing water fail to cooperate on issues related to quality, quantity and infrastructure.

The world is now facing three major conflicts in international river basins (Johnson, 2014). It fears that the construction of several dams on the Chinese side of the Brahmaputra River will reduce water flow in the Indian and Bangladeshi parts of the river. The conflict continues to grow as there is no international treaty on this issue. Another major example of a similar conflict is in the region of the Nile between Ethiopia and Egypt. Because Ethiopia is constructing the huge Renaissance dam, the annual water flow of the Nile in Egyptian territory will supposedly be reduced. Again, the construction of the Illisu Dam on the Tigris River by Turkey is not considered beneficial by Iraq, giving rise to yet another conflict.

Major river basins and aquifers across different countries are under threat of over-exploitation and severe pollution. Table 2.6 illustrates some major basins and aquifers under threat.

Table 2.6 Major basins and aquifers of the world under threat.

Country or basin	Present status
Israel	Israel which is basically a water-stressed country, shares the River Jordan with Jordan and also withdraws water from the Sea of Galilee. Groundwater is so highly depleted that Israel has to buy water from Turkey.
Ogalla Aquifer	The Ogalla Aquifer is a major geological formation in the USA that covers about 450,000 sq. km of eight states. The saturated thickness of the aquifer varies from 60 m to 300 m and is being over-exploited. Water level has declined more than 30 m between 1940 and 1990. In spite of restrictions in groundwater mining, the annual rate of decline in water level is about 82 cm (Kromm, 2016).
Euphrates and Tigris Basin	Apparently this basin has no water crisis. But being international rivers, their water is shared by Turkey, Syria and Iraq. Under present growth conditions, water crisis may develop by 2020. The river basin has political tensions (Kliot, 1993) and trans-boundary management problems, which have been aggravated by the recent civil war in Syria.
Lake Chad	Lake Chad is situated in the borders of Chad, Niger and Cameroon in West Africa. It is the source of fresh water for agriculture in these countries. Lake Chad, which is a large aquatic ecosystem, is now under threat (UNEP 2008). Its area has been reduced from 25,000 sq. km to 1350 sq. km in just 30 years since 1963 (the World Bank, 2014).
Nile Basin	The Nile is one of the largest international river basins. It is the lifeline of two major countries, Ethiopia and Egypt. Water crisis in this region has developed due to recent management plans of Ethiopia, which is a threat to Egypt's economic growth.
Yellow river	The Yellow River, China's second-largest river is under the threat of severe over-exploitation, mainly because of irrigation. Since the early 1970s, the river has occasionally dried up in its downstream region. According to a WWF report, the river is under threat due to pollution, hydropower and intensive water extraction for human consumption, agriculture and industrial use.
The Ganga Basin	The Ganga is both a rain-fed and a glacier-fed river. It is the lifeline of northern and eastern part of India. The Ganga Valley is one of the largest alluvial plains in the world. The thick alluvium is also a major freshwater aquifer which is hydraulically connected to the river and its tributaries. Widespread exploitation of groundwater, withdrawal of water for irrigation, drinking and industrial purposes, pollution from industrial, agricultural, urban and religious sectors are some of the activities, leading to the gradual death of this river.
Murray-Darling Basin	The Murray-Darling Basin in Australia is suffering severe water shortage due to climate change, and water abstraction. The annual flow of the river has been reduced by three-quarters to cater to water for large dams.

2.2.5   *Physical water risk in business*

Will Sarni (2011) has warned that physical water risk for business may develop from the impact of direct use and disruption in supply chain. It means that there are two possible sources of physical business risks; first, water scarcity at the local source and second, the supply chain disruption by water shortage.

2.2.6   *Disruption in the supply chain*

Disruption in the supply chain impacts business seriously because companies do not have any direct control over the supply chain when the source is from another country. There are several examples of such disruption and potential threats from climate exposure. A report published by CDP (2015) on the performance of supply chains in different countries reveals that the chains in the US, China and Italy are 'vulnerable'. India and Canada are neglecting their performances under climatic exposure while Brazil is attempting to manage it somehow (CDP, 2015). Agricultural supply chains are highly complex (Wilcox, 2015) because agriculture production is dependent on local water supply which may be disrupted, even by a temporary draught. Sugar, beverages and agri-businesses are always vulnerable to supply chain disruption. As for example, the case of Anheuser-Busch, the world's largest brewer of beer may be mentioned. This company suffered a severe supply chain disruption in 1991 which was very complex. See Figure 2.6.

2.2.7   *Failure to meet basic water needs*

There are business risks in some developing countries where poor people are deprived of basic water and sanitary requirements. According to Morrison and Gleick (2004), water risk is increasingly higher in developing

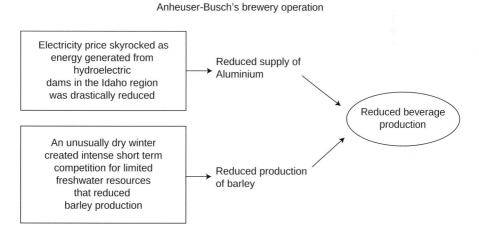

Figure 2.6 Example of supply chain disruption in business.

countries than in the developed ones. If a factory is installed in a water-stressed area of a country, the business will face major challenges in securing water. Primarily, the resistance will come from the local people, especially when the source is traditionally used by them. In places where the government has failed to provide proper water supply and basic services to the people and where the poorer people live in unhealthy conditions, industries may find the environment hostile to them. But this may be an opportunity for the corporation to secure 'social licence' or local community's acceptance by extending water-related services to the people. The responsibility of the business should be extended to make provisions for secured drinking water to the community and to develop appropriate water and sanitation infrastructures.

2.3   Global water trends

Let us also consider the current global trends that impact water resources.

- There will be an additional '2.5 billion people to the planet in the next 40 years' (Godfray *et al.*, 2010; and WWF, 2013) and therefore water requirements for both people and business will be increased.
- As global climate change is a crucial factor, it will result in greater weather variability (WWF, 2013), thereby giving rise to more water-related disasters, like drought and flood.
- Rise in the consumption and patterns of consumption as a consequence of higher income of people will result in higher need for agriculture production, biodiesel production and meat production, and will ultimately raise the overall water footprint.
- Regulatory trends show that laws for industrial water allocation are becoming more stringent and restricted.

These global trends are heading towards a future that has a complex social inter-relationship in the context of water shared by different sectors. A single organisation or institute cannot address the issues to satisfy all sectors. A collective action involving people and organisations from different corners of society and effort initiated by the government may lead to a sustainable solution to water crisis. The corporation has the opportunity to lead the initiative towards a sustainable development goal.

2.4   Business models

Industries, decision makers, politicians and the scientific community are under continuous pressure to find a sustainable solution to existing water crises. Several models for assessment of trends and for framing modalities to cope with the trends have been formulated. There are several models for the water scenario projections. One of the most talked about is the business as usual model.

2.4.1 Business as usual model

In this model, it is assumed that our society in all sectors will ignore the actual water crisis and allow the growth of population, agriculture and industry without initiating any effective measure. The consequences of this model is that, by 2050, developing countries, where the industrial growth and trend of population growth is also alarmingly high will experience high water stress (FAO, 2009), which in turn will affect the GDP. The forecast under this model is given in Table 2.7.

Apart from GDP, other factors that will be affected are water quality and water cost. In a recent study in Ghana by Adombire, Adjewodah and Abrahams (2013) on the business as usual scenario, it is forecast that the effect will be observed in declining water quality, rise in the cost of 'water production' and decrease in usable water availability in urban sector.

The business as usual model is unable to achieve sustainability because by 2050 more than 50% of the world population will be under stress. So, a business model must be conceptualised to lead the business to its desired goal. Turner (2015) has shown that in spite of capital investment in resource extraction, there is a possibility of global collapse because of depletion of non-renewable resources and increased pollution that may reach the threshold limit.

2.4.2 Alternative model

Bocken *et al.* (2014) proposed that a business model should create value by exploring new opportunities, deliver values to market and generate measurable social and ecological values. Bocken *et al.* also proposed a business model archetype (Figure 2.7) which addresses technological, social and organisational aspects of a business. Technology is applied in a business to achieve maximum efficiency in its use of material and energy and to create value from waste and apply renewable resources. In his business model,

Table 2.7 Water stress and GDP in the business as usual model in 2050 compared to 2010.

Water stress level	< 20%	20% – 40%	>40%
Population under stress	32%	16%	52%
GDP under risk by water stress level	30%	25%	45%

Adapted from Growing Blue (2016).

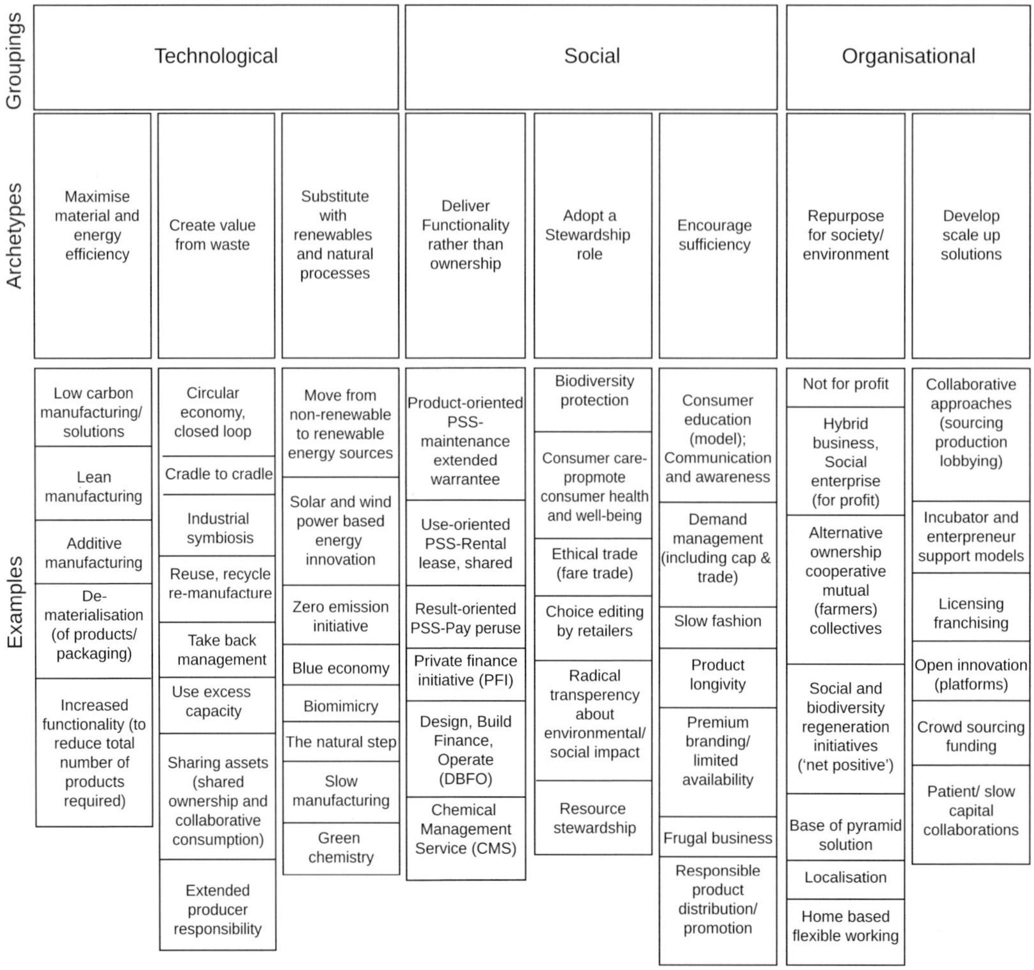

Groupings	Technological			Social			Organisational	
Archetypes	Maximise material and energy efficiency	Create value from waste	Substitute with renewables and natural processes	Deliver Functionality rather than ownership	Adopt a Stewardship role	Encourage sufficiency	Repurpose for society/environment	Develop scale up solutions
Examples	Low carbon manufacturing/ solutions	Circular economy, closed loop	Move from non-renewable energy sources	Product-oriented PSS-maintenance extended warrantee	Biodiversity protection	Consumer education (model); Communication and awareness	Not for profit	Collaborative approaches (sourcing production lobbying)
	Lean manufacturing	Cradle to cradle	Solar and wind power based energy innovation	Use-oriented PSS-Rental lease, shared	Consumer care-propmote consumer health and well-being	Demand management (including cap & trade)	Hybrid business, Social enterprise (for profit)	Incubator and entrepreneur support models
	Additive manufacturing	Industrial symbiosis	Zero emission initiative	Result-oriented PSS-Pay peruse	Ethical trade (fare trade)	Slow fashion	Alternative ownership cooperative mutual (farmers) collectives	Licensing franchising
	De-materialisation (of products/ packaging)	Reuse, recycle re-manufacture	Blue economy	Private finance initiative (PFI)	Choice editing by retailers	Product longevity	Social and biodiversity regeneration initiatives ('net positive')	Open innovation (platforms)
	Increased functionality (to reduce total number of products required)	Take back management	Biomimicry	Design, Build Finance, Operate (DBFO)	Radical transparency about environmental/ social impact	Premium branding/ limited availability	Base of pyramid solution	Crowd sourcing funding
		Use excess capacity	The natural step	Chemical Management Service (CMS)	Resource stewardship	Frugal business	Localisation	Patient/ slow capital collaborations
		Sharing assets (shared ownership and collaborative consumption)	Slow manufacturing			Responsible product distribution/ promotion	Home based flexible working	
		Extended producer responsibility	Green chemistry					

Figure 2.7 Business model archetypes. Source: Boken *et al.* (2014). © Elsevier Ltd. Printed with permission from Elsevier Ltd.

society is a functional element in economic development, and the business will fulfill its stewardship role. To reach these goals, an organisational restructuring is also essential.

2.5 Integrated water resource management

According to Biswas (2016), 'the concept of integrated water resources management (IWRM) has been around for some 60 years. It was rediscovered by some in the 1990s.' In recent times, this model has been aligned with the recommendations for sustainable development goals and is in practice mainly in agricultural sectors. The components of IWRM are so varied and versatile that they cover almost all aspects of water resource management and can be a paradigmatic business model for industries. By definition, IWRM is a coordinated approach for abstraction and management of freshwater resources. The IWRM paradigm, of course, assumes that all users in the common water resource pool are interdependent and function in harmony to achieve ecosystem sustainability.

'Integrated Water Resources Management is a cross-sectoral policy approach' (Global Water Partnership, 2010) and is totally different from traditional ideas that considered industry as an entity isolated from the other two sectors, namely agriculture and domestic use. IWRM calls for an ecosystem-based approach in water management where industries can participate as leaders.

2.5.1 History of IWRM

IWRM has a long history of development. Along with the advancement of civilization, the role of water in human history has undergone a series of successive transformations, which, according to Hassan (2011) and Mayfield (2010) have involved the following stages.

In the beginning of civilisation after a long history of nomadic life, fishing and fowling, some societies, about 10,000 years ago adopted agriculture, which in no time demanded irrigation and water management through digging of canals, diverting water flow and digging of wells. If we look at the ancient civilisations of the Indus Valley, Mesopotamia and Egypt we find an efficient network of canals and water storage systems. This was followed by the development of water distribution systems and by the appearance of water-lifting technologies; thus began the age of mechanised water lifting with devices (e.g. water wheels and the Archimedes wheel). In the steam-powered Industrial Revolution, distribution of water became an industry itself. A typical scene immortalising the Industrial Revolution is shown in Figure 2.8.

In the latter half of the Industrial Revolution, methods of modern water management started evolving. The idea of IWRM emerged from a cooperative approach prevalent from 1960 to 1970.

To sum up, the evolution of IWRM can be visualised as (a) the sectoral approach, which lasted from the middle of the Industrial Revolution to the 1950s; (b) the cooperative approach in the 1960s and 1970s, (c) management-oriented IWRM in 1980s; and finally (d) goal-oriented IWRM from the 1990s to the present.

2.5.2 Principles of IWRM

Mayfield (2010) has identified eight principles which are guiding factors in water stewardship (printed with permission from Dr. Colin Mayfield):

   i 'Water source and catchment conservation and protection are essential.
   ii Water allocation should be agreed to between stakeholders within a national framework.

Figure 2.8 Chelsea waterworks, 1752. Two Newcomen beam engines pumped Thames water from a canal to reservoirs at Green Park and Hyde Park. Source: https://commons.wikimedia.org/wiki/File:Chelsea_Waterworks_1752.jpg.

  iii Management needs to be taken care of at the lowest appropriate level.
  iv Capacity building is the key to sustainability.
  v Involvement of all stakeholders is required.
  vi Efficient water use is essential and often an important "source" in itself.
  vii Water should be treated as having an economic and social value.
  viii Striking a gender balance is essential.'

The essence of the eight principles is an inclusive approach to water management, failing which management plans are likely to fall prey to internal conflict and collapse. This approach is thus vital to assessment of the water balance of a watershed. IWRM is a concept that integrates nature with both society and business ensuring, sustainable development of all. The conceptual framework showing the intersection of factors in IWRM is shown in Figure 2.9.

In an IWRM initiative several factors should be taken into considerations. Eleven key considerations are listed (Mayfield, 2010):

  i 'Collecting information on the water resource from the local environment. This involves: (a) collection of data on the dynamic water resource and estimation of available water for various uses, (b) assessment of quality of both surface and groundwater and (c) preparation of thematic maps. To develop the database a programme for systematic data collection programme should be taken up.
  ii IWRM involves technological innovations and application for retrofit of appliances and use of innovative technologies that helps in conservation of water.

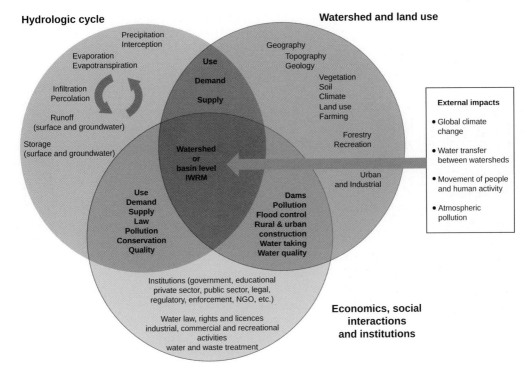

Figure 2.9 Intersection of factors in IWRM (Mayfield, 2010). Printed with permission from Dr. Colin Mayfield.

iii Human resource development is essential for IWRM because its success depends on the informed participation of stakeholders. One objective of IWRM is to engage people from various sections of the society who will actively implement the IWRM programmes.

iv IWRM depends upon how accurately data on water withdrawal, consumption and loss is collected and assessed. This is continuous process that generates further scopes for water conservation.

v IWRM is a sequence of activities that are periodically assessed, calibrated and improved. These activities are conducted in phases. Obvious and low cost options are always preferred and should be adopted.

vi IWRM is an integrated process where stakeholders are involved in planning and practice.

vii IWRM will achieve compatibility of industrial process with quality and quantity of water. For example industrial processed must avoid using potable water wherever applicable. For example potable water is not required for rinsing.

viii Reuse of water should be encouraged and a systematic plan for reusing water in maximum possible ways should be adopted. For example spent water from rinsing may be reused for gardening or cooling.

ix Conduct water audit to assess the water loss and to determine the water balance in the industrial process. This identifies how and where water conservation measures are required.

x Water footprint assessment and life cycle assessment of water consumption of a product or service is an integral part of IWRM. This should be conducted as a part of sustainability assessment of the industry in terms of water availability.

   xi IWRM also works on gender issues and has a goal to provide easy access of water for women.'

Printed with permission from Dr. Colin Mayfield.

Under an IWRM programme, an industry has to act beyond the fence and consider the entire spectrum of water users; it cannot be done in isolation. In other words, big business needs to adopt IWRM to remain in business.

2.6 Sustainable development goal for business sector

On 25 September 2015, in an initiative taken up by the United Nations member countries, a set of goals to end poverty, protect the planet, and ensure prosperity for all as part of a new sustainable development agenda (Nino, 2016) was adopted. Each goal has specific targets to be achieved over the next fifteen years. In 2000, the United Nations adopted the Millennium Development Goals, which were a useful framework for coordinated global action for future development. After about fifteen years from that time, the positive role of the business community is recognised in this effort by governments and civil society and included in the goals when the UN embarks upon an inclusive process to determine a Post-2015 Development Agenda (United Nations, 2017). Before that, in 2014 the Stakeholder Forum, a civil society group conducted a series of exercises to identify the role of business in sustainable development. The outcome emphasised that business is the key driver of the society to 'create the jobs and livelihoods and provide the necessary technical resources to create and deploy new solutions to the sustainable development challenges the international community is facing today.'

The Sustainable Development Goal (SDG) has a set of 17 goals and 169 targets to be achieved by 2030. Out of these goals, some are directly related to water and business, whereas Goal 6 and Goal 9 are related to water stewardship (ICSU, 2015) and propagate the message of sustainable business through improving quality, quantity, access and affordability of fresh water for society. It also emphasises a cooperative and inclusive involvement of industries, civil society and government to achieve the goals and fulfil the targets. Goals 6 and 9 are detailed in the Table 2.8.

2.7 Conclusion

During the 21st century, business and industry are going to play vital roles in global water management. The participation of industries in that direction may not be uniform across the globe, but it is expected that several innovative models will come up because they are necessary for businesses

Table 2.8 Sustainable development goals and targets in the water sector.

Goals	Targets
Goal 6: Ensure access to water and sanitation for all	• By 2030, achieve universal and equitable access to safe and affordable drinking water for all
	• By 2030, achieve access to adequate and equitable sanitation and hygiene for all and end open defecation, paying special attention to the needs of women and girls and those in vulnerable situations
	• By 2030, improve water quality by reducing pollution, eliminating dumping and minimizing release of hazardous chemicals and materials, halving the proportion of untreated wastewater and substantially increasing recycling and safe reuse globally
	• By 2030, substantially increase water-use efficiency across all sectors and ensure sustainable withdrawals and supply of freshwater to address water scarcity and substantially reduce the number of people suffering from water scarcity
	• By 2030, implement integrated water resources management at all levels, including through trans-boundary cooperation as appropriate
	• By 2020, protect and restore water-related ecosystems, including mountains, forests, wetlands, rivers, aquifers and lakes
	• By 2030, expand international cooperation and capacity-building support to developing countries in water- and sanitation-related activities and programmes, including water harvesting, desalination, water efficiency, wastewater treatment, recycling and reuse technologies
	• Support and strengthen the participation of local communities in improving water and sanitation management
Goal 9: Build resilient infrastructure, promote sustainable industrialization and foster innovation	• Develop quality, reliable, sustainable and resilient infrastructure, including regional and trans-border infrastructure, to support economic development and human well-being, with a focus on affordable and equitable access for all
	• Promote inclusive and sustainable industrialization and, by 2030, significantly raise industry's share of employment and gross domestic product, in line with national circumstances, and double its share in least developed countries
	• Increase the access of small-scale industrial and other enterprises, in particular in developing countries, to financial services, including affordable credit, and their integration into value chains and markets
	• By 2030, upgrade infrastructure and retrofit industries to make them sustainable, with increased resource-use efficiency and greater adoption of clean and environmentally sound technologies and industrial processes, with all countries taking action in accordance with their respective capabilities
	• Enhance scientific research, upgrade the technological capabilities of industrial sectors in all countries, in particular developing countries, including, by 2030, encouraging innovation and substantially increasing the number of research and development workers per 1 million people and public and private research and development spending
	• Facilitate sustainable and resilient infrastructure development in developing countries through enhanced financial, technological and technical support to African countries, least developed countries, landlocked developing countries and Small Island Developing States*
	• Support domestic technology development, research and innovation in developing countries, including by ensuring a conducive policy environment for, inter alia, industrial diversification and value addition to commodities
	• Significantly increase access to information and communications technology and strive to provide universal and affordable access to the Internet in least developed countries by 2020

*Small Island Developing States (SIDS) are small coastal countries that share similar challenges like limited resources, remoteness, vulnerability to natural disaster, excessive dependence on international trade, and fragile environments. Eighteen such countries have been recognised by the United Nations.

Source: 'Sustainable Development Goals' by Nino (2016), © (2016) United Nations; reprinted with permission from the United Nations.

to survive in water-stress situations. Location specific business models will develop to mitigate business risks that will increase societal value as well as build up the hope for a sustainable future.

Bibliography

Abdirahman, A. (2015) *Somalia: At least 10 people killed in tribal fighting*. Available at https://horseedmedia.net/2015/01/20/somalia-at-least-10-people-killed-in-tribal-fighting/(accessed: 9 March 2016).

Adombire, I.M., Adjewodah, P., and Abrahams, R. (2013) *Business as usual (BAU) scenario information and analysis covering the Pra and Kakum river basins prepared for nature conservation research centre*. Available at http://forest-trends.org/documents/files/Ghana_BAU_Scenario_assessment.pdf (accessed: 2 April 2016).

Biswas, A.K. (2016) Integrated water resources management: A reassessment, *Water international*, 29(2): 248–256. doi: 10.1080/02508060408691775.

Boccaletti, G., Stuchtey, M., and Olst, M. van (2010) *Confronting South Africa's water challenge*. Available at http://www.mckinsey.com/business-functions/sustainability-and-resource-productivity/our-insights/confronting-south-africas-water-challenge (accessed: 27 February 2016).

Bocken, N.M.P., Short, S.W., Rana, P., and Evans, S. (2014) 'A literature and practice review to develop sustainable business model archetypes', *Journal of cleaner production*, 65: 42–56. doi: 10.1016/j.jclepro.2013.11.039.

Carmody (ed), L. (2010) *Water in China*. Available at http://www.sustainalytics.com/sites/default/files/water_in_china-_issues_for_responsible_investors_feb2010.pdf (accessed: 2 April 2016).

Catholic Online (2016) *Severe drought in Brazil destroying herds and lives—Americas—international—news - Catholic online*. Available at http://www.catholic.org/news/international/americas/story.php?id=48869 (accessed: 2 April 2016).

CDP (2015) *Supply chain sustainability revealed: A country comparison supply chain report 2014–15*. Available at https://www.cdp.net/CDPResults/CDP-Supply-Chain-Report-2015.pdf (accessed: 2 April 2016).

CIA (no date) *The world factbook*. Available at https://www.cia.gov/library/publications/the-world-factbook/fields/2201.html (accessed: 27 February 2016).

Coonan, C. (2006) *China secretly executes anti-dam protester*. Available at http://www.independent.co.uk/news/world/asia/china-secretly-executes-anti-dam-protester-427401.html (accessed: 2 April 2016).

FAO (2003) *Review of world water resources by country*. Food and Agriculture Organization of the United Nations. Rome, Italy. ISSN 1020-1203

FAO (2009) *How to feed the world in 2050*. Available at http://www.fao.org/fileadmin/templates/wsfs/docs/expert_paper/How_to_Feed_the_World_in_2050.pdf (accessed: 8 May 2016).

Friedman, T.L. (2016) *Postcard from Yemen*. Available at http://www.nytimes.com/2013/05/08/opinion/friedman-postcard-from-yemen.html?_r=0 (accessed: 2 April 2016).

Gassert, F., Reig, P., Luo, T., and Maddocks, A. (2013) *Aqueduct country and river basin rankings*. Available at http://www.wri.org/resources/data-sets/aqueduct-country-and-river-basin-rankings (accessed: 2 April 2016).

Gleik, P.H. (2000) 'The changing Water Paradigm, A look at Twenty First Century Water Resource Development', *Water international*, 25(1): 127–138.

Global Water Partnership (2010) *What is IWRM?*. Available at http://www.gwp.org/the-challenge/what-is-iwrm/(accessed: 30 April 2016).

Godfray, H.C.J., Beddington, J.R., Crute, I.R., Haddad, L., Lawrence, D., Muir, J.F., Pretty, J., Robinson, S., Thomas, S.M., and Toulmin, C. (2010) *Food security: The challenge of feeding 9 Billion people*. Available at http://science.sciencemag.org/content/327/5967/812.full (accessed: 2 April 2016).

Government of India Ministry of Water Resources (2011) *Strategic plan for ministry of water resources*. Available at http://www.tgpg-isb.net/sites/default/files/document/strategy/Water.pdf (accessed: 2 April 2016).

Growing Blue (2016) *What is the future of water? | water in the future | water in 2050*. Available at http://growingblue.com/water-in-2050/(accessed: 2 April 2016).

Hassan, F. (2011) *Water history for our times*. Available at http://unesdoc.unesco.org/images/0021/002108/210879e.pdf (accessed: 17 May 2016).

ICSU (2015) *Review of targets for the sustainable development goals: the science perspective (2015) — ICSU*. Available at http://www.icsu.org/publications/reports-and-reviews/review-of-targets-for-the-sustainable-development-goals-the-science-perspective-2015 (accessed: 20 March 2016).

Indonesia water crisis assessment (no date) Available at https://sites.google.com/site/isat380eindonesia/(accessed: 24 February 2016).

James, I. (2016) *The costs of Peru's farming boom*. Available at http://www.desertsun.com/story/news/environment/2015/12/10/costs-perus-farming-boom/76605530/(accessed: 2 April 2016).

Javedanfar, M. (2013) *The water threat in Iran—al-monitor: the pulse of the middle east*. Available at http://www.al-monitor.com/pulse/originals/2013/11/water-scarcity-frustrates-iranians.html (accessed: 2 April 2016).

Johnson, P. (2014) *Three international water conflicts to watch*. Available at http://www.geopoliticalmonitor.com/three-international-water-conflicts-watch/(accessed: 2 March 2016).

Kliot, N. (1993) *Water resources and conflict in the middle east*. United Kingdom: Taylor & Francis.

Kromm, D.E. (2016) *Ogallala aquifer—depth, important, system, source*. Available at http://www.waterencyclopedia.com/Oc-Po/Ogallala-Aquifer.html (accessed: 2 April 2016).

List of countries by area (2016).Wikipedia. Available at https://simple.wikipedia.org/wiki/List_of_countries_by_area (accessed: 8 March 2016).

Lucas, M. (2013) *New Brunswick fracking protests are the frontline of a democratic fight*. Available at http://www.theguardian.com/environment/2013/oct/21/new-brunswick-fracking-protests (accessed: 8 May 2016).

Maddocks, A., Shiao, T., and Mann, S.A. (2014) *3 Maps help explain São Paulo, Brazil's water crisis*. Available at http://www.wri.org/blog/2014/11/3-maps-help-explain-s%C3%A3o-paulo-brazil%E2%80%99s-water-crisis (accessed: 2 April 2016).

Mayfield, C. (2010) *Introduction to IWRM*. Available at http://www.yemenwater.org/wp-content/uploads/2013/03/Introduction-to-IWRM-_-History-and-Introduction-Capacity-bui.pdf (accessed: 14 May 2016).

Morrison, J., and Gleick, P. (2004) *Freshwater resources: managing the risks facing the private sector*. Available at http://www.pacinst.org/wp-content/uploads/sites/21/2013/02/business_risks_of_water3.pdf (accessed: 12 March 2016).

Nino, F.S. (2016) *Sustainable development goals - United Nations*. Available at http://www.un.org/sustainabledevelopment/sustainable-development-goals/(accessed: 2 April 2016).

ORASECOM (2006) *Mandate - Orange-Senqu river commission (ORASECOM).* Available at http://www.orasecom.org/about/mandate.aspx (accessed: 27 February 2016).

Pacific Institute (2013) *Water conflict - the world's water.* Available at http://worldwater.org/water-conflict/(accessed: 2 April 2016).

Peek, K. (2014) *Where will the world's water conflicts erupt? [Infographic].* Available at http://www.popsci.com/article/science/where-will-worlds-water-conflicts-erupt-infographic (accessed: 2 March 2016).

Reig, P., Maddocks, A., and Gassert, F. (2013) *World's 36 most water-stressed countries.* Available at http://www.wri.org/blog/2013/12/world%E2%80%99s-36-most-water-stressed-countries (accessed: 2 April 2016).

Revision World (2007) *Water conflict.* Available at http://revisionworld.com/a2-level-level-revision/geography/global-challenges-and-issues-0/water-conflict (accessed: 2 April 2016).

Sarni, W. (2011) *Corporate water strategies.* United Kingdom: Taylor & Francis.

Shiklomanov, I.A. (2000) Appraisal and assessment of world water resources, *Water International*, 25(1): 11–32. doi: 10.1080/02508060008686794.

Singh, G. (2009) *Water wars strike ahead of predictions.* Available at http://ecolocalizer.com/2009/05/16/water-wars-strike-ahead-of-predictions/(accessed: 2 April 2016).

Turner, G. (2014) 'Is Global Collapse Imminent?', in *MSSI Research Paper No. 4.* Melbourne: Melbourne Sustainable Society Institute, the University of Melbourne.

UNEP (2008) *Lake Chad: Almost gone—vital water graphics.* Available at http://new.unep.org/dewa/vitalwater/article116.html (accessed: 20 March 2017).

UNESCO (2015) *National commissions for UNESCO: annual report, 2014; 2015.* Available at http://unesdoc.unesco.org/images/0023/002326/232649E.pdf (accessed: 26 February 2016).

United Nations, (2017) *Millennium development goals and post-2015 development agenda.* Available at: http://www.un.org/en/ecosoc/about/mdg.shtml (accessed: 24 March 2017).

UN-Water (2013a) *MEX pagebypage.* Available at http://www.unwater.org/fileadmin/user_upload/unwater_new/docs/Publications/MEX_pagebypage.pdf (accessed: 18 September 2016).

UN-Water (2013b) *UN-water country brief year.* Available at http://www.unwater.org/fileadmin/user_upload/unwater_new/docs/Publications/BGD_pagebypage.pdf (accessed: 18 September 2016).

Water Politics (2007) *Brazil turns to big projects to fix water crisis.* Available at http://www.waterpolitics.com/2015/04/07/brazil-turns-to-big-projects-to-fix-water-crisis/(accessed: 22 February 2016).

Water conflict (2016) Wikipedia. Available at https://en.wikipedia.org/wiki/Water_conflict#cite_note-21 (accessed 25 October 2016).

Wilcox, M. (2015) *A farm-level view on supply chain water risk.* Available at https://www.greenbiz.com/article/farm-level-view-supply-chain-water-risk (accessed: 2 April 2016).

The World Bank (2014). *Restoring a disappearing giant,* Available at https://www.worldbank.org/en/news/feature/2014/03/27/restoring-a-disappearing-giant-lake-chad (accessed 21 March 2017).

Worldometers (2016) *Population by country (2016).* Available at http://www.worldometers.info/world-population/population-by-country/(accessed: 19 February 2016).

WWF (2013) *Water stewardship, brief perspectives on business risks and responses to water challenges.* Available at http://awsassets.panda.org/downloads/ws_briefing_booklet_lr_spreads.pdf (accessed: 12 March 2016).

3

Understanding Water

3.1 Introduction

When we talk about ancient civilisation, we visualise great river valleys like the Nile and the Indus River where these civilisations flourished. Water has always been a pivotal factor in determining the location and development of any civilisation. Biodiversity is best showcased in water-rich areas, such as rainforests and river basins; likewise, history shows that the most populated areas of the world have also developed in water-rich regions.

We cannot know for sure the exact period of the arrival of water on earth or whether it has co-evolved with the evolution of the solar system, but it is certain that water has always played the most vital role in nurturing life on earth and modifying terrestrial landscapes.

Water is essential to us, both for living and economic growth. For the sake of economic growth, large investments are made for water infrastructures at great cost to the environment. For economic activities, water is extracted from lakes, rivers or from aquifers. Recently, water-deficient countries are using purified seawater for both agriculture (as in Oman) and industry (as in Saudi Arabia).

Water, as all human users are aware, is a shared resource with high social, cultural, economic and environmental value. Water serves many parallel functions (Figure 3.1) which are essential for:

- Human health;
- Aquatic ecosystems;
- Biomass production;
- Conveying of nutrients and commodities (and also pollutants);
- Geomorphological processes shaping the earth; and of course, for
- Industries (which is the main topic of this book).

Industrial Water Resource Management: Challenges and Opportunities for Corporate Water Stewardship, First Edition. Pradip K. Sengupta.
© 2018 John Wiley & Sons Ltd. Published 2018 by John Wiley & Sons Ltd.

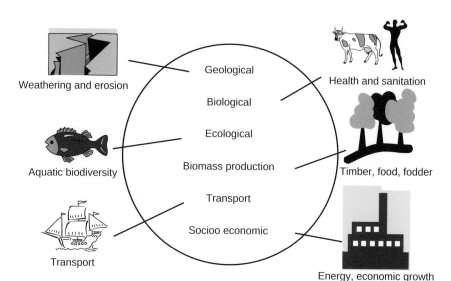

Figure 3.1 Water has many parallel functions.

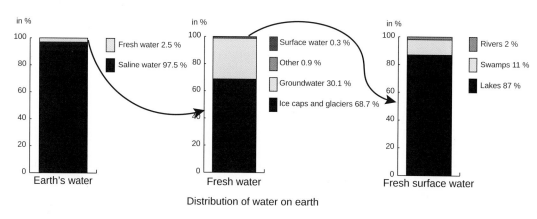

Figure 3.2 Distribution of water on earth. Source: Adapted from R.J. Oosterbaan via Wikimedia Commons.

The earth is essentially a water-planet on which more than 70% of the surface is covered by water. Despite being a water-planet, the availability of fresh water is astonishingly low compared with the total water; only 2.5% of the total water on earth is fresh water. Out of this 2.5%, 68.7% is trapped in glaciers and ice caps of the polar regions (Figure 3.2); the rest is rain, atmospheric water and other terrestrial sources.

In other words, only a very small amount of water is directly accessible to humankind; in addition, accessible water is not evenly distributed throughout the world. This fresh water is shared by animals, plants, humans and the other components of nature for maintaining the ecosystem. Some places in New Zealand and India experience more than 6,000 millimetres of rainfall annually, whereas, the Atacama Desert of Peru, arguably the driest desert, seldom experiences any rainfall.

Agriculture consumes a major part of the freshwater resources of the earth, but consumption amounts and rates vary from region to region. In most countries in the world, the major user of water is the agricultural sector, and second comes industry. In the USA, Canada and countries in Europe, industrial consumption of water is more than 50% of the total withdrawal. In contrast, agriculture in developing countries consumes more water, which often rises up to 70% of the fresh water, whereas industry consumes only about 22% of fresh water. But these statistics are ever-changing. It is claimed that industrial water withdrawal is generally responsible for inflicting hardship upon marginal people. Industries withdraw much more water than they consume, which makes industries guilty of wasting a life-sustaining resource. In Figure 3.3, the percentage of freshwater utilised by industries against withdrawal is shown on a world map.

Although water sustains life, the vast majority of consumers are surprisingly ignorant about the hydrological cycle and how it is affected by human interference. This section discusses the fundamentals of water science or hydrology, specifically targeting managers and decision makers. For example, managers and decision makers must be scientifically informed to select a landfill site or to judge the suitability of a particular water resource in terms of its potential (to contaminate or extract), or to evaluate the expertise of hired help where necessary, and especially where water assessment, accounting and disclosure are called for. What is really important to know about water is shown in Figure 3.4.

3.2 Hydrological cycle

As emphasised earlier, water is a dynamic resource. The same water molecules are continuously circulating between the land and the seas, and move into the atmosphere by evaporation; they form clouds and come down as precipitation, finally returning to the sea via rivers and groundwater. This endless circulation of water is known as the *hydrological cycle* or *water cycle*. See Figure 3.5

Water plays a fundamental role in the climate system through the hydrological cycle. Evaporation of millions of cubic kilometres of water into the atmosphere requires an enormous amount of energy that is supplied in the form of solar energy. When water vapour in the atmosphere condenses to water droplets, it releases the latent heat of vaporisation absorbed from solar radiation at the rate of 532 calories per gram vapour; the total energy transfer involving millions of cubic kilometres of water vapour can be calculated, but the number is inconceivable in terms of workaday experience. Likewise, there are other energy transfers in the hydrological cycle involving ice-water phase transformation. Although fresh water is in limited supply, it also mediates in climate change to some degree through H_2O phase transfer in lakes, snow covers, glaciers, wetlands and rivers (Shearer, 2014). The hydrological cycle is an immensely powerful engine running on solar energy.

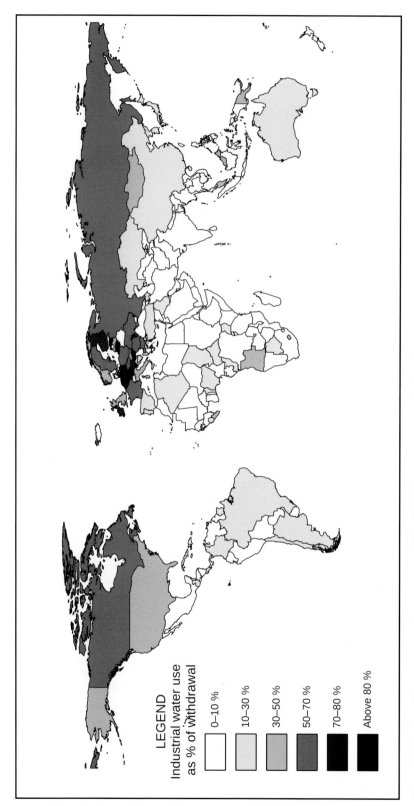

Figure 3.3 Industrial water uses by country. AQUASTAT data from FAO (no date).

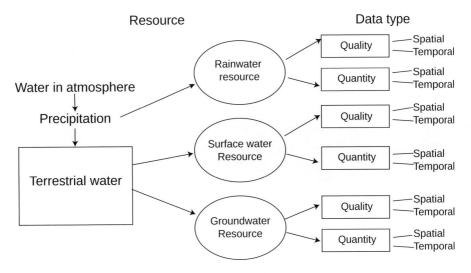

Figure 3.4 Water knowledge matrixes.

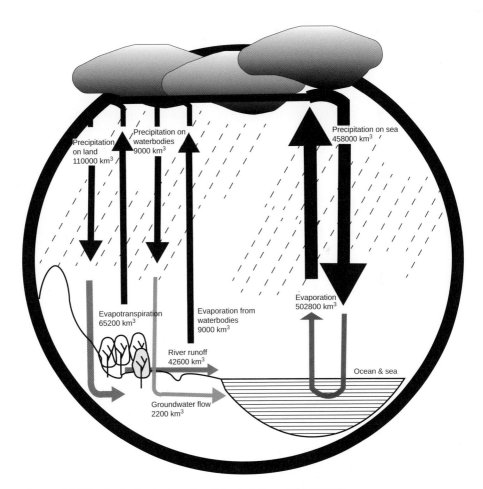

Figure 3.5 The hydrological cycle. Adapted from Oki (2006).

Of all the water on Earth, only about 0.35% is in the atmosphere at any given time, yet this small percentage controls formation of vital climatic elements, namely clouds, fog, rain, snow and hail. If precipitation were distributed evenly, every place on earth would receive 857 mm rainfall per year (Muriel Martin Elementary School, 2016). About 77% of all precipitation falls into the sea. The amount of evaporation from the Mediterranean Sea was first calculated by Edmund Halley (1656–1742), the English astronomer, who found that the amount of evaporation from the sea is equal to the inflow from rivers.

There are six important processes that make up the water cycle (Arnaiz, 2016). These are:

1 **Evaporation:** Evaporation is the process in which a matter is transformed from its liquid state to vapour state; in nature, this phase transformation involves vaporisation of water due to solar energy. Suspended and dissolved materials in water are left behind during evaporation; H_2O in vapour form is thus cleaner than the terrestrial water.

2 **Condensation:** When water vapour reaches the upper atmosphere, it is condensed due to the release of latent heat of vaporisation. Condensed water vapour occurs in the atmosphere as very minute water droplets which remain suspended in the air, forming what we call clouds. Formation of such condensed droplets near ground level is fog. Condensation takes place only around a nucleus, which in nature takes place around suspended dust particles in air.

3 **Precipitation:** In appropriate weather conditions, minute droplets in clouds come in contact with each other and form larger and heavier droplets that can no longer remain in suspension and precipitate as rain; sometimes precipitation occurs as hail or snow when condensation takes place in sub-zero-temperature clouds.

4 **Runoff:** The flow of rain and melted snow as moving sheets of water, down the slope is known as runoff. Runoff has two parts namely, surface runoff and sub-surface runoff or underground flow. A part of the sub-surface runoff remains as soil moisture and the other part percolates vertically downwards and accumulates in the aquifer. It is called groundwater.

5 **Surface Storage:** A part of the rainwater is stored in depressions as ponds, lakes, marshes and various types of surface bodies of water. These act as the storehouse of surface water.

6 **Evapo-transpiration:** Plants draw water from the soil and release it to the atmosphere through a process called plant transpiration. Water also evaporates directly from the soil, bodies of water and rivers. Collectively, this phenomenon is called evapo-transpiration. Bodies of water and large forests transpire millions of cubic kilometres of water into the atmosphere, which has a direct relationship to the amount of annual rainfall; which in turn means that deforestation is likely to adversely affect rainfall.

The water cycle originates from oceanic evaporation and completes when all terrestrial runoff goes back to the sea. Groundwater too flows, but at a slower rate and generally seeps out as natural springs or baseflows that feed rivers and bodies of water.

Every component of the water cycle or hydrological cycle discussed so far is important and has a deep impact on the ecology of a region. Human intervention in the cycle can have damaging repercussions. Constructing dams for the withdrawal of a large amount of water impacts the ecological flow. Large-scale withdrawal from the system gives rise to an imbalance which may not be perceivable in the short term, but will invariably have long-term implications in water availability in the region. Similarly, if a large part of a forest is cleared and turned into barren land or a built-up area, the affected natural evapo-transpiration can have a serious regional climate impact.

Due to an advancement of hydrological sciences, each component of a water cycle is now measurable on both a local and a global scale. There are innumerable measurable parameters that participate in the hydrological cycle. Important among them are precipitation, stream flow, groundwater, evapo-transpiration and soil moisture; all of which, therefore, require regular assessment and monitoring. In the following pages, these components and modalities of hydrological measurement are discussed in detail.

3.2.1   Water cycle and ecosystems

The water cycle nourishes the ecosystem by providing supplies of fresh water every year. The development, growth and sustainability of an ecosystem are controlled by the water cycle, while the water cycle itself is naturally controlled by geographical factors, as follows:

1 Geology controls the direction and volume of the terrestrial flow of water and the sediment load the streams carry. Groundwater storage and flow are controlled by the configuration of sub-surface geological formations.
2 Geographical location (i.e. latitude) influences the water cycle because the duration of sunshine is the deciding factor controlling evaporation.
3 The nature of land cover, biodiversity and human control also influences the water cycle.

Three components—industry, society and the ecosystem—participate in the functioning of the water cycle. These three components withdraw water and give it back as return flow, evaporation and transpiration. Industrial activities involving water withdrawal, discharge and storage gradually change the hydrological environment of a region. The effect of the change is most conspicuous in the (a) degradation of water quality, (b) reduction of flow in rivers, (c) reduced population of aquatic life forms and reduced biodiversity and (d) the depleting trend of groundwater table. Industrial impact on hydrological cycle is shown in Figure 3.6.

3.3   Water on land

When it rains or snow melts, water flows towards the lowest altitude of the basin. Rain lands first on trees, open land, or on rivers, lakes and seas; when it falls on a body of water, it only adds to the volume of the existing water. Rain has a different impact when it falls on land. Water falling on a plant or a tree trickles down the leaves, twigs and branches and eventually

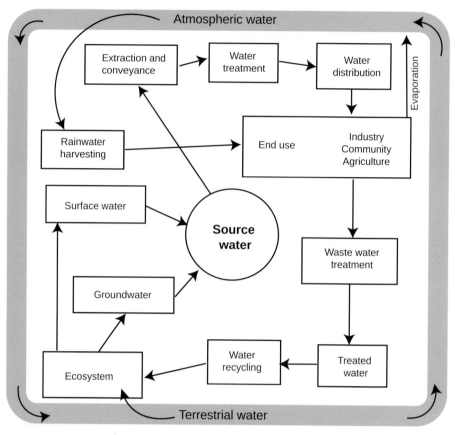

Figure 3.6 Functions of industry and ecosystem within water cycle. Adapted from Better World Solutions (2014).

falls on the ground. In the process, a small amount of water is captured by the tree, and a part of it is lost to evaporation.

The water cycle is a dynamic system which is in a state of continuous flux owing to major processes, namely evaporation and precipitation. After falling on land surfaces, rainwater takes four major courses: (1) part of rain is absorbed by the soil, (2) part is evaporated, (3) part of it percolates down to be stored as groundwater, and (4) the remaining part flows down-slope as surface runoff and drains into streams. Distribution of rainwater on land is shown in Figure 3.7.

3.3.1 Soil water

Soil has a unique property of holding water, but it also holds water for plants and organisms in the soil. Only plants have the ability to generate high enough osmotic pressure (up to 33 bars) to draw water from the root zone and send it to the leaves. Cohesive force between soil particles and water prevents human beings from extracting usable water. See Figure 3.8.

When all pore spaces of the soil are filled with water it becomes saturated. A saturated soil slowly releases the amount of water it cannot hold against gravity, and the released water flows downwards, which is called vertical

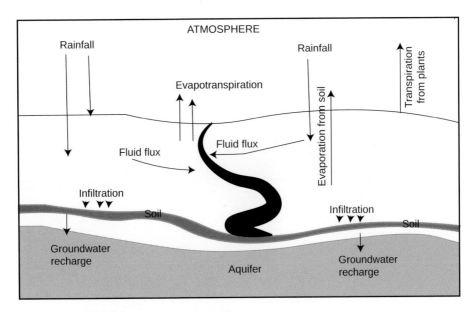

Figure 3.7 Distribution of water on land.

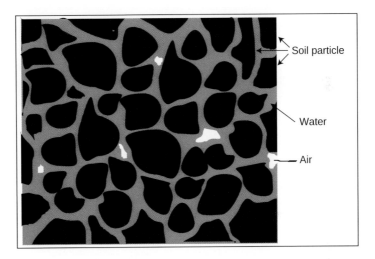

Figure 3.8 Water embedded in soil.

infiltration or percolation. Percolation in turn continues until the soil can hold water against gravity and attains a state defined as field capacity. Field capacity varies from soil to soil. For example, field capacity is about 15% to 25% for sandy soils, 35% to 45% for loam soils, and 45% to 55% for clay soils. The term 'wilting point' refers to the condition of soil when plants cannot draw any more water from the soil (dry condition).

Soil moisture

Soil moisture is defined as the amount of water present at a depth of one metre of soil, which is expressed either in percentage or in the unit of depth

Table 3.1 Available soil moisture and total available water of some important soil types.

Soil type	Soil moisture (%)	Total available water (mm/m)
Medium sand	6.8	68
Fine sand	8.5	85
Loamy sand	11.3	113
Sandy loam	7.9	79
Sandy clay loam	10.2	102
Loam	18.1	181
Silt loam	19.8	198
Clay	22.6	226

Adapted from Cornell University (2010) and Soil (2016).

(in mm of water depth). For example, a soil horizon of one metre thickness containing, say 15% of soil moisture can also be expressed as 150 mm/m (Brouwer, Goffeau and Heibloem, 1985).

Water holding capacity

Water holding capacity is the amount of water held between field capacity and the wilting point; this property depends on porosity and soil thickness. Available soil moisture and total available water of some important soil types are shown in Table 3.1.

The soil horizon does not extend much below the surface; depending on geological and geomorphological conditions it generally ranges from a few centimetres to a couple of metres.

3.4   Stores of water

The surface of the earth is neither flat nor smooth. There are many types of depressions formed both by natural and anthropogenic activities. Depressions on the earth's surface and other geological structures provide space for water accumulation. In a complete year, water is stored in all available storehouses, natural or otherwise, until the dynamism of the water cycle transfers it from one medium to another, say, for example, in the transfer of water from a river to a lake or vice versa. Major storehouses of water on earth are

- **Ice caps and glaciers:** 68.7% of fresh water on earth is trapped in the polar ice caps and glaciers, 'and if all land ice melted the seas would rise about 70 meters' (USGS, 2016).
- **Sea and ocean:** Ocean and seas are the biggest storehouse of water, hosting a great variety of marine biodiversity. Oceanic currents control global weather through circulation of heat energy between the equator and the poles.
- **Soil:** The amount of total soil-water is 16,500 cubic kilometres accounting for about 0.001% of the total water on earth. Estimation of soil moisture is

crucial to agriculture and crop planning. In a given area of land measuring one hectare with soil thickness of 1 metre, the soil moisture is measured at say 30% on a particular day; the amount of total water content of the soil then turns out to be 300 mm × 1 hectare = 0.3 hectare-meter (ham).

- **Surface water storage:** Surface water includes rivers, lakes and freshwater wetlands, which is subject to (a) evaporation, (b) sub-surface seepage, (c) withdrawal use and (d) discharge into the streams. The only replenishment of surface water is through precipitation or release of groundwater. Surface water has always been central to the development of human civilisation and its availability. Nature and distribution have been decisive factors in the formation of different cultures. Ancient cities like Tenochtitlan (in Mexico) and Tiahuanaco (in Bolivia) were built on the banks of great lakes. Surface water storages act as heat sinks and regulate the climate; if the local culture is not mindful of this balance it affects the local climate which may eventually lead to obliteration of the culture. Throughout the world, there are many large bodies of water which are being abused and threatened by various human activities. The Aral Sea and Lake Chad are two examples of severe degradation due to human abuse. The Aral Sea, one of the largest lakes in the world was once navigable by large ships but is now facing extinction. The water level has been sinking steadily since 1960 because of water diversion from the Amu Deriya and Sir Deriya rivers that feed the Aral Sea. 'By 2007, it had declined to 10% of its original size' (*Aral Sea*, 2016). The satellite images of the Aral Sea in 1989 and 2014 are shown in Figure 3.9. Similarly Lake Chad in Africa is facing extinction due to overexploitation of water by surrounding countries.

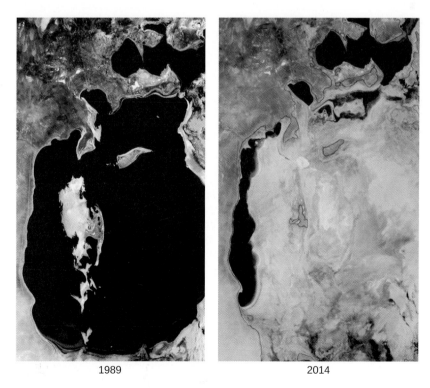

1989   2014

Figure 3.9 Aral Sea by NASA. Collage by Producer Cunningham. Public domain through Wikimedia Commons.

3.5   Surface runoff

To quote from USGS (2015a), 'when rain hits saturated or impervious ground it begins to flow overland down slope. It is easy to see if it flows down your driveway to the curb and into a storm sewer, but it is harder to notice it flowing overland in a natural setting. During a heavy rain you might notice small rivulets of water flowing downhill. Water will flow along channels as it moves into larger creeks, streams, and rivers.'

The amount of rainwater that tends to flow as surface runoff in a watershed or catchment is controlled by several meteorological and geographical factors.

3.5.1   Meteorological factors affecting runoff

The following meteorological factors affect runoff:

- **Type of precipitation (rain, snow, sleet, etc.):** Snow and sleet take considerable time to melt and to be manifest in surface runoff. Rain, on the other hand, causes almost immediate surface runoff in a catchment, provided that the rate of precipitation is higher than the combined effect of evaporation, initial abstraction and absorption by soil. Three major precipitation-related factors that control the rainfall runoff are (1) rainfall intensity, (2) rainfall amount and (3) rainfall duration. Rainfall intensity is defined as the ratio of the total amount of rain (rainfall depth) to the duration of rain (Critchley and Siegert, 1991). It is expressed in depth per unit time, usually as mm per hour or inches per hour.
- **Direction of storm movement:** Direction of storm movement influences the drainage or runoff of a stream and exercises a significant influence on the peak flow and time to peak flow (Shearman, 1977).
- **Soil moisture conditions:** Dryness of soil has an impact on the manifestation of storm runoff. If the soil is dry, rainwater will first be absorbed by the soil and the rest will go into overland flow. On the other hand, if the soil is saturated (called antecedent soil moisture condition) due to a previous rainfall, soil will capture less, or no water at all, causing instant runoff.
- **Evaporation and evapo-transpiration:** These two climatic factors determine how much water will be lost to the atmosphere due to evaporation and transpiration. Potential evapo-transpiration (PET) determines the amount of evapo-transpiration from soil and evaporation from open bodies of water. If sufficient water is available, PET equals evaporation from lake or actual evapo-transpiration (AET) from soil. Depending on these factors, the amount of water lost to evaporation and transpiration reduces the actual surface runoff.

3.5.2   Physical factors affecting runoff

- **Land use:** Land use affects runoff. Land covered by vegetation lowers the rate of runoff, whereas urban land use patterns cause rapid surface discharge. In general, runoff is between 50% and 90% of precipitation in areas of human settlement, whereas in forested areas the runoff may be as low as 5% of precipitation.

- **Drainage area:** The drainage area of a river basin determines the volume of runoff that will ultimately generate the stream flow. The other factors that determine stream flow are slope and elevation. Slope of the basin is important because in high-slope regions runoff will be faster and larger than in areas with low or gentle slope. It is therefore extremely important to study the catchment gradients to manage runoff and prevent erosion in the catchment.
- **Drainage pattern:** Drainage patterns are the matrix of streams that combine with other smaller streams to form a larger one. This network determines the length of the path that water will have to traverse to reach the mouth of the basin. If the total path is long, runoff will take more time to reach the mouth of the river basin.
- **Ponding:** Ponds, lakes, reservoirs, sinks, potholes and so on capture a good amount of precipitation and reduce or delay the downstream flow of water.
- **Latitude:** The duration of sunshine is a determining factor in estimating potential evapo-transpiration (PET). The duration of sunshine in a particular place depends upon the time of the year, length of a day and, most important, on the latitude of the place.
- **Altitude:** Three major rivers, namely the Ganga, Indus and Brahmaputra constitute the lifeline of India, and all three of them originate from Himalayan glaciers. It can thus easily be appreciated just how important a role glaciers and high altitude snow play in surface runoff.

The relation between rainfall intensity, runoff and soil is too complex for deterministic modelling; this complex relationship is therefore understood on the basis of empirical formulae. The formulae have been validated by hydrologists over decades of observation and have also been validated in laboratory experiments. The data acquired from these empirical equations have long been used in crop planning and irrigation design. It needs emphasising that such data are also integral to planning major industries and are routinely used as baseline data for storm-water management in major industries.

The amount of runoff against rain varies in space-time. For a given amount of rainfall, the Thor Desert in India will not have the same runoff effect as in the rain forest of Indonesia. Only about one-third of the precipitation on land is delivered as surface runoff in its journey back to the ocean; the remaining part undergoes evaporation or transpiration, or infiltrates into groundwater aquifers.

3.5.3 Human activities can affect runoff

Urbanisation, civil construction, dams, reservoirs and building any impervious infrastructure affect both surface and sub-surface runoff. In urban areas storm rain flows more rapidly than in areas under vegetation or forest cover. Withdrawal of water from streams or aquifers also affects runoff and reduces the flow. In short, human activities fundamentally affect runoff. See Figure 3.10.

The ultimate effect of surface runoff is the formation of a stream. The size and shape of a stream network depends on the size, geology and geomorphology of the basin. There are several reasons for which the flow in a river may be impaired. In Table 3.2, the mechanism that causes changes in stream flow are compared.

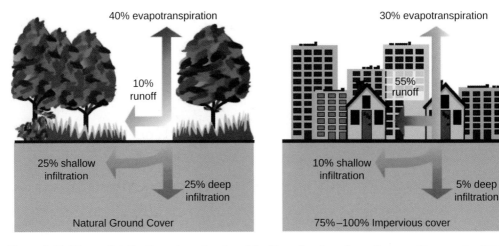

Figure 3.10 Water distributions in natural and built up land surface. Source: Stormwater (2016). Original figure by U.S. Environmental Protection Agency, Washington, D.C., via Wikimedia Commons.

Table 3.2 Mechanisms that cause changes in stream flow.

Natural mechanisms	Human-induced mechanisms
• Runoff from rainfall and snowmelt	• Surface-water withdrawals and trans-basin diversions
• Evaporation from soil and surface bodies of water	• River-flow regulation for hydropower and navigation
• Transpiration by vegetation	• Construction, removal and sedimentation of reservoirs and storm-water detention ponds
• Groundwater discharge from aquifers	• Stream channelisation and levee construction
• Groundwater recharge from surface-bodies of water	• Drainage or restoration of wetlands
• Sedimentation of lakes and wetlands	• Land-use changes, such as urbanisation, that alter rates of erosion, infiltration, overland flow or evapo-transpiration
• Formation or dissipation of glaciers, snowfields, and permafrost	• Wastewater outfalls
	• Irrigation wastewater return flow

Source: USGS (2015b).

3.6 River and river basin

Understanding a river system is very important for water managers, especially when the source of water is a stream or river valley or when industrial discharge is released into a stream. In both cases, the stream or river basin is affected by the industrial activity and has a long-term effect on the water cycle. Industries prefer proximity to rivers for three reasons: (1) the convenience (low-cost, higher traffic safety, higher carrying capacity and much reduced probability of goods damage) of waterway transport; (2) easier and cheaper way of pumping water; and (3) easy disposal of waste and effluents. Environmental activism and laws are forcing a change in industry attitudes.

Figure 3.11 A river valley. Photo courtesy Debasish Guha.

Howrah Station, one of the busiest and most crowded railway stations of the world, draws its water from the Hooghly River (Ganga) and after waste water treatment discharges better-quality water back into Hooghly River.

Rivers are naturally flowing pathways of water on the land surface (Figure 3.11). Barring, of course, the few underground rivers flowing through caves and caverns in a limestone country; the Camuy River, one of the largest underground rivers in the world is located in a limestone cavern near Puerto Rico. A river may flow into another river, lake, sea or ocean, or may disappear in parched land before reaching its destination. Movement of water from upstream to downstream is measurable and is one of the most defining elements of the river basin called the river discharge.

Water in a river may come from three sources;

1 **Snow or glacier:** There are several rivers on the earth which originate from glaciers, one of which is Ganga, the holy river of India. Snow or glaciers generally supply water during the summer season. Glacial rivers are essentially perennial, that is, they carry water round the year, even during non-monsoon months.
2 **Rainfall in the river catchment:** Rainfall in the river catchment supplies water to the stream through direct runoff, baseflow and overland flow.
3 **Groundwater:** Spring and effluent seepage also contribute to river flow where the water table is above the river bed.

3.6.1   Stream order

Streams are classified according to their hierarchical position in the river network of a river basin. The first stream that originates from the highest upstream region as a spring or a little pool or a glacial snout is called a first-order stream. Two first-order streams combine to form a second-order stream, and two second-order streams unite to form a third-order stream and so on. Stream order is used for understanding the stream network of a river basin.

3.6.2   Drainage basin, catchment and watershed boundaries

As defined in the *Federal Standard for Delineation of Hydrologic Unit Boundaries,*

> A hydrologic unit is a drainage area delineated to nest in a multi-level, hierarchical drainage system. Its boundaries are defined by hydrographical and topographic criteria that delineate an area of land upstream from a specific point on a river, stream or similar surface waters. A hydrologic unit can accept surface water directly from upstream drainage areas, and indirectly from associated surface areas such as remnant, non-contributing, and diversions to form a drainage area with single or multiple outlet points. Hydrologic units are only synonymous with classic watersheds when their boundaries include all the source area contributing surface water to a single defined outlet point (USDA, no date).

A river catchment is an area surrounded by relatively higher terrain within which all precipitation flows into a river system and ultimately drains out through a 'single point at a lower elevation' (ABC Science, 2008). Big catchments cover several thousand square kilometres, but there are small catchments as well occupying only a few square kilometres, while the forces operating within the catchment are identical irrespective of shape and size of the catchment. There are several terms to describe a catchment, such as catchment basin, river basin, drainage basin, watershed and so on. According to the free dictionary, 'a watershed is defined as a ridge of high land dividing two areas that are drained by different river systems.' See Figure 3.12.

3.6.3   Classification of river basin and hydrological unit

According to the USDA, 'Watershed boundaries define the aerial extent of surface water drainage to a point. The intent of defining hydrologic units (HU) for the Watershed Boundary Dataset is to establish a baseline drainage boundary framework, accounting for all land and surface areas.' The United States Geological Survey created a hierarchical system of hydrologic units originally called regions, sub-regions, accounting units, and cataloguing units. Each unit was assigned a unique Hydrologic Unit Code (HUC).

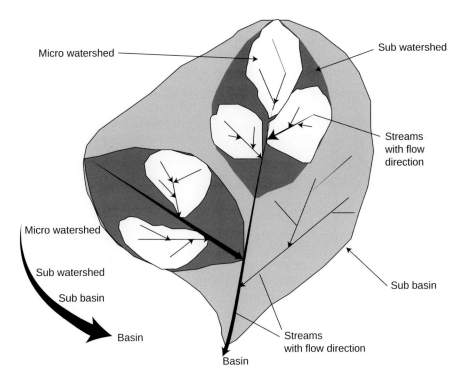

Figure 3.12 Concept of basin hierarchy.

Table 3.3 Classification of river basin.

Name	Level	Average size (sq. miles)
Region	1	177,560
Sub region	2	16,800
Basin	3	10,59
Sub basin	4	700
Watershed	5	227

Source: USDA (no date)

USGS classification of hydrological units are given in Table 3.3

The Indian classification of river basins is a little different from that of the USA. Catchments smaller than a minor river basin or a sub basin (as in the case of the US classification) are called watersheds. See Table 3.4.

River basins do have their own network that follows a natural drainage hierarchy. Mini, small or micro watersheds deliver water into larger sub watersheds and sub watersheds drain into a larger basin or sub basin.[1]

A river basin is not simply a piece of land but a complex eco-hydrological unit consisting of various landscapes, hills, valleys, river courses, lakes, sediments, aquifers and cultural elements associated with it, such as villages, towns, cultivated fields, industries, and so on. A river catchment can be divided into an upper catchment (Figure 3.13) where the erosion by the river

Table 3.4 Indian classifications of river basin and watershed.

Sr. No	Type of river basin/watershed	Area covered
1	Major river basin	>20,000 sq. km
2	Medium basin	20,000 to 2,000 sq. km
3	Minor (Coastal areas) basin	< 2,000 sq. km
4	Macro watershed	400 to 1000 ha
5	Sub watershed	200 to 400 ha
6	Mini watershed	40 to 200 ha
7	Small watershed	10 to 40 ha
8	Micro watershed	0 to 10 ha

Source: SANDRP (2014), and *Classification of watershed* (2015).

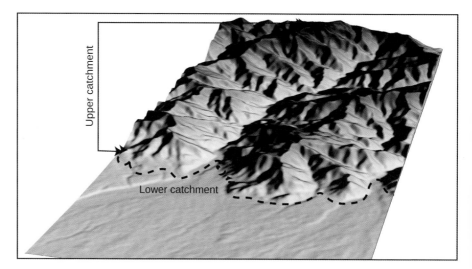

Figure 3.13 Upper and lower catchment of a river.

dominates and the lower catchment where sediment deposition is the major activity. This division into upper and lower catchment does not apply to small rivers with high altitudes, which constitute almost exclusively an erosional domain of a large river system. For instance, the Alakananda River, which is an upper catchment tributary of the Ganga, does not have any lower catchment.

3.7 Industrial impact on river flow

The collective impact of human activity causes large-scale despoliation of a river basin. But the impacts in upper and lower catchments are radically different, largely because damages in the upper catchment have a cascading effect and become even more damaging in the lower catchment of the same river basin. Negative consequences of industrial activity on river basins are listed below (adapted from Govorushko, 2007).

3.7.1 Temporal and spatial control over river flow

Dam and reservoir construction for power generation radically modifies the river configuration by flow redistribution both in space and time. Flood control dams associated with hydel-projects and irrigation canals are used to control the flow volume and duration. For example, in a trans-boundary river basin (like the Ganga which flows through India and Bangladesh[2]) the daily flow rate into Farakka feeder canal, or into the Bangladesh part of the river, is regulated by an international commission/treaty, thereby bringing about a fundamental alteration in the natural system.

3.7.2 Water direct withdrawal

Water direct withdrawal is a common practice among industries located in the vicinity of rivers. These industries source their water from the rivers, causing water shortages in the downstream region.

3.7.3 Physical disturbance of riverbeds

A river flows through a valley that it creates. In the lower catchment, valley width increases with the maturity stage of the river. A mature river valley is wide and consists of floodplains and terraces. Terraces are older floodplains and a raised surface from the river bed. Industries situated above the flood plain and on the upper terraces often construct check dams and low transverse bars within the river to facilitate water withdrawal. Sometimes, even in seasonal rivers baseflow is also withdrawn by constructing infiltration galleries along the river bed. These constructions and mining of sand, minerals, and so forth from the river bed cause serious physical disturbance and affect normal flow of river.

3.7.4 Pollution

Industries not only cause physical damage to the river system by 'water direct withdrawal' and water infrastructures but also pollute the river qualitatively. Industries located near rivers generally dispose their wastewater into the rivers or municipal sewer system (UNESCO–WWAP, 2006), which in turn discharges in the downstream part of the river. The discharged water is sometimes treated following national guidelines, but tragically such instances of treated water returning back into the river are rare.

3.7.5 Water clogging

Littering river beds with industrial waste is by no means uncommon, especially in the context of small-scale industries. The commonest example of clogging comes from river bed brickfields in the Indian sub-continent (Figure 3.14). Fish nurseries, which are also constructed on river beds, do not cause littering (for their own good), but they do cause serious effects in the natural flow of rivers.

Since time immemorial, rivers have sustained civilisation throughout the world, but river pollution is a recent phenomenon, and pollution levels have now reached a point where rivers have started hitting back by causing serious health hazards. To prevent this problem from getting worse, industry will have to make up the damage it is doing to the river system, taking the government into their confidence. For example, the pollution of Yellow River of China has reached a level where its water is not even fit for industrial use. The Chinese Government launched an intensive campaign in 2008 to reverse the abysmally poor environmental condition of Yellow River.

A simple guideline for industries is prescribed below. The following tasks can be taken up to protect a river (adapted from Montana River Action, no date)

- Collect or prepare all information regarding history, geography, culture, politics and natural resource information of the river catchment and prepare different thematic maps to comprehend the problems and opportunities in protecting the river.
- Protect the riparian forests, rain forests, wetlands, springs and the floodplain from degradation and loss of soil and biodiversity.
- Involve the community in taking action.

Figure 3.14 A brickfield on the floodplain of Padma River, Bangladesh.

- Moderate water withdrawal, river training and diversion of flow.
- Allow the river to play freely over its floodplains.
- Promote appropriate land use involving stakeholders.
- Control invasive species of plants and animals that can impact ecological balance.
- Control discharge of polluting agents into river.

3.8   Surface water management

Surface water management is a part of the water resource management of a river basin or catchment with the objective to maximise the availability of water from surface bodies of water and rivers and to minimise the flooding problems that impact economic growth. Such management is generally carried out by the government or by local authorities. Industries with their vision on water stewardship should participate in surface water management programs for their own benefit. A surface water management program may be a complementary activity under a source-water assessment program or any activities under a CSR program. It may also be a prerequisite of an industry prior to commissioning. In most countries, there are frameworks which guide the program involving different sections of stakeholders. DEFRA (2010) has prepared a document on surface water management plans (SWMP). According to that document, a SWMP is a plan which outlines the preferred surface water management strategy in a given location. This framework often has the scope to accommodate different organisations to work together and develop a shared understanding of the most suitable solutions.

Under the framework described in the document published by DEFRA (2010), the scope of the work may also be extended in industrial ecosystems. Industries may involve themselves in surface water management programmes on their own with the following objectives:

- Develop a storm-water management plan and flood control.
- Initiate a programme to develop local bodies of water, small streams and ponds by creating structures like check dams, *bunds*[3] and other rainwater-harvesting and flood-controlling structures.
- Manage surface water to abstract stream flow for industrial use.
- Manage surface water for future capital investment, drainage maintenance, land-use planning.
- Managing the drainage of industrial effluent and mining discharges.

The circumstances in which an industry should conduct a SWMP depend on the geographical location and size of the company. Some examples where such managements are required are:

- Large plantations located in rain-fed areas or in a flood-prone area of a basin need to take up such management activities.
- Large tea gardens and coffee plantations are generally located in hilly areas where there is ample rainfall. In these areas, large amounts of overland flow can cause damage to the crop and can cause landslides.

- Industries that occupy a large land parcel in the strategic areas in the flood plains of a river basin are also vulnerable to flood and require appropriate management strategies to control the runoff.

3.8.1   Key component of a SWMP

A surface water management plan should be well structured and supported by local environmental directives. Some of the key components are as follows.

Identify the need for a SWMP study

The main drivers of SWMP are (adapted from DEFRA, 2010):

a **Legal requirements:** Environmental and water laws of some countries ask industries to assess the risk of flood in the industrial location as a requirement under environmental impact assessment (EIA).
b **Flooding history:** If the location of the industry has a history of flooding, the study may help to assess the risk and vulnerabilities.
c **Planning for new development:** Future extension of the industry or development of township or changing the land use may be properly planned.
d **Environmental protection:** According to US-EPA (US-EPA, no date), surface water management for environmental protection includes activities to limit the discharge of pollutants into the water, and regulate storm-water discharge in accordance with the local laws and also the management of runoff and run-on in connection with waste disposal.

Establish partnership

It is always necessary to identify who are going to be the partners of the endeavour. Partners should belong to the stakeholders, who generally come from the government side (water resource authority, river basin authority, environment and forest department, rural and urban development authorities, navigation and canal authorities, etc.), industrial side and local community. There are other stakeholders belonging to the local resilience forum, riparian owners, farmers' union, local contractors, and so on.

Scope the SWMP study

Clearly identify the stakeholders before you share the process and possible outcomes. Clarify the strategic relationship among the stakeholders and with the leadership. Provide the investment strategies and the deliverables from individual members of the group

- **Establishing an engagement plan:** This covers an identification of the actual workforce from the stakeholders and distribution of the tasks.
- **Identifying the availability of information:** Identification of sources of information, availability of data, maps, tables etc., along with quality of data, relevance and applicability.

Table 3.5 Key components of strategic assessment.

Criteria	Description
Purpose	To obtain a general understanding of surface water flooding within a local authority area.
Scale	Local catchment
Inputs (data and information)	• Historic flood incident data
	• Nationally developed surface water flooding map
	• Ground model data
	• Information from Strategic Flood Risk Assessments (SFRAs), Catchment Flood Management Plans (CFMPs) and Shoreline Management Plans (SMPs) of a country
Process	To combine historic flood incident data with the existing maps, or other simple analysis techniques, to identify areas more vulnerable or susceptible to surface water flooding
Output	Objectives and prioritised list of locations for further assessment. Mapping which identifies the hotspots or settlements more vulnerable to surface water flooding
Benefits	Further assessment is targeted in locations which are more vulnerable to surface water flooding.

Source: Adapted from: DEFRA (2010) © Crown copyright 2010.

- **Identifying the level of assessment:** This sets the spatial and temporal scale of the assessment such as identifying the catchment boundary, length of study, number of years of historical data collection, periods of assessment.
- **Undertake a strategic assessment:** Finally, undertake a strategic assessment based on the specific objectives of the study. The strategic assessment is oriented and limited by the objectives of the work. Key components of a strategic assessment are shown in Table 3.5.

3.9 Groundwater

A good share of terrestrial fresh water is trapped in ice caps and glaciers. If the frozen reserves are discounted, about one-third of the freshwater reserve occurs as groundwater (Siebert *et al.*, 2010). Groundwater therefore is a major source of water for industries, agriculture and domestic consumption. Industries (including extractive industries, e.g. mines and oil fields) draw vast quantities of groundwater for industrial processes and for extending water-related services to the employees or community. In fact, industries located in major groundwater basins are almost entirely dependent on groundwater. For example, recent industrial development in Chittagong, Bangladesh is highly dependent on groundwater.

Use of groundwater in industries is a global practice; but the quantity of water drawn and consumed varies widely. Figure 3.15 shows the 2010 factsheet (NGWA, 2016) of groundwater withdrawal in different sectors from fifteen major countries.

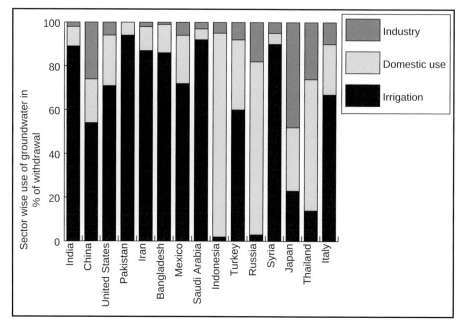

Figure 3.15 Sector-wise and country-wise groundwater use expressed as percentage of withdrawal in 2010. Data source: NGWA (2016).

3.9.1 Groundwater hydrology (hydrogeology)

A working knowledge of hydrogeology begins with a very common natural phenomenon which is familiar to everyone, namely percolation of rainwater into ground. The downward movement of rainwater first rests a while in the uppermost part of the soil horizon, that is, the root zone, and then resumes its downward journey until field capacity is attained; the second phase of the downward journey continues until it reaches a level where it begins to accumulate. The zone of accumulated water in the lower reaches of the soil horizon is called an aquifer. By definition, an aquifer is a layer of sediments or fractured or porous rocks which can yield water. It needs pointing out in this context that this property of yielding water is the defining element of an aquifer; for example, a water-saturated clay horizon that does not yield water is not considered as an aquifer, no matter how much water is stored in the clay horizon. The aquifer material is mainly of two types: (1) unconsolidated sediment and (2) fractured or porous rocks; both types of aquifers may occur at the same geographical location in which, it goes without saying, that the former always overlies the latter.

By and large, aquifer materials are unconsolidated sediments or lithified sediments, that is, sedimentary rocks. Other than aquifer material of sedimentary origin, there occurs below the soil horizon a zone of metamorphic or igneous rocks altered by weathering, which is underlain by relatively unaltered but fractured rocks that also act as the aquifer of that region.

I have already introduced the definition of groundwater and an aquifer. The discussion can now take up the groundwater environment that controls the storage and yield of an aquifer. Yield and storage are functions of aquifer parameters, which are determined experimentally. Without going through the complex mathematical details of groundwater hydrogeology, some salient terms, their definitions and applicability are discussed in this chapter. Corporate managers will get some idea as to how the terms are applied in a hydrogeological report and the implications of these terms in the context of taking appropriate measures for groundwater conservation, or even in a relatively simpler decision of installing a tube well.

3.9.2   Fundamentals concepts

A groundwater reservoir is a domain of multiple hydrogeological dimensions and is classified under several zones. According to Heath (2004), 'All water beneath the land surface is called underground water' and the water that is stored or flows over the land surface is called surface water. Underground water occurs in two zones. The uppermost zone, which is aerated, is called the unsaturated zone, that is, containing both water and air; this is underlain by a water-saturated zone (i.e., aquifer). The underground water domain may be divided into four parts:

1 **The soil zone:** The zone which extends up to a maximum depth of one metre or two and is crisscrossed by the roots of plants.
2 **The intermediate zone:** Defined as the zone between the aerated soil and the water-saturated zone.
3 **The capillary zone (capillary fringe):** The lower part of intermediate zone contains some water held in place by capillary suction, hence called the capillary zone. As capillary suction works against gravity, the hydraulic pressure here is negative.
4 **Saturated zone:** The capillary fringe is underlain by a saturated zone where water is stored as groundwater. The hydraulic pressure of groundwater increases with increasing depth; the water table is the upper bounding surface of the saturated zone where the hydraulic pressure is equal to atmospheric pressure.

These four zones are schematically shown in Figure 3.16, of which the top three zones are collectively known as the 'vadose zone'.

3.9.3   Aquifer and confining beds

Unconsolidated sediments are of various types, depending on grain size and porosity. A good aquifer must have enough inter-granular space to accommodate water, and the pores must be interconnected to allow the movement of water (Heath, 2004). Very fine sediments (i.e. clay, grain size less than or equal to 0.002 mm) do not form aquifers, even though they can hold a large quantity of water. Clay does not allow movement of water, because the water remains adsorbed on the surface of the clay particles. Table 3.6 shows the grain size of different types of sediments and their nomenclature according to the US standard.

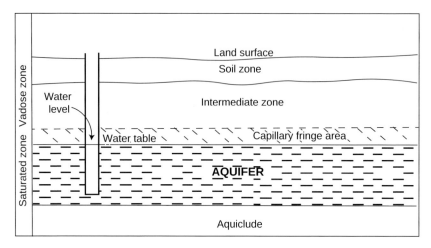

Figure 3.16 Zones of underground water. Adapted from Heath (2004).

Table 3.6 Grain size of some aquifer materials

Nomenclature	Grain size in mm	Aquifer type
Very coarse sand	1 to 2	Good
Coarse sand	0.5 to 1	Very good
Medium sand	0.25 to 0.5	Good
Fine sand	0.125 to 0.25	Good to moderately good
Very fine sand	0.050 to 0.125	Poor

Like rivers, groundwater also flows under the force of gravity and moves from a state of higher to lower hydraulic head. Movement of groundwater in a particular aquifer, unlike rivers, occurs in different directions following the hydraulic pressure gradient of the water table. This inevitably leads to the question as to how rivers and aquifers interact. There are two types of interactions:

1 Where the aquifer intersects the river valley feeding the river, called 'effluent seepage' (Figure 3.17a)
2 Where water from river beds seeps underneath, feeding the aquifer, called 'influent seepage' (Figure 3.17b).

This hydraulic connectivity between river and aquifer is an important component of the water cycle and an environmental phenomenon that maintains the ecological flow of region.

From the standpoint of groundwater storage, all rocks, consolidated or otherwise, are either aquifers or confining beds. Confining beds are those rocks which act as a barrier between the atmosphere and the aquifer or between two aquifers. Confining beds do not allow water to pass through or move within the bed.

Groundwater storage in aquifers takes place under two different conditions: (1) confined aquifers, and (2) unconfined aquifers. See Figure 3.18. Unconfined

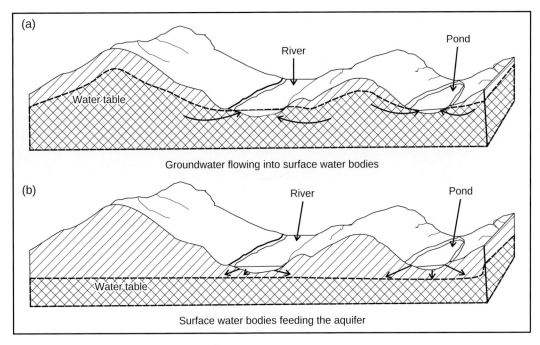

Figure 3.17 Interconnectivity between groundwater and surface water. Adapted from Khattak (2014).

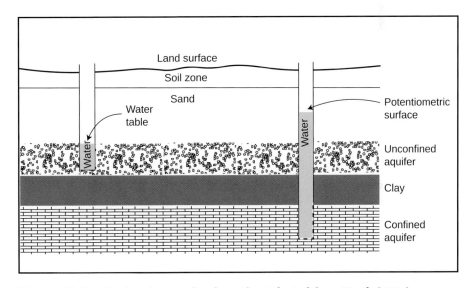

Figure 3.18 Confined and unconfined aquifer. Adapted from Heath (2004).

aquifers are those in which precipitation can vertically percolate to the saturated zone. The aquifer is generally partially filled, and the hydraulic pressure of the upper surface is equal to the atmospheric pressure. In confined aquifers, the whole aquifer remains filled with water and is overlain by an impervious layer which does not allow vertical percolation of rainwater into the aquifer

Figure 3.19 An auto-flowing well.

(Heath 2004). The hydraulic pressure of a confined aquifer at any point is higher than the atmospheric pressure because of the added pressure of the overlying rocks. It is well known that water seeks its own level, so when a well is drilled in a confined aquifer the water level rises above the aquifer and 'stands at a level where the hydraulic pressure is equal to the atmospheric pressure or the level of potentiometric surface' of the aquifer (Heath, 2004).

This spontaneous rise of water from a confined aquifer is called the artesian condition; it may occur in this way at places where the potentiometric surface is above the ground level; a well drilled in such situations becomes auto flowing. See Figure 3.19.

The main parameters that govern flow, yield and recharge capacity of an aquifer are (1) porosity, (2) specific yield, (3) head and gradient, (4) hydraulic conductivity, (5) storage coefficient and (6) transmissibility. These are briefly discussed below.

Porosity

Porosity is an attribute of an aquifer that denotes the capacity (Heath, 2004) of the aquifer to accommodate water. It is defined by the ratio of actual volume of the material to the volume of total pore space in that material; porosity of a material is:

$$\text{Porosity (n)} = \frac{\text{Total volume of void space in the aggregate}}{\text{Total volume of aggregate}} = \frac{Vv}{Vt} \quad (3.1)$$

Table 3.7 Porosity, specific yield and specific retention of some aquifer materials.

Material	Porosity	Specific yield	Specific retention
Soil	55	40	15
Clay	50	2	48
Sand	25	22	3
Gravel	20	19	1
Limestone	20	18	2
Sandstone (semi consolidated)	11	6	5
Granite	0.1	0.09	0.01
Basalt (young)	11	8	3

Adapted from Heath (2004).

where, porosity 'n' is expressed in percentage or decimal function; and 'Vv' and 'Vt' are volume of void space in the rock and total volume of the rock respectively. Porosity is of two types, primary and secondary. Primary porosity is the original porosity of the rock. Secondary porosity is generated in lithified rock affected by fracturing. Porosity depends on the grain size, sorting of grains and extent of lithification. For example, a typical aquifer medium may consist of (1) unconsolidated sand which has primary porosity only, (2) sandstone which has both primary porosity and secondary porosity, and (3) crystalline rocks, e.g. granite, have secondary porosity only.

Specific yield and specific retention

Specific yield and specific retention are two hydrogeological parameters of an aquifer which determine how much water is actually available. Specific yield and specific retention are different from porosity but these three aquifer parameters are interrelated by the equation:

$$n = S_y + S_r,$$ (3.2)

where, 'S_y' is specific yield and 'S_r' is specific retention, and 'n' is the porosity as mentioned earlier. Table 3.7 shows general values of porosity in relation to specific yield and specific retention of some selected aquifer materials.

Heads and gradient

Hydraulic head or piezometric head is a specific measurement of liquid pressure above a geodetic datum. It is usually measured in units of length. Gradient is the pressure difference between two points on the piezometric surface or water table, which governs the flow of groundwater. Quantified understanding of hydraulic head and pressure gradient are essential for maximising pumping efficiency. For measuring the head and gradient of

Water level measurement

Depth of water level in an unconfined aquifer or the depth of piezometric surface is measured from land surface, and the value should be determined relative to a common datum plane. The most widely used datum plane is sea level or mean sea level. By subtracting the depth of water level from the altitude of the measuring point, the reduced level (RL) or the total head at the well is obtained. 'As out of the three components of total head, that is elevation head, pressure head and velocity head the velocity component is very low and ignored; the total head represents two components: elevation head and pressure head. Groundwater moves in the direction of low total head from a higher total head' (Heath, 2004, p. 10).

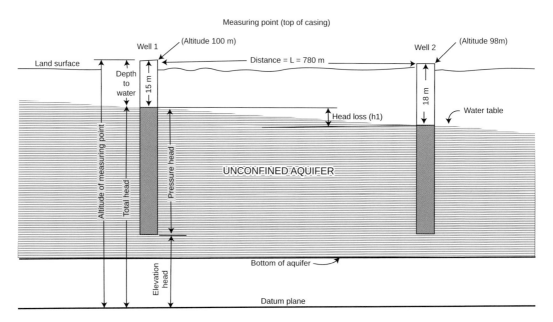

Figure 3.20 Hydraulic head and hydraulic gradient of groundwater. Adapted from Heath (2004).

a groundwater system, the parameters that are measured in the field are (1) depth of water level from the ground level and (2) the 'reduced level' of the water level with respect to the datum surface. These are illustrated in Box 3.1.

The rate of groundwater flow depends on hydraulic gradient, which is defined as the change in head per unit distance. In Figure 3.20 'Well 1' and 'Well 2' are two observation wells, and it is assumed that water flows from 'Well 1' to 'Well 2'. The horizontal distance between the wells is 'L' and the head loss is expressed as 'h_l'. The hydraulic gradient is expressed as h_l/L.

Hydraulic conductivity (Heath, 2004):

The capacity of a porous medium to transmit water is called hydraulic conductivity. *Coefficient of permeability* is another term for hydraulic conductivity. Groundwater moves from higher total head to lower total head. According to Darcy's Law:

$$Q = KA(dh/dl) \tag{3.3}$$

Where 'Q' is the quantity of water flow per unit of time; K is the hydraulic conductivity that depends on the size, texture of medium and the dynamic characteristics of the fluid (i.e. density and kinematic viscosity of water), 'A' is the cross-sectional area of a unit prism of aquifer, at a right angle to the flow direction, and dh/dl is the hydraulic gradient.

In short, hydraulic conductivity can be defined as the volume of water (Q) that will move in a unit of time (commonly, a day) under a unit hydraulic gradient (such as a meter per meter) through a unit area (such as a square metre). The unit of hydraulic conductivity 'K' is m/day.

Methods of determining hydraulic conductivity

There are two broad methods of determining hydraulic conductivity of a porous media (e.g. soil, aquifer material). These are (1) hydraulic methods based on Darcy's Law and (2) correlation methods.

Hydraulic methods are based on field or laboratory experiments that come from supposedly certain flow and boundary conditions, where Darcy's Law is applicable for computation of experimental data. The hydraulic field methods can be divided into small-scale and large-scale methods, where small-scale field methods serve for fast testing of many locations. In small-scale methods, the experiments are conducted on small auger holes, pits and piezometers, where rate of percolation of water through the pit or hole is determined. Pumping tests are conducted in the large-scale method.

Correlation methods are based on the empirical approach. The 'K' value can be obtained from a predetermined relationship between soil property (e.g. grain size distribution, texture.) which can be easily determined.

The method of determining hydraulic conductivity is shown in Figure 3.21

Transmissivity

Transmissivity is a quality of an aquifer that denotes the capacity of the aquifer to transmit water of the prevailing kinematic viscosity. In the case of transmissivity or transmissibility, the total thickness of the aquifer is taken into consideration. It is also defined as the rate of flow of water through the entire thickness of the aquifer. The value of transmissivity derives from the equation below:

$$T = Kb \tag{3.4}$$

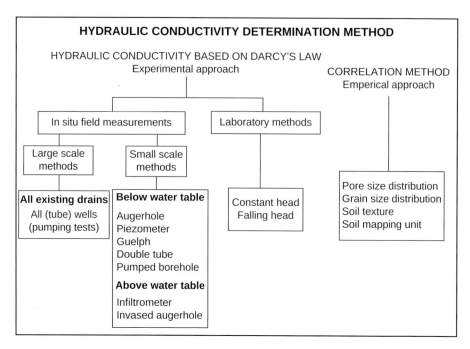

Figure 3.21 Methods of determining hydraulic conductivity. Adapted from R.J. Oosterbaan via Wikimedia Commons.

where 'T' is transmissivity, 'K' is hydraulic conductivity and 'b' is aquifer thickness; the unit of transmissivity is m²/day.

Flow of water

The amount of water moving through a cross section of the aquifer per unit gradient and viscosity also derived from Darcy's Law, that is,

$$Q = KA(dh/dl) \tag{3.5}$$

where, 'Q' is the flow and 'A' is cross-sectional area of the aquifer. It needs pointing out that in this context A is the product of width (W) and thickness (b), i.e.:

$$A = bW \tag{3.6}$$

Replacing 'A' by 'bW' in equation (4) we get

$$Q = KbW(dh/dl) \tag{3.7}$$

Expressing transmissivity (T) as Kb we obtain

$$Q = TW(dh/dl) \tag{3.8}$$

It means that flow through a medium of cross-sectional area A is the product of three components: transmissivity, width of the aquifer and the hydraulic gradient.

Storage coefficient

The storage coefficient (S) of an aquifer defines the ability of the aquifer to take into storage or release water per unit change of head per unit surface area (Figure 3.22). Storage coefficient applies to a unit prism of the aquifer (Heath, 2004).

$$S = \frac{\text{volume of water}}{(\text{unit head change})(\text{unit area})} = \frac{m^3}{(m)(m^2)} = \frac{m^3}{m^3} \quad (3.9)$$

Cone of depression

Until water is withdrawn, the water level of an aquifer rests in equilibrium. What occurs is only the natural flow in response to natural hydraulic gradient. When withdrawal starts, the geometry of the water level of the aquifer begins to change at the point of pumping, which is called 'drawdown'. The zone of aeration is the region between the earth surface and water table. See Figure 3.23.

The head of the well goes below the existing water table or potentiometric surface (for confined aquifers), and as a result the rate of flow of water from the surrounding part of the aquifer increases and water flows towards the pumping well. This rate of flow will continue to increase until the rate of flow equals the rate of withdrawal. As a result, a cone of depression is formed surrounding the pumping well.

In an unconfined aquifer, the cone of depression expands slowly because the storage coefficient is equal to specific yield. As dewatering continues from the aquifer, transmissivity decreases resulting in drawdown of both well and the aquifer.

In a confined aquifer, in contrast, the effect of drawdown is manifested in the potentiometric surface which is above the upper surface of the aquifer. The effect of dewatering is not manifested in the aquifer as such. Due to the small value of the storage coefficient, the cone of depression expands rapidly.

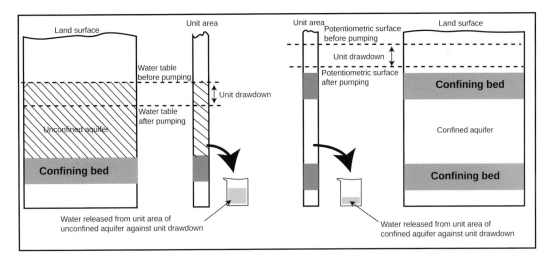

Figure 3.22 Storage coefficients. Adapted from Heath (2004).

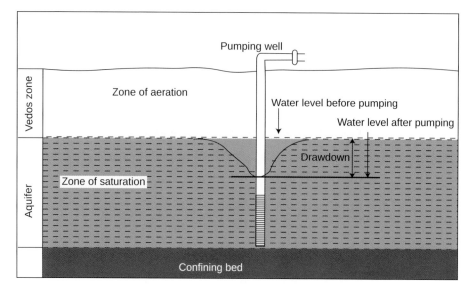

Figure 3.23 Drawdown.

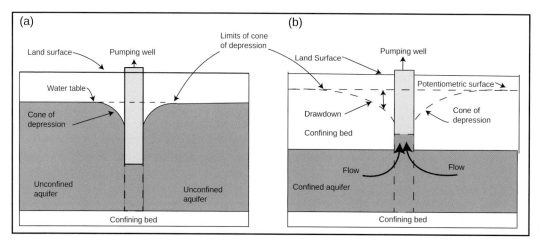

Figure 3.24 Cone of depression: (a) unconfined aquifer, (b) confined aquifer. Adapted from Heath (2004).

Study of the cone of depressions is important for two reasons: (1) it reflects the effect of pumping on the water level of surrounding wells and (2) the effect of change in flow direction in the aquifer. See Figure 3.24.

Groundwater velocity

Groundwater velocity can be derived using both Darcy's law and the velocity equation. Darcy's law states:

$$Q = KA(dh/dl) \tag{3.10}$$

And the velocity equation is:

$$Q = vA$$

where 'Q' is the rate of flow, 'K' is hydraulic conductivity, 'A' is the area of cross section at a right angle to the direction of flow, 'v' is the velocity and 'dh/dl' is the hydraulic gradient.

Combining these two equations we get

$$vA = KA(dh/dl)$$

Or

$$v = K(dh/dl) \qquad (3.11)$$

This is Darcy's velocity, which is the rate of flow in a porous medium for a given value of K. In other words, for a given hydraulic gradient water will move faster in an aquifer with higher hydraulic conductivity. Porosity also determines the effective area of the flow path in the sense that velocity is inversely proportional to porosity. By taking porosity into account the velocity of water in the aquifer is then given by the equation:

$$groundwater\ velocity\ (v) = \frac{hydraulic\ conductivity\ (K) \times hydraulic\ gradient\left(\dfrac{dh}{dl}\right)}{porosity\ (n)} \qquad (3.12)$$

3.9.4   Groundwater system

Both aquifers and confining beds perform the hydrological functions of storing and transmitting water in which the aquifer plays a more important role of storing fresh water (barring exceptional instances of saline and brackish water aquifers).

Groundwater flows from an area of high total head to one of low total head. Determining such a flow vector is difficult because groundwater is not a visible resource, unless observed and measured in wells. To estimate the storage, recharge and flow in both qualitative and quantitative terms, some experimental or investigational work must be done. It is very important to conduct flow analysis of a groundwater basin to demarcate the recharging and discharging areas to ensure better groundwater management.

Recharging and discharging areas

The recharging area is a part of the land surface that constitutes the natural intake zone in a groundwater basin; water that enters the aquifer then moves towards the lower hydraulic head. This movement of water towards the lower hydraulic head also recharges confined aquifers

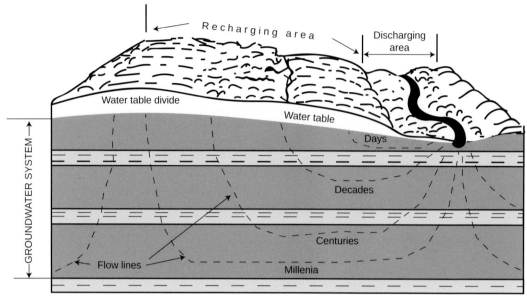

Figure 3.25 Recharge area and discharge area. Adapted from Heath (2004).

located below an impervious confining layer. Recharge and rate of movement of water in confined aquifer is rather slow and it may take several days to several years to move from the recharging area to the discharging area depending on the depth and distance of the aquifer from the recharging zone.

Discharging areas are defined as the areas where the water table meets the land surface (Sengupta, Majumdar, and Roy, 2002). According to Heath (2004), 'Recharge involves unsaturated movement of water in the vertical direction; in other words, movement is in the direction in which the hydraulic conductivity is generally the lowest. Discharge, on the other hand, involves saturated movement, much of it in the horizontal direction—that is, in the direction of the largest hydraulic conductivity.' See Figure 3.25.

3.9.5 Essential studies in groundwater

In addition to the hydraulic parameters of an aquifer discussed above, for a fuller understanding it is essential to undertake the following studies.

The geometry of the aquifer

The geometry of the aquifer is defined by its depth, thickness, overlying and underlying geological formations and its areal extent, which are then shown on a map. This information becomes available only after collection of detailed hydrogeological data from the terrain.

Water Table and piezometric level

The water table and piezometric-level data are obtained from dedicated observation wells that are regularly monitored.

Aquifer type(s)

As has already been discussed, an aquifer may be confined or unconfined or semi-confined or leaky. The first two types are already described in this chapter. A leaky or semi-confined aquifer is a water-bearing formation with both upper and lower surfaces bound by aquitards, or one boundary is an aquitard and the other is an aquiclude. In terms of hydraulic character, a semi-confined aquifer lies in between confined and unconfined aquifers (Kovalevsky, Kruseman, and Rushton, 2004). An aquitard is defined as a layer that acts like a confining bed but has a capacity to transmit water at a very slow rate in some conditions; whereas aquicludes do not transmit water at all.

Groundwater storage

This is defined as the amount of water stored in an aquifer, which is a function of the volume of the aquifer and the effective porosity or specific yield. Groundwater storage is an important measurable parameter of an aquifer system essential to effective management. The storage volume at any point of time t is defined as:

$$\text{Volume} = \text{Area} \times \text{ST} \times S_y \tag{3.13}$$

Where ST = saturated thickness of the aquifer, and 'S_y' is the specific yield.

Groundwater budget

Groundwater budget is defined by the equation:

$$\text{Recharge} = \text{Discharge}$$

The details of recharge and discharge are shown in Table 3.8

Table 3.8 Possible sources of water entering and leaving a groundwater system under natural conditions.

Inflow (recharge)

1. Areal recharge from precipitation that percolates through the unsaturated zone to the water table.
2. Recharge from losing streams, lakes, and wetlands.
3. Recharge from return flow.

Outflow (discharge)

1. Discharge into streams, lakes, wetlands, saline-water sources (bays, estuaries, or oceans).
2. Groundwater evapo-transpiration.
3. Springs (outflow).

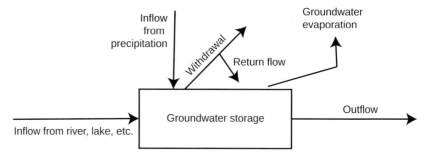

Figure 3.26 Groundwater budget.

Table 3.9 Withdrawal mechanisms for different types of aquifer

Types of well	Withdrawal mechanism
Dug well, small diameter (< 1m)	Rope and bucket, hand pumps, small electrical or fluid fuel pumps
Large diameter (> 1m)	Animal-driven water wheel, electrical or fluid fuel pumps
Tube well (Shallow aquifer)	Centrifugal pump
Tube well (deep aquifer) low-duty	Low-duty submersible pump
Tube well (deep aquifer) heavy-duty	Heavy-duty submersible pump

A schematic representation of the water budget with human intervention is shown in Figure 3.26.

Groundwater withdrawal data

Groundwater is generally withdrawn from the aquifer by constructing different types of wells and using various devices. Table 3.9 shows the common practices in different types of wells.

Water quality data

Quality of groundwater is vital to development and management needs. Water samples are collected from wells and analysed in the laboratory to determine the concentration of physical parameters (e.g. turbidity, colour, odour) and chemical parameters (e.g. As, F, Cr content) which are hazardous to human health and, on a larger scale, the ecosystem itself. Details of chemical parameters and their permissible limits that determine the quality of water is discussed in Chapter 6.

3.9.6 Relation between groundwater withdrawal and stream flow

Breault *et al.* (2009) have shown that 'groundwater withdrawals can affect stream flow in several ways.' In undisturbed hydrogeological situations, groundwater will discharge to the streams (Figure 3.27a). In instances

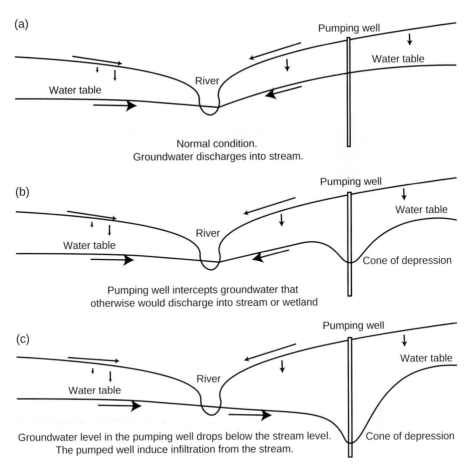

Figure 3.27 Effect of pumping on stream flow. Adapted from Breault *et al.* (2009).

where water is pumped out from the aquifer, and if the quantum of water withdrawal is high enough to create a cone of depression, then it can give rise to two situations: (1) even after formation of a cone of depression, the water table remains above the stream level and (2) where the drawdown lowers the water table below the stream level. In the former situation, groundwater will still flow to the stream but at a reduced rate (Figure 3.27b). In the second case, the flow direction is reversed, and the stream will contribute water to the aquifer, that is, induced infiltration (Figure 3.27c); if the rate of induced infiltration exceeds the rate of stream flow, then the stream may lose baseflow, which is an extreme, but by no means an unrealistic, example of the adverse effect of groundwater withdrawal on streams. A study conducted in the Maurice River (Corzine, 2009) shows that due to excessive withdrawal of groundwater from an unconfined aquifer in the river basin, stream baseflow has depleted by 93%. It needs clarifying in this context that the aforementioned discussion holds good for unconfined aquifer systems only.

3.9.7   *Groundwater withdrawal in the recharging zone*

Human activities are encroaching into the recharging zones of an aquifer system in direct proportion to population growth. The vast Indo-Gangetic plain contains numerous recharging and discharging zones relating to aquifer systems beneath the Indo-Gangetic alluvium. Some part of the upper catchment can be clearly identified as a recharge zone. Human activity and extension of irrigated agriculture into the recharge zone have resulted in (1) groundwater depletion in discharge zones and (2) reduced recharge flow into the confined aquifers.

3.9.8   *Hydrogeological investigation*

One of the major constraints in the hydrogeological investigation is heterogeneity of geological formations. This may be reflected in the spatial arrangement and continuity or discontinuity of aquifer mass (Kresic, 2007). So, in a groundwater basin a single experiment or test may not yield the desired result; moreover, it may lead to flawed decision making. To understand different dimensions of a groundwater regime, a series of investigations or experiments are required, which are collectively called hydrogeological investigation. The major field-based experiments are (a) hydrogeological surveys, (b) aquifer performance tests or pumping tests, (c) periodical monitoring of groundwater levels, and (d) periodic chemical analysis of groundwater samples, which are discussed below.

Hydrogeological survey

A hydrogeological survey is a set of field investigations carried out for compiling hydrogeological maps and evaluating the hydrogeological condition of an area. Data required for a hydrogeological survey are (1) mapping of surface geology and (b) mapping of thickness and extent of sub-surface water-bearing formations. In addition, aquifers are evaluated in terms of factors like geomorphology, hydrology, climate and anthropological factors that affect the underground water system and its recharge.

To understand the geometry of the sub-surface formations, two methods are applied: invasive drilling and non-invasive geophysical methods. MacDougall *et al.*, (2002) state that the intrusive technologies are cost-intensive and requires space and time, but these methods define the different sub-surface geological horizons more accurately. The most common intrusive method is direct drilling with the help of a drilling rig or by hand drilling (details of drilling methods are discussed in Chapter 7). Non-intrusive techniques include geophysical methods (e.g. vertical electro-resistivity sounding); resistivity sounding[4] is also important for differentiating between saline and freshwater aquifers. Figure 3.28 shows how a geophysical survey is being conducted in order to understand the aquifer disposition of a proposed industrial area.

Figure 3.28 A hydrogeologist conducting VES. Photo Source: Earthcare Products and Services, Kolkata, India.

Hydrogeological surveys vary according to the scale and purpose of the survey. A low-resolution hydrogeological survey which covers a large area (e.g. a district, river basin or even a country) is the responsibility of the concerned government; whereas, industries typically undertake high-resolution mapping pertaining to the area of their interest. From the acquired hydrogeological survey data, industries compile a hydrogeological report for (a) compliance with the statutory requirements, and (b) for preparing a database for strategic planning for groundwater management of the basin.

Aquifer performance tests (APT) or pumping tests

An aquifer performance test is a method to determine the aquifer parameters (storativity, transmissibility, specific yield and so on). In an arrangement for a pumping test, a pumping well and one or more observation wells (located at different distances from the pumping well) are constructed. Water is pumped out in the pumping well, and the water level is measured continuously at specified time intervals from the observation wells, manually or with a data logger.

The measurements continue even after the pumping is stopped until the water levels in the observation wells return to the original level. Time-series curves are plotted with the time-vs.-drawdown data and matched with aquifer-specific 'type curves'. From this experiment, the storativity and transmitivity values of the aquifer are determined.

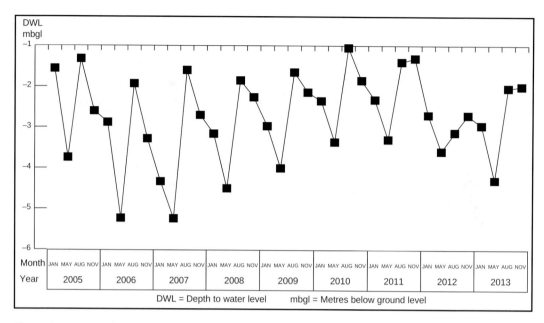

Figure 3.29 Groundwater level hydrograph of Sachin, Gujarat. Based on data published in India WRIS.

Water level measurement

Groundwater level fluctuates naturally even if there is no extraction. During the pre-monsoon period water levels go down to their lowest position, and during the monsoon the water level rises up to its maximum height. The difference of level between pre- and post-monsoon periods is called the annual water level fluctuation. Hydrogeologists take measurements from wells and record the water level in a time series and plot the data in a graph, which is called a 'water level hydrograph'. The time gap for water level hydrograph data collection may be hourly, daily, monthly or seasonal depending on the objective of the investigation. The method of water level measurement is discussed is Chapter 8, 'Water Resource Assessment'. Based on time series water level data of a monitoring well in Sachin, Gujarat the water level hydrograph is shown in Figure 3.29.

The data that can be derived from water level measurements are:

- The gradient of water table or piezometric surface of a confined aquifer
- The direction of flow of groundwater
- The determination of recharge and discharge areas
- Both annual and seasonal water level fluctuation
- Quantification of groundwater recharge and discharge

Chemical analysis of groundwater

Groundwater quality assessment 'includes field analysis of physical characteristics, as well as laboratory analysis for nutrients, major inorganic ions, and selected trace metals' (USGS, no date). After completion of a hydrogeological map, the number and locations for water sample collection are selected; each such selected spot should represent an area of homogeneous hydrogeology. In a small river basin of 50 to 100 sq. km area with a homogeneous aquifer, 10 to 15 samples usually suffice; but, in a multi-aquifer region, especially with varying human activity and multiple pollution sources, more extensive sampling is required.

Water samples for laboratory analysis collected from tube wells, open wells and springs are stored in 500 ml plastic bottles free from contaminants. Some parameters, like temperature, specific conductivity, dissolved oxygen and pH are generally determined in the field itself with portable instruments. A sample groundwater analysis report prepared under baseline study is given below in Table 3.10.

3.9.5   Groundwater management

Groundwater management is especially relevant to both underground and open-cast mines, which are major sources of groundwater pollution. Management of wastewater discharge from mines is therefore a major concern. However, to deliver efficient water stewardship in the mining sector, it is vitally important, as has already been emphasised earlier, for corporate managers to have a working knowledge of hydrogeology.

Schmoll (2006) wrote that groundwater management measures fall into three broad categories: control, assessment and good practice. Additional categories, such as, site selection and CSR may be identified as the drivers for groundwater study. Business schemes for water management for which groundwater knowledge is essential are given in Table 3.11.

Other important factors of groundwater management are:

- 'Class' of aquifer (see Table 3.12).
- The geological character of the aquifer, which, among hydrologists, is referred to as the 'aquifer management unit'.
- Stage of development of aquifer management unit, i.e. the percentage of annual renewable groundwater abstracted annually.

Research on groundwater has advanced by leaps and bounds with continual advancements in investigation technology and computer-aided modelling tools; a corporate manager will benefit from these applications and thus can select the appropriate technology or method of investigation.

3.10   Conclusion

This chapter covers basic concepts of hydrology, so that corporate managers can get an overview of how to deal with development and conservation of water resources. In essence, the chapter should make it clear that

Table 3.10 Report of chemical analysis of groundwater in Panagarh, West Bengal, India.

Parameter	Unit	Location						
		Chandpur	Dharala	Panagarh	Randiha	Sukdal	Pondali	Khandari
Total hardness as CaCo3	mg/l	84.84	210.80	234.32	278.76	250.48	105.04	121.20
Ca	mg/l	12.29	58.29	59.91	71.25	72.87	29.15	32.38
Cl	mg/l	15.31	11.49	15.31	36.37	32.54	21.06	28.71
Ph	–	7.02	7.18	7.28	7.38	7.34	7.02	7.02
Mg	mg/l	5.89	15.71	20.62	24.54	16.69	7.85	9.82
TDS	mg/l	132.0	316.0	346.0	346.0	336.0	176.0	175.00
SO4	mg/l	4.27	11.73	5.69	10.27	6.67	2.13	4.56
Fl	mg/l	0.92	0.73	0.85	0.97	0.85	0.81	0.78
No3	mg/l	2.90	4.30	2.26	4.86	3.92	1.27	2.26
Fe	mg/l	0.30	0.26	1.28	0.24	0.24	0.37	0.26
Al	mg/l	BDL	BDL	BDL	BDL	BDL	BDL	BDL
Boron	mg/l	0.77	0.70	0.28	0.48	1.00	0.57	1.14
Na	mg/l	10.40	19.50	20.90	15.10	14.00	18.20	16.20
K	mg/l	1.90	1.70	2.30	1.40	1.30	0.92	2.00
Hg	mg/l	BDL	BDL	BDL	BDL	BDL	BDL	BDL
As	mg/l	BDL	BDL	BDL	BDL	BDL	BDL	BDL
Pb	mg/l	BDL	BDL	BDL	BDL	BDL	BDL	BDL
Cd	mg/l	BDL	BDL	BDL	BDL	BDL	BDL	BDL

BDL = below detection level.

Source: Sengupta (2015), unpublished report.

Table 3.11 Business schemes for groundwater management.

Sl No	Business scheme	Application of groundwater knowledge
1	Source-water assessment and protection	• Selecting the appropriate aquifer in a multiple-aquifer region and assessing its 'safe-yield' and 'safe-distance'. • Taking up appropriate source-water monitoring and protection plans.
2	Water treatment	• Since water treatment planning depends on the quality of groundwater, the application begins with a selection of appropriate technology; e.g. mitigation of salinity and excess iron content problems requires very different technologies. • Identification of source of contamination is also vital to water treatment.
3	Groundwater development	Selecting appropriate methods of drilling, well design and technology of withdrawal.
4	Water conservation	Planning for artificial recharge of groundwater is essential where natural recharge is compromised by human interference.
5	CSR	Groundwater knowledge is vital to CSR. Sinking of community tube wells under CSR programmes are known to have ended as failures because of absence of groundwater knowledge in some cases.
6	Waste disposal	Selecting proper location for land fill so that groundwater is not contaminated.

Table 3.12 Classification of aquifer for groundwater development.

Class of aquifer	Low density development	High density development
Small/medium unconfined	Low transaction costs in developing and institutional arrangements (example Sub-Saharan aquifer).	High-density development may lead to drying up of aquifer, causing interference between wells and depletion of water quality.
Small/medium confined	Transaction cost is high. Management issues are generally uncalled for.	High transaction cost, high pressure on aquifer. Institutional management and surveillance required (example: Sundarban Delta).
Large/extensive unconfined	Higher transaction costs. Example: the Guarani aquifer system of Brazil.	Very high transaction cost. Lowering of water table. Management intervention and monitoring essential.
Large/extensive confined	Initial cost may be high, but institutional arrangement is optional since abstraction is low.	High infrastructure cost and pressure on aquifer is also high. Proper institutional management and monitoring are required.

Adapted from Kemper (2007).

groundwater is a resource that cannot be directly observed; consequently, it needs both intrusive and non-intrusive studies. It also needs emphasising in this context that both theoretical understanding and field experience are crucial to understanding hydrology; neither suffices by itself. I hope the chapter will empower corporate managers in planning sustainable development.

Notes

1. There is practically no such accepted classification based on river hierarchy. Here the concept is introduced to understand the management boundaries of a river basin.
2. The Bangladesh part of the distributary of the Ganga River is called the Padma River
3. *Bund* is a common name for certain types of water harvesting structures used in Indian subcontinent.
4. A reference on resistivity methods is available at https://archive.epa.gov/esd/archive-geophysics/web/html/resistivity_methods.html.

Bibliography

ABC Science (2008) *ABC catchment detox.* Available at http://www.abc.net.au/science/catchmentdetox/factsheet/(accessed: 10 April 2016).

Aral Sea (2016) in *Wikipedia.* Available at https://en.wikipedia.org/wiki/Aral_Sea (accessed: 10 January 2016).

Arnaiz, N. (2016) *The water cycle.* Available at https://prezi.com/fyccsivcqkvk/the-water-cycle/(accessed: 10 April 2016).

Better World Solutions (2014) *Roadmap with proven integrated and sustainable water resource approach—BetterWorldSolutions—The Netherlands.* Available at https://www.betterworldsolutions.eu/roadmap-with-proven-integrated-and-sustainable-water-resource-approach/(accessed: 8 June 2016).

Breault, R.F., Zarriello, P.J., Bent, G.C., and Masterson, J.P. (2009) *Effects of water-management strategies on water resources in the Pawcatuck river basin, southwestern Rhode island and southeastern Connecticut.* Available at http://pubs.usgs.gov/circ/circ1340/pdf/circ1340_pawcatuck508.pdf (accessed: 10 February 2016).

Brouwer, C., Goffeau, A., and Heibloem, M. (1985) *Irrigation water management: Training manual No. 1: Introduction to irrigation.* Available at http://www.fao.org/docrep/r4082e/r4082e00.htm (accessed: 10 January 2016).

Classification of watershed (2015) Available at http://www.agriinfo.in/?page=topic&superid=8&topicid=76 (accessed: 6 May 2016).

Cornell University (2010) *Northeast region certified crop adviser (NRCCA) study resources.* Available at https://nrcca.cals.cornell.edu/soil/CA2/CA0212.1-3.php (accessed 6 July 2016)

Corzine, J.S. (2009) *New Jersey Geological Survey: Potential rate of stream-base-flow depletion from groundwater use in New Jersey. New Jersey department of environmental protection.* Available at http://www.state.nj.us/dep/njgs/pricelst/tmemo/tm09-1.pdf (accessed: 18 May 2016).

Critchley, W., and Siegert, K. (1991) '3.Rainfall-runoff analysis', in *Water harvesting, A Manual for the Design and Construction of Water Harvesting Schemes for Plant Production.* Rome: FAO.

DEFRA (2010) *Surface water management plan technical guidance*. Available at: https://www.gov.uk/government/uploads/system/uploads/attachment_data/file/69342/pb13546-swmp-guidance-100319.pdf (accessed: 5 March 2016)

Drainage basin (2016) in *Wikipedia*. Available at https://en.wikipedia.org/wiki/Drainage_basin (accessed: 10 April 2016).

FAO (no date) *AQUASTAT database, database query results*. Available at http://www.fao.org/nr/water/aquastat/data/query/results.html? (accessed: 10 January 2016).

Govorushko, S.M. (2007) *Effect of human activity on rivers*. Available at http://www2.dsi.gov.tr/english/congress2007/chapter_2/37.pdf (accessed: 10 February 2016).

Heath, R.C. (2004) *Basic groundwater hydrology prepared in cooperation with the North Carolina Department of Natural Resources and Community Development*. Available at http://pubs.usgs.gov/wsp/2220/report.pdf (accessed: 10 February 2016).

Kemper, K. (2007) Instruments and institutions for groundwater management. *The agricultural groundwater revolution: opportunities and threats to development*. Wallingford, Oxfordshire, U.K. CABI, pp. 153–172.

Khattak, O. (2014) *Learning geology*. Available at http://geologylearn.blogspot.in/2015/12/groundwater-flow.html (accessed: 10 February 2016).

Kovalevsky, V.S., Kruseman, G.P., and Rushton, K.R. (eds.) (2004) *Groundwater studies: An international guide for hydrogeological investigations; IHP-VI series on groundwater*, Vol. **3**. Available at http://unesdoc.unesco.org/images/0013/001344/134432e.pdf (accessed: 10 February 2016).

Kresic, N. (2007) *Hydrogeology and groundwater modeling*. 2nd edn. Boca Raton, FL: Taylor & Francis.

Macdougall, K., Fenning, P., Cooke, D., Preston, H., Brown, A., Hazzard, J., and Smith, T. (2002) *Non-intrusive investigation techniques for groundwater pollution studies*. Available at https://www.gov.uk/government/uploads/system/uploads/attachment_data/file/290964/sp2-178-tr-1-e-e.pdf (accessed: 10 February 2016).

Montana River Action (no date) *Methods to protect rivers and streams*. Available at http://www.montanariveraction.org/methods-to-protect-rivers.html (accessed: 10 February 2016).

Muriel Martin Elementary School (2016) *Water cycle diagram*. Available at http://www.mmem.spschools.org/oldsite/grade5science/weather/watercyclediagram.html (accessed: 10 April 2016).

NGWA (no date) *Facts about global SM groundwater usage*. Available at http://www.ngwa.org/Fundamentals/Documents/global-groundwater-use-fact-sheet.pdf (accessed: 20 March 2017).

Oki, T. (2006) Global hydrological cycles and world water resources, *Science*, **313**(5790): 1068–1072. doi: 10.1126/science.1128845.

SANDRP (2014) *Know our rivers: a beginners guide to river classification*. Available at https://sandrp.wordpress.com/2014/11/15/know-your-rivers-a-beginners-guide-to-river-classification/(accessed: 6 May 2016).

Schmoll, O., Howard, O. and Chilton, J. (eds)(2006) *Protecting groundwater for health: Managing the quality of drinking-water sources*. London: IWA Publishing

Sengupta, P. K. (2015) *Hydrogeological study of Panagarh area*, Burdwan, West Bengal.Unpublished report.

Sengupta, P.K., Majumdar, P.K. and Roy, A. (2002) Determination of spring line as a tool for water resource planning and development. *Analysis and practice in water resources engineering for disaster mitigation*. Kolkata: Indian New Age International (P) Limited, Publishers, 162–167.

Shearer, J. (2014) *Water properties hydrologic cycle aquatic ecosystems.* Available at http://www.thinktrees.org/my_folders/2014_envirothon/aquatic_ecology.pdf (accessed: 10 January 2016).

Shearman, R.J. (1977) The speed and direction of movement of storm rainfall patterns with reference to urban storm sewer design/Vitesse et direction de la distribution de l'intensité de la pluie en ce qui concerne le dessin des égouts-déversoirs urbains, *Hydrological Sciences Bulletin,* **22**(3): 421–431. doi: 10.1080/02626667709491735.

Siebert, S., Burke, J., Faures, J.M., Frenken, K., Hoogeveen, J., Döll, P., and Portmann, F.T. (2010) Groundwater use for irrigation – a global inventory, *Hydrology and Earth System Sciences,* **14**(10): 1863–1880. doi: 10.5194/hess-14-1863-2010.

Soil (2016) Available at https://en.wikipedia.org/wiki/Soil (Accessed: 19 October 2016).

Stormwater (2016) in *Wikipedia* Available at https://en.wikipedia.org/wiki/Stormwater (accessed: 8 October 2016).

UNESCO–WWAP (2006) *Water: A shared responsibility the United Nations world water development report.* Available at http://unesdoc.unesco.org/images/0014/001454/145405e.pdf (accessed: 10 April 2016).

USDA (no date) *Information about hydrologic units and the watershed boundary dataset.* Available at http://www.nrcs.usda.gov/wps/portal/nrcs/detail/national/water/watersheds/dataset/?cid=nrcs143_021616 (accessed: 9 May 2016).

USGS (2015a) *Surface runoff - the water cycle, from USGS water-science school.* Available at https://water.usgs.gov/edu/watercyclerunoff.html (accessed: 10 January 2016).

USGS (2015b) *Streamflow—the water cycle, from USGS water-science school.* Available at http://water.usgs.gov/edu/watercyclestreamflow.html (accessed: 10 January 2016).

USGS (2016) *Glaciers and icecaps, the USGS water science school water-information site.* Available at http://water.usgs.gov/edu/earthglacier.html (accessed: 15 September 2016).

USGS (no date) *Groundwater quality.* Available at http://mi.water.usgs.gov/pubs/WRIR/WRIR00-4120/pdf/gwq.pdf (accessed: 16 May 2016).

Water distribution on earth (2016) in *Wikipedia.* Available at https://en.wikipedia.org/wiki/Water_distribution_on_Earth (accessed: 8 October 2016).

4

Corporate Water Stewardship

4.1 Introduction

Corporate Water Stewardship, CWS for short, is a recent concept that aims at institutionalising fair distribution of water for all, from business magnates to, say, endangered frogs. Over the last few decades, the corporate world is beginning to understand (after losing enormous sums of money in litigation and compensation) how ecological services degraded by industrial effluents can affect the business itself. After being penalised in courts of law, large corporations have merely become wary of the cost of restoring degraded ecological services, but they still do not get the big picture. They are unconcerned about how species extinction affects ecological balance irreversibly, and the fact that healthy ecosystems boost business still remains incomprehensible to them. Running a profitable business on the one hand and preserving the ecosystem on the other require educating the corporate world. CWS educates business magnates so that endangered frogs do not go out of business; if they do, profit curves will dip sooner or later. The bottom-line of CWS is simple: preserve ecological balance or perish.

Global business is under the threat of losing out to the challenges of water scarcity and climate uncertainties. The problem is further compounded by the elusiveness of sustainable solutions. The challenges are big and many, but at the same time, these challenges (IUCN, 2006) can create new business opportunities, new technologies, products, markets, new businesses (Baue,1998) and new revenue streams. These challenges also demand that the corporate assess the ecosystem services that nature provides and to explore possible local actions that may lead to an enhancement of these services. A fair amount of trade-off is involved in the management of changes. To maintain the water resources of the environment along with services demanded by society, industries should operate with a global perspective,

Industrial Water Resource Management: Challenges and Opportunities for Corporate Water Stewardship, First Edition. Pradip K. Sengupta.

abiding by the legal framework of the country. These strategic acts and services by the industrial sector collectively come under the umbrella of Corporate Water Stewardship.

The term 'Corporate Water Stewardship' is defined as the approach towards the management of water-related business risks. When the same source of water is shared by a number of stakeholders, apart from the company itself, the activities of the company to protect the water resource in a sustainable way are incumbent upon the corporate governance. Water-intensive industries, especially where the public water governance is inadequate to cover the business viability, desperately need to apply corporate stewardship to play a large role in achieving water sustainability.

4.2   Why water stewardship?

Given the dismal track record of environmental protection, the title question and its answer may seem rhetorical to the cynic, because media hullabaloo about curbing water pollution has been going on for years, and yet a critically endangered species, say tree frogs in densely forested parts of Amazon Basin, is lucky if it is not extinct. The long shadow of polluted waters penetrates deeply. Remoteness is no protection against ravages of industrial pollution; the cynic has a point because tokenism lies at the root of the problem. All governments have hitherto allowed one-dimensional economic growth in which ecosystems stressed by polluted water are shielded from public attention by tokenism, while people residing around polluted waters suffer in silence. It is ironical that poison-arrow frogs are dying in toxic waters leaving the bow-and-arrow wielding ancient tribes high and dry, who have lived in harmony with nature for millennia. Elected governments have routinely turned a blind eye to industries by passing regulatory laws that are only a form of tokenism.

Industries are familiar with trade union strikes, but nothing strikes industrial growth more effectively than retaliation from nature. Water stewardship is designed to indoctrinate the government-industry-media nexus with the idea of preserving the balance of nature—that is, healthy coexistence—is essential to economic growth. The European Water Partnership (2016) has identified the benefits of water stewardship that:

- Leads to partnership development on river-basin level;
- Improves efficiency of water use and thereby economic benefit by creating leaders in water management;
- Generates public acceptance of production sites and business sectors;
- Gives positive incentives for other operators in the sector;
- Balances risk and economic performance; and
- Reinforces communication with investors, customers, and supply chain and also, between actors on a regional level.

In Figure 4.1 the drivers of corporate water stewardship are illustrated.

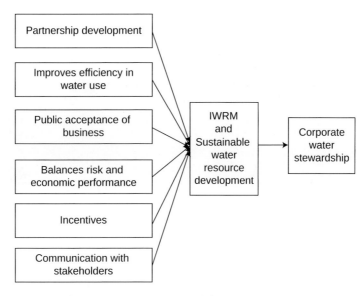

Figure 4.1 Drivers of corporate water stewardship.

4.2.1 Partnership development

Partnership is the essence of water stewardship; it builds a joint endeavour along with stakeholders and peers. The main objective is to generate the framework of a unified approach to managing water risks in a basin and consequently sharing experience, knowledge and best practices with partners. This framework needs an alliance or consortium of business institutions who will build up the partnership and a common strategy to mitigate business risks. The partnership framework has several steps in its life cycle, which are (1) identification of opportunity, (2) identification of partners, (3) alliance formation, (4) negotiation and agreement, (5) implementation and management, (6) reporting and disclosure (Figure 4.2). Leading companies around the world have already developed a number of consortiums, such as the CEO Water Mandate and the Alliance for Water Stewardship, which are platforms for sharing best practices and reporting water performances.

4.2.2 Improve efficiency

A water stewardship programme in an industry must have activities and management reforms that improve water efficiency in the value chain. A successful water stewardship initiative will improve performance in the supply chain and, in the process, in the market. In other instances, where laws protect the resource, corporate water stewardship is an opportunity to continue business under water-stressed conditions. The processes reduce conflicts, enhance social acceptance of the company and protect shareholders' value. Corporate stewardship is not only an accepted paradigm but a way of functioning that is embedded in the sustainability of the economy.

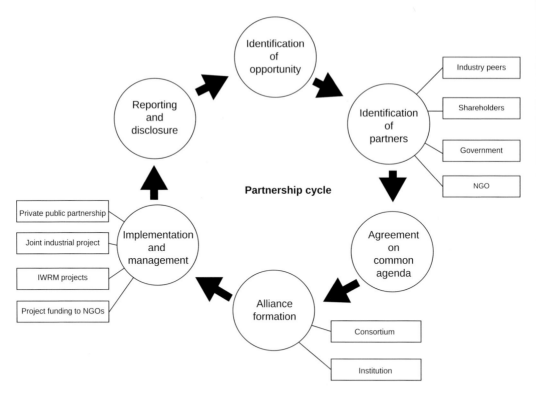

Figure 4.2 The partnership framework in water stewardship. Adapted from Simon (2011).

The necessity of establishing corporate water stewardship should be considered against the backdrop of the current global water scenario.

4.2.3   Public acceptance

Consumers worldwide are now becoming more and more concerned about the environment, and they are ready to pay a premium price for goods which are branded eco-friendly. Customers and the public will accept a business in their local environment only if it is water-efficient and shows evidence of overall improvement in the water environment of the area.

4.2.4   Incentives

It is a common belief that industries are reluctant to manage water in a conservative way because they do not get enough incentives to do so. For the same reason, they are not inclined to invest in water stewardship initiatives. Challenging this idea, Schulte and Morrison (2014) clearly explained that the benefits of water stewardship are (1) secured water availability (2) fewer water risks, (3) improved water-use efficiency and (4) reduction of pollution; in addition, there are four market benefits: (a) immunity from market failure, (b) reduced cost of water, (c) secured business and (d) enhanced brand value. These are the most sought-after incentives of a business.

Figure 4.3 Redefined business case for water stewardship. Source: Rozza *et al.* (2013), printed under CC BY 3.0 licence.

4.2.5 Balancing risk and economic performance

Companies should appreciate that their performance maintains a balance of risk, sustainability and economic performance (Figure 4.3). These three components are mutually reinforcing (Rozza *et al.*, 2013), and each component has its own value.

How sustainability, risk and productivity work together is best defined in a framework where corporations engage in proactive programmes involving community, sources of water (river, reservoir, aquifer, etc.) and water-management facilities. See Figure 4.4.

According to Barton, *et al.* (2011), water across a large multinational business empire requires 'robust governance and management systems' that manage risk and sustainability. Risk management is crucial to achieving a state of harmony with the society and the ecosystem. Fundamental tasks under risk management are listed in Table 4.1.

Apart from risk management and sustainability issues, the corporate sector will improve its performance under water stewardship when legal, political, technological aspects are considered simultaneously along with social partners who will help the corporate sector in overcoming challenges (Pedley, Pond and Joyce, 2011). These social partners are marginal people on the receiving end of water pollution.

4.2.6 Reinforces communication

Investors and other stakeholders are always keen to learn about water performance of a business empire because water crisis is one of the major causes of market failure. Water stewardship will enable investors to remain informed about the water profile of the company. Since water stewardship covers a large area of operation beyond the fences, it will build up communication with local people, government, community organisations, market and consumers.

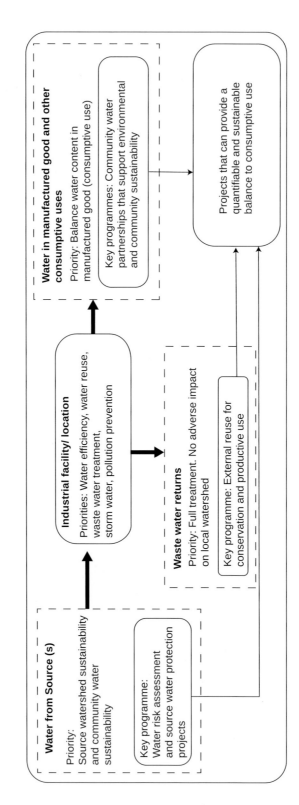

Figure 4.4 Framework for achieving a sustainable water balance. Source: Rozza *et al.* (2013). Printed under CC BY 3.0 licence.

Table 4.1 Risk management tasks.

Risk management tasks	Conditions for engagement and tasks
Operational and technical interventions	A natural primary focus of a company's approach to water management is in specific operational and technical interventions which include: • Reduce non-revenue water and water wastage; • Pilot projects to determine if treated wastewater could be used for agriculture or fisheries; • Reduction of real or physical water losses associated with infrastructure; • Reduce the high level of water losses in both the network and on properties; • Water recycling reuse in facilities; and • Application of water harvesting technologies.
Measuring corporate impacts	Measuring impact is a complex task, and it takes regular data generation. The tasks involve collection of data and translating the data matrix into a measure of local impact. The required tasks are: • Collect data on operational water use and wastewater discharges; • Water footprint assessment of the facility and supply chain; • Greywater footprint assessment; and • Periodical source-water assessment.
External factors	External factors are often potential risks, and mitigation measures are complex and nonlinear. The risks may generate from local regulatory and economic conditions, political uncertainty, climate change and the impacts from other water users. The primary activity to combat such situations is to develop resilience through watershed-based collaborations that effectively engage other stakeholders to improve the shared management of water.
Scenario planning	Scenario planning is a task within the overall planning where the hydrological and water use scenario is constructed and modelled to estimate the impact for different assumed conditions. To generate a reliable scenario, several types of baseline data are required to be collected. These are: • Historical hydrologic records; • Flood and drought records; • Changing climatic conditions; • Land-use and population records; • Water-use data and past records;
Understanding value chain impacts	• Mapping the social, environmental and economic impacts in the value chain; • Identifying and assessing risk along the value chain; and Identifying the scope of risk management. • It is also important to quantify the dormant risks and scoping remediation in the supply chain, and in customers' use of products.
In harmony with sustainability issues.	Major sustainability issues are: • Connectivity between water and energy consumption; • Water and biodiversity linkage; and • Human rights and sanitation issues. Companies should appreciate the local hydrological situation and should remain prepared to manage trade-offs between cross-cutting issues.

Adapted from Jhunjhunwala *et al.* (2014) and Barton *et al.* (2011).

4.3   Aspects of water stewardship

The domain of water stewardship is extensive and goes far beyond efficient water use (WWF, no date). It covers the collaborative activities of all stakeholders including the government, NGOs, civic body organisations and common people where the private sector plays the lead role in a proactive manner. There are five major aspects of corporate stewardship. These are (1) legal, (2) social, (3) technological, (4) environmental and (5) economic. The basic framework of water stewardship can function only under a full consideration of all of these aspects. See Figure 4.5.

4.3.1   Legal aspect

Industries often become embroiled in legal battles over environmental issues. Regardless of the outcome of the court case, which is a long drawn-out process, it is bad publicity. Both legal expenses and bad publicity can be avoided if industries *suo moto* comply with the laws of the country including all cross-sectoral statutes and norms. And it is not just the law of the land; industries located in the jurisdiction of an international river basin authority are subject to international laws, which require industries to apply for water allocation and obtain a licence to withdraw water. Following the licence, industries are required to follow guidelines on water quality of

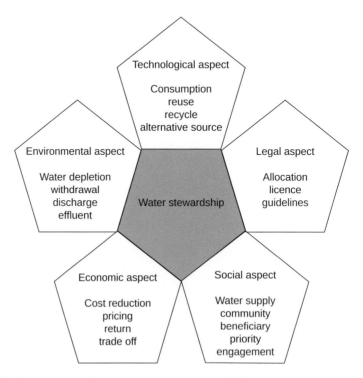

Figure 4.5 Five aspects of corporate water stewardship.

effluent discharge, waste disposal, and to declare their water budget (daily or annual). Industries in countries like USA and UK are under legal obligation to collect water-related data for the basin where they operate.

4.3.2   Environmental aspect

Industries that are focused solely on profit will sooner or later pay for their brinkmanship. More than the letter of the law, it is the spirit of the law by which industries must abide to save themselves from incurring environmental wrath. To consider environmental aspects beyond the obligations of local water laws, an industry should minimise the impact on the ecosystem through understanding the catchment water dynamics. A proactive and vigilant industry can always keep track of water risks, sustainable withdrawal, use and discharge of water. If, for example, an industry adopts a body of water or the recharging area of an aquifer and keeps it free from anthropogenic damage, it not only protects long-term ecological services but also their own business interests.

4.3.3   Social aspect

Social aspects of WSI bring stakeholders together for a joint action that includes sharing of knowledge and experience, capacity building and fostering innovation. Social performance of water stewardship demands transparency and integrity from all who are engaged in collective actions (UN Global Compact, 2015).

4.3.4   Technological aspect

Science and technology are life-supporting systems but the tunnel vision of profiteers sometimes gives science and technology a bad name. Technology has great potential in the context of business (Advameg, 2016), and it can be applied in all stages of the value chain for reducing production costs, developing new products, capturing new consumer markets, improving consumer services and improving management efficiency. Efficiency in water use can be achieved through an adaptation of appropriate technology in different sectors of the value chain. These technological interventions facilitate activities like (1) remediation of water loss in withdrawal and conveyance, (2) enhanced water efficiency in supply chain, and (3) improved purification and treatment facilities. Maximum water efficiency can be achieved by technological improvements in the manufacturing process. The process of adopting an appropriate technology has several steps. See Figure 4.6.

Major technological aspects are:

- **Identify technology need and scope:** This is the primary consideration in water management. Industries should identify the business cases where technological interventions can bring about the desired changes.

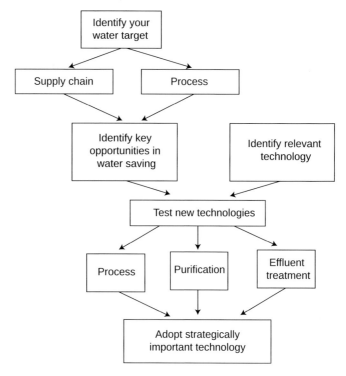

Figure 4.6 Identifying relevant technologies.

- **Invest in technology:** Industries should invest in technologies so that there are sufficient water savings and payback of the investment. Expecting instant recovery of investment costs from the installation of new technologies is a fundamentally flawed understanding of water stewardship. Fulfilment of the basic objectives of water stewardship begins to show on the profit curve only after the projected target period. For example, investment in a water recycling plant reduces water cost equal to the 25% of investment in the recycling system; the investment can be recovered in four years.
- **Identify relevant technology:** Conventionally, product-specific technologies are routinely procured from the market. Innovative technology, on the other hand, emerges from brainstorming among employees and is fine-tuned by testing. As opposed to conventionality, sensitised industries that engage R & D teams to explore innovations invariably fare better in both profit making and effluent cleansing. For example, water pollution from the use of wet dyes is a major problem in textile industries. To mitigate this problem, a Netherlands-based company has introduced a waterless textile-dyeing machine that uses carbon dioxide instead of water; the technological innovation not only saves water but also recycles excess dyes (Ecouterre, 2016).
- **Adopt strategically important technology:** New technologies are 'critical to the firm's ability to compete in the product market' (Zhu and Weyant, 2003) and sharpen an oligopolistic cutting edge for business. Additionally, they have the effect of reducing the impact on the environment. In terms of the highly polluting big businesses of yesteryear, oligopoly seems like the repugnant stranglehold on the market, but in the Third Millennium oligopolistic business comes from value addition by technologies like rainwater harvesting and recycling treated effluents.

4.3.5 *Economic aspect*

Catering to social responsibilities and environmental laws at the cost of lowered profit is not a viable business policy. Corporate Water Stewardship reincarnates the 'invisible hand' of Adam Smith in a new avatar, in the sense that water stewardship activities are designed to promote business markets as well as protecting ecological services.

4.4 Challenges in water stewardship

Water stewardship is never smooth sailing. Water managers have to face scores of both, visible and unforeseen challenges at every turning point of the endeavour.

4.4.1 *Legal challenges*

Legal challenges in water stewardship arise both from national governance or international water and environment related frameworks applicable at trans-boundary locations. Some of the legal challenges are:

Uncertainty

One of the major challenges in legal compliance is uncertainty. It is often seen that a 'consent to operate' from the licensing authority is challenged in a court of law. The motivation from which legal frameworks are drawn may well have hidden political agendas. In some countries, partisan politics play an 'inappropriate role' (Meidinger, 2000) from the very outset, in that, the licensing process itself is open to venality.

Bureaucratic delay

The licensing authority is sometimes overtly dependent on tardy bureaucracy. In addition to the cost of fixed assets, the licensee, then, has to factor in the cost of procedural delay in its capital outlay; for instance, loss incurred after land acquisition due to delayed water allocation. This red tape necessitates engaging consultants and lobbyists to deal with bureaucrats, a headache that the corporate sector has to live with. Deeply embedded local customs are handed down through generations as legacies in another barrier that the licensee has to surmount, especially in tribal areas. In electoral democracies, the government allows some latitude to numerically strong ethnocentric communities where the licensee has to assess the trade-off between industrial demand and ethnocentricity on their own. In an electoral democracy, the licensing authority cannot be counted upon for help.

Integrity issues

Corruption and impaired integrity in the public sector are two major challenges in water stewardship because they not only hamper equitable distribution of water but also deprive the legitimate user from his share of water. Some of the most impacting integrity issues in the praxis of water governance are as follows (Water Integrity Network, 2015):

- Collusion or bribes over contracts;
- Officials turning a blind eye to transgressions;
- Favouritism, clientelism, cronyism and nepotism;
- Illegal fees to water companies for standard services; and
- Fraud, embezzlement and theft.

The effects of corruption are long lasting and result in:

- Destabilisation of the water sharing process;
- Adverse effect on partnership;
- Reputational risk;
- All of which ultimately leading to unsustainability.

Right of way

A right of way is the legal right or easement granted by the owner of land to pass through or to lay pipelines to transport water from the source to the factory (Water and Sanitary Sewer Easement Information, 2010). The process of acquiring a right of way from the owner is very complicated and often takes a long time.

4.4.2   Challenges in the value chain

The OECD, WTO and World Bank group (2014) have jointly published a report that states, 'The growth of global value chains (GVCs) has increased the interconnectedness of economies and led to a growing specialisation in specific activities and stages in value chains, rather than in entire industries.' Under the GVC, industries are pursuing business in a system where the supply chain is fragmented and industries procure raw materials from producers located in different corners of the globe. From the point of view of macroeconomics, the fragmented supply chain is assumed to be interconnected. But microeconomic forces acting on a local scale can disrupt this interconnectedness. Consider, for instance, pharmaceutical companies in Europe sourcing chemicals from Indian producers; when under water crisis, Indian producers reduce production or even stop it in a severe water crisis, the pharmaceutical industry in Europe inevitably suffers loss. The vulnerability of the global value chain to local conditions is best exemplified by the cascading effect of water-stressed areas in the supply chain. This vulnerability is not industry specific; it works across the full spectrum of interconnected businesses operating in the world. Larson *et al.*, (2012) mentions a telling example. In 2011 drought in Texas, India, Pakistan and

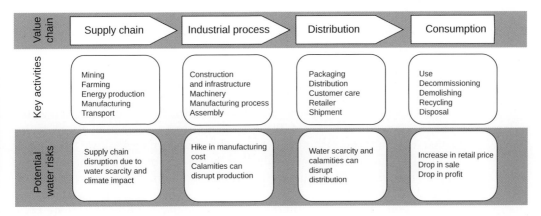

Figure 4.7 Water challenges in the value chain. Adapted from Hwang, Waage and Stewart (2007) and the Value Chain Map 22 (no date).

Brazil caused an escalation in cotton prices affecting the profit curve of Gap and Levi Strauss, two well-known garment companies. This illustrates that water crisis is an emerging crisis, and it can be an indiscriminate killer. See Figure 4.7.

4.4.3 Watershed Challenges

Watershed challenges are defined as the challenges in identifying the boundary of IWRM activities which arise from an overlapping of natural, social and political boundaries. The challenges are as follows:

Boundary challenge

Defining a watershed boundary is complicated even if the hydrology is well established and the aquifer system is also well defined. The challenge arises from the appropriate scale selection for the watershed. As defined in Chapter 3, a watershed size can vary from a small rivulet straddling a few hectares to the enormous Amazon basin. The task of the water stewardship manager is delineating watershed boundaries that will be effective. Indeed, the inherent nature of watersheds unfolds 'a number of different boundary options' (Cohen and Davidson, 2011), and in such cases the selection of a hydrological boundary calls for GIS-based information systems reinforced by natural and social priorities. Enlisting the services of experts from universities and concerned government departments will be helpful in the delineation exercise.

Political and accountability challenge

Political conditions influence the selection process of watershed boundaries. The legal framework follows the political boundary (except in regions subject to international laws of water sharing). Parts of political boundaries may be based on river courses or water divides, but watershed boundaries do not conform to political boundaries. The challenge is compounded when watershed boundary delineation falls foul of 'asymmetries between watersheds and administrative boundaries (municipal, state and national) scales' (Cohen and Davidson, 2011) because of vested interest groups. The crux of the watershed delineation problem is that the vast majority of those claiming access to water want to avoid the responsibility of protecting the watershed. So, who will accept the responsibility of protecting the carrying capacity of a given watershed? Since the advent of agriculture, the size of human settlement has been controlled by the size of watersheds delineated by nature; population booms and increased human longevity have disturbed the natural balance. The carrying capacity of nature has now fallen prey to politics of vested interests. At the cost of sounding rhetorical, I will say this that nature will happily support requirements of civilisation if modernity supports nature.

Problem-shed challenge

A problem-shed is defined as a 'geographic area that is large enough to encompass the issues but small enough to make implementation feasible' (Griffin, 1999). In fact, many of the problems may have originated from outside the chosen watershed boundary and actions for mitigation within the watershed may be defeated by the scale of the problem. For example, if the designed capacity of a storm discharge system in an upstream reservoir turns out to be inadequate (e.g. during cloudbursts), it can cause flooding in the lower reaches of the watershed; the problem becomes endemic to the area and cannot be mitigated at the local watershed level. Problems, economic or ecological, arising out of trans-boundary transfer of water from one watershed to another cannot be mitigated by activities within one watershed. According to Cohen and Davidson (2011), the 'problem-shed challenges are extendable to social issues' when the concept of a watershed boundary is inconceivable to the affected people.

4.4.4   Social challenges

In *The Millennium Development Goal 2000* (MDG, 2000) adopted by the United Nations there are eight goals in which 'Goal 7' is to 'ensure environmental sustainability'. Since all aspects of sustainable development were not covered under MGD 2000, the United Nations adopted another set of goals designated as the Sustainable Development Goal (SDG). The *SDG Report 2015* clearly states that more than 40% of people (United Nations, 2015) are affected globally by water scarcity and the situation is worsening. SDG 6 aims at attaining sustainability in water resource management. In a document jointly released by the WWF, CEO Water Mandate and Water Aid (Ha, 2015), SDG 6 has been elaborated to align with the corporate water stewardship initiatives.

The sixth target of SDG or 'SDG 6' calls for strengthened community participation. This means that in the functioning of business at a local level, multi-stakeholder representation is essential (Ha, 2015). While delivering community level initiatives, the company may encounter several social challenges:

Integrity

In the context of 'Corporate Water Stewardship' (CWS) the importance of this particular issue cannot be overemphasised. There is a deep-rooted divisive perception among marginal people that the rich do not care for the poor; this mistrust is rooted in centuries of monarchy, followed by centuries of feudalism, and as if that was not a long enough period of misfortune, the harried lot then fell into the lap of capitalism (including state capitalism). CWS has a formidable job on its hands and is doomed to failure if it cannot take the poor on board in its implementation plans. While performing water stewardship tasks in the community, the initial unfavourable response of the community may be overcome through integrity performances as explained by the UN Global Compact (2015). The corporate strength in integrity issues rests broadly on (1) clear objectives and tangible outcomes of the water stewardship Initiatives (WSI), (2) integrity and capacity of participants, (3) transparent, responsive and inclusive decision-making process.

Human rights issues

In 2010, the United Nations General Assembly and Human Rights Council formally recognised the 'human rights to water and sanitation' (Ha *et al.*, 2015). In this document, a complete guideline for companies to deal with human rights issues in water and sanitation has been incorporated. According to this document, the dimensions of human rights in water and sanitation are (1) availability, (2) accessibility, (3) quality, (4) acceptability and, (5) affordability. To deal with human rights issues, the corporate sector must assure the community of the availability of affordable safe water. That means business activity should, in no way, prey upon the water service already available to the community naturally, traditionally or through public utilities. If such services are inadequate by human rights standards, corporate-sector interests will be best served by supplementing that service. The CEO Water Mandate guideline may be referred to deal with this issue.

Priority issues

Society has social priorities that may not align with the corporate priorities in water appropriation. In the developing countries, people generally do not consider the true value of water (University of Cambridge, 2013); instead, they ask for short-term benefits which conflict with business interests. This affects the long-term goal of WSI. The goal of WSI is up-scaling the business by mobilising government, business and civil society towards improved water management. If the civil society feels its priorities are not given adequate weight, WSI loses its social acceptance and the essence of stewardship is defeated.

Gender issue

Women are more attached to water starting from procurement to discharging household tasks. In water-scarce areas as in Rajasthan, India, women walk several kilometres to bring water to their homes. All water-related development programmes should, therefore, have a gender-sensitive approach that can have a positive impact. Involving women in the design and implementation of interventions leads to effective new solutions to water problems; helps governments to avoid poor investments and expensive mistakes (UN Water, 2015) It always makes sense that if gender issues are well addressed in WSI, it will yield maximum social benefits.

Bargaining

Bargaining in water issues is commonplace. When an industry tries to abstract water from an underdeveloped area, it often becomes difficult for it, even with a legal licence, due to resistance from local people. The only way out in such cases is to negotiate with local people and politicians on financial terms.

Local politics

Local politicians often try to manipulate the corporate programme to consolidate their vote bank. For example, an industry willing to install a number of drinking water tube wells in a neighbouring village selects the sites based on hydrogeological and socioeconomic data. Politicians may try to capture the programme to harvest maximum political benefit and ask the industries to relocate tubewell sites according to their political priorities.

4.4.5   *Market challenges*

It goes without saying that the whole point of business is making money from the market. A break-even situation is not a driver of growth; consequently, while delivering water stewardship tasks an industry always focuses on profit, growth and brand value. The effectiveness of WSI shows up in the money saved, which in the present context, is the equivalent of money earned from the market. The European Union claims that EU companies saved around €600 billion annually (Orr, 2014) from the water stewardship initiatives.

Another market-oriented consideration is that water appreciated through WSI brings in new opportunities in investment, new job avenues and new market possibilities. CDP has reported that 75% of companies who submit annual water reports to them have achieved remarkable success in water savings through WSI.

4.5 Developing a corporate strategy in water stewardship

Developing a corporate water strategy (UNESCO, 2015) is the primary task in an efficient water stewardship programme. Basic challenges in this process are (a) appreciating legal formalities, (b) specific profiling in the value chain, (c) market analysis, and (d) best practice in water management. Although these four components entail distinctly separate activities, they eventually converge on a clear plan for mainstreaming a wholesome corporate water strategy.

Davies (2011) while reviewing Will Sarni's book on 'Corporate Water Strategies' stated that water strategy is a 'call to action for corporations to move towards water stewardship' and a call to engage all stakeholders in those initiatives. Sarni's book offers deep insight into water management, which if adopted in earnest, will ensure water sustainability through fixing goals and following certain steps. Martin (2011) has identified six components in developing a corporate water strategy. As adapted from Martin, the steps are shown in Figure 4.8.

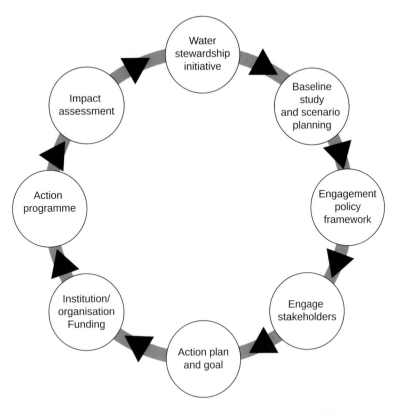

Figure 4.8 Steps in developing corporate water strategy. Adapted from Martin (2011).

4.5.1   Understand and recognise sustainability

Water sustainability is the main driver behind the building of a corporate water strategy. A company must understand what sustainability means and appreciate the application of the concept at every point in its value chain. Companies must assess the areas where the company has a large impact on the social and natural environment or where social and natural issues impact the business. The investment decision of the company must be guided by the sustainability concept.

4.5.2   Develop an engagement framework

Every business has stakeholders coming from different sectors of society playing a pivotal role in achieving success. Business, on the other hand, needs to build up relations with stakeholders and engage them in various business-related functions. Some of the functions are:

   i Water risk management: In a situation of water crisis, participation and cooperation of stakeholders hold the key to lessening water risk. If stakeholders are indifferent to the fact that water is a shared resource, the management plan falls apart.
  ii Regulatory compliance: Regulatory compliance needs the engagement of external stakeholders who work as non-financial assurance providers coming from academics, consultants and experts.
 iii Baseline data generation: Companies are required to engage hydrologists and sociologists to prepare baseline data prior to industrialisation.
  iv Innovation: Saving water saves money. Companies need to look for people externally or from their own staff who are innovative, thereby lending a cutting edge to business through water-saving practices.
   v Strategy: Developing a corporate water strategy calls for hiring experts when the company does not have in-house expertise.

Exploring engagement opportunities is essential to effective water stewardship. This becomes possible after a company has explored the area of operation (i.e. basin/watershed) and has developed a water policy based on ground truth. It can then plan for achieving the desired goal; the objectives being mitigating water risks, operational risks and regulatory risks, and of course ensuring social acceptance of the scale of operation. There are several actors in the engagement policy who may have limited relevance and opportunities in framing water policies. The company should prioritise the requirements, the scale of engagement and identify the key actors in the task. Table 4.3 summarises relevant actors, their water policy relevance, and how engagement opportunities differ depending on scale.

4.5.3   Identification of stakeholders

Stakeholders are people who have an interest in or are affected by an effort (Rabinowitz, 2015). Understandably, therefore, stakeholders are a heterogeneous lot representing different sectors of society with variable interests in relation to industrial water-resource development plans. Water stewardship efforts will encounter stakeholders who can be classified into different categories based on their level of participation in the WSI.

Stakeholders are characterised under three categories:

1 Primary stakeholders: People under this category are (1) direct beneficiaries, and at the opposite pole, (2) those adversely affected by the project. For example, in a large irrigation project people in the command area are benefitted, but a section of people are uprooted from their native settlements. In this case, both groups are primary stakeholders.

2 Secondary stakeholders: Secondary stakeholders are close to primary stakeholders but are not directly affected or benefitted. But, they care for the interest of primary stakeholders. For example, an NGO rallying against the eviction of people is a secondary stakeholder. Intermediaries in an effort are also secondary stakeholders.

3 Key stakeholders: These are the people or a group who may belong to either or neither of the above groups but take the positive or negative role in an effort. For example, while politicians argue for a bill or government bodies implementing a water project they act as key stakeholders.

The task of a WSI is to identify the stakeholders in a project and classify them according to their involvement and projected benefit from the effort. The applicability of key agencies, their importance in water strategy, and how engagement opportunities differ by scale are summarised in Table 4.2.

4.5.4 Engagement risks

The engagement framework may require remaining cautious over policy and regulatory capture which they may have to cope with. Policy and regulatory capture is a dangerous trend and has the potential to undermine the interests of other stakeholders and deny the cause of water stewardship. The source of imbalances in power and influence is given in Table 4.3.

4.5.5 Collective action framework

Water is a shared resource and hence competition for water intensifies pressure on both business and the ecosystem which, in turn, generates social dissent. Under such consequences, business can perform well if a framework of collective action (Greenwood, Willis, and Ross, 2012) is developed taking all competitors under a joint action programme. Business will, however, have to compromise or trade off with the stakeholders. All water level risks and sustainability issues are manifested at the catchment level. Therefore, collective action programmes should be taken up in the catchment level in alignment with IWRM programmes. The parties involved in collective action are other companies or groups of companies, supply chain operators, stakeholders, academia, and the government. At times, programmes for collective action extend to regional, national or international level.

Table 4.2 Engagement across different scales of water policy.

Scale	Typical entities engaged	Common water policy relevance	Engagement opportunities
Corporate/ Internal Operations	• Facility and sustainability managers • Line workers • Environmental health and safety personnel • Legal counsel	• Meet regulatory standards • Reduce demand on water sources • Minimize pollution • Assess and mitigate social and environmental impacts	• Assess risks and opportunities • Communicate expectations for meeting or exceeding regulatory or legislative compliance • Share water data to help benchmarking and impact assessment • Share innovative technologies and practice
Local (Municipality/ Community)	Local water providers • City planners • Community councils and committees • Community-based organizations • Related government departments	• Set water rates and distribute water • Establish and amend building, plumbing and planning codes • Set local priorities and manage in the public interest • Service delivery to underserved areas	Change building codes and planning processes to consider non-structural water treatment • Encourage community engagement in water management and planning • Share investment in service-delivery infrastructure development and O & M
Regional (Catchment/ Watershed)	Regional water providers • River basin authorities • River basin boards and commissions • Catchment stakeholder forums • Research institutions and universities • Local and international NGOs	• Set water rates • Develop quality goals and corresponding parameters for each body of water • Integrate water services • Provide a meaningful and legitimate forum for public participation • Developing contextually specific responses to shared risks	User fees that recover full capital and O & M costs and encourage efficiency • Create catchment-based planning units that integrate data and create economies of scale • Support transparent decision making and oversight for accountable water governance • Participatory decision-making and sign-off of allocation decisions and conflict resolution
National	National governing bodies (e.g. legislature and parliament) • National agencies (e.g. water management and infrastructure development) • National NGOs (e.g. for environmental protection) • National water boards and water research councils	• Legislation (e.g. national frameworks like the South African Water Law) • Water allocation processes (e.g. water rights framework) • Enforceable and enforced standards (e.g. contaminant limits) • Monitoring networks (e.g. water quality testing) • National and regional water and land development and planning	Establish polluter-pays and beneficiary-pays principles • Avoid inappropriate subsidies for water infrastructure or services • Require policies for integrated approach to water management • Establish enforceable water-quality standards that protect human and ecosystem health • Monitor institutional performance

(*Continued*)

Table 4.2 (Continued)

Scale	Typical entities engaged	Common water policy relevance	Engagement opportunities
International	• Bilateral development partners (e.g. government entities like DFID, USAID, Danida, JICA, etc.) • Multilateral development partners (e.g. UN Agencies such as WHO) • International financial institutions (e.g. World Bank, IFC and regional development banks) • International NGOs and networks of smaller NGOs or researchers • Multi-stakeholder initiatives	• International law (e.g. trans boundary management) • Standards (e.g. drinking water) • Human rights (e.g. to water) • Financing (e.g. for large infrastructure projects) • Research and development • International policy and methodologies	• Establish international law and standards around water supply and quality concerns to enhance certainty and reduce risk • Advance best management practices • Finance water infrastructure and wastewater treatment • Invest in or directly support innovation for improved SWM • Unlock progress in SWM through international advocacy

Source: Morrison *et al.* (2010); © Pacific institute, published with permission from the Pacific Institute.

There are four types of collective action (Figure 4.9), which are:

1 Informative: Where the stakeholders share information and knowledge to build an environment of trust and mutual dependence.
2 Consultative: Consultative status of engagement is required for getting new ideas; sharing experience and conducting water-related accounting and assessment for taking informed decisions. Companies often arrange consultative meetings with stakeholders, academia, government and other companies for sharing ideas and experience.
3 Collaborative: Collaboration with other organisations who have common objectives and the willingness to share responsibilities. A good example is a watershed development programme launched by ITC in its supply chain area in collaboration with local NGOs.
4 Integrative: In this type of engagement, everybody works harmoniously in accordance with corporate decisions and functions as a team in the face of challenges.

4.6 Goals and commitments

Water stewardship aims to reduce water stress and water risk and to ensure social well-being. These goals are aligned with the goals of IWRM and SDG 6. But each company can set its own stewardship goals. For example, Coca-Cola has declared that by 2020 it will recycle the water wasted in its

Table 4.3 Sources of policy and regulatory capture.

Source of imbalance in power	Description	Example
Knowledge	Where the knowledge and expertise available to some stakeholders and/or preferential access to information or the ability to generate information prejudices one position or perspective over others. Related to epistemological concerns and the favouring of some knowledge 'types' over others.	Policy makers tend to value quantitative, 'scientifically' derived evidence over qualitative, 'local' perspectives, despite the arguably equal value of pluralistic knowledge types. Information and technical expertise is more accessible, and can be generated or mobilized by those with financial or other forms of power.
Financial	Where financial considerations are permitted to dominate other public-interest criteria to the detriment of some stakeholders.	Protectionist policies to favour large economic interests. Rejection of enforcement action where the offender is a significant employer and source of tax revenue.
Political	Where political influence and reach are exercised to further political ends—by a member of the executive or those seeking favour with members of the executive.	Used to win favour of local electorates, to curry political power or favours by business, and to encourage patronage. For example, influence by state governors or local councillors in planning decision-making processes to ensure favourable outcomes for select interests.
Access	Where, by design or accident, some stakeholders or stakeholder groups are better represented or have greater access to decision makers and thus are more able to influence outcomes.	Easy access to government offices by telephone or in person, compared to other stakeholders from outside commercial centres or who are seen as less important. Preferential or unbalanced membership of stakeholder forums, particularly where resource outlay is involved.
Opportunity	Where individuals and teams working for government are offered future enhanced professional opportunities in exchange for favouring certain positions and perspectives.	'Revolving door' employment within corporate concerns who offer employment, training, or career development opportunities to public-sector workers. For example, some mines in Africa offer well-paid career development secondment opportunities to regulators who 'perform well'.
Social and psychological	Where some stakeholder positions are favoured over others because of slick and persuasive presentation as modern, advanced, progressive, etc.	Some public-sector institutions may be easily swayed by the confident presentation of certain positions, attributing the trappings of financial power— luxury vehicles and offices— to 'success' to be emulated and supported.
Logistical capabilities	Where the resources or logistical capabilities, such as ability to access field sites, tends to favour certain stakeholders and their perspectives over others.	Remote sites where field observations and investigations are necessary to support one perspective over another are only accessible to those with 4x4 vehicles, airplanes, or other specialist equipment.
State protection	State capture occurs when state parastatals, ministries, agencies, or local government receive preferential treatment at the expense of other stakeholders.	Non-enforcement of regulatory conditions against state-owned or state-managed facilities or operations (such as wastewater treatment works).

Source: Morrison et al. 2010. Published with permission from the Pacific Institute.

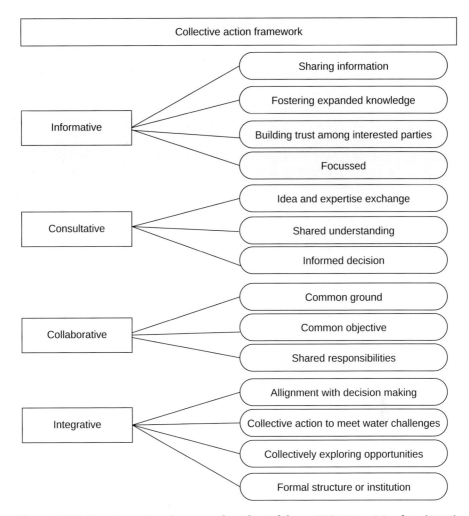

Figure 4.9 Collective action frame work. Adapted from CEO Water Mandate (2013).

manufacturing processes and treat all wastewater before discharging it to the environment. Similarly, BASF has also set its goal to reduce organic pollutants and nitrogen in water by 80% and heavy metals by 60% by 2020. Walmart has focused its goal on the use of renewable energy which in turn saves water. Some specific goals are suggested below that companies can achieve within a specific time frame:

- Completion of baseline study;
- Development of regulatory compliance efficiency;
- Completion of water accounting;
- Optimisation of water use;
- Reduction of emission of hazardous chemicals;
- Expansion of IWRM to the watershed level;
- Develop institutional capacity;
- Ensure water, sanitation and hygiene (WASH) facility to the community;
- Train stakeholders in IWRM; and
- Implement tasks under CWS.

4.7 Establish systems and processes

Once the goals are established, the system and process to implement the action components become crucial. This needs development of proper executive decisions that will guide the process. A committed leadership is required to steer the whole process. In the 2012 Report of Sustainability Leaders (VOXGlobal, 2012), the key to the success of sustainability has been identified as personal skills of the leadership in the company. The four major qualities that a corporate leader should have are:

1 Interpersonal skill to implement changing processes with knowledge-driven expertise.
2 Corporate vision in balance with ecological services.
3 Excellent communication skills to motivate people within the company and its environs to align with water stewardship programmes.
4 Collaborate with other companies, institutions, government, politicians, NGOs and colleagues.

4.8 Opportunities in water stewardship

Complying with the government norms in water management, despite being a binding condition, is not enough to deliver social responsibilities; companies need to be proactive. Water management has to be a clearly visible part of the corporate strategy, for it shows up on the social radar. There is a sharp 'distinction between firms that are compliance driven and merely aim to meet legal requirements, and those that adopt more proactive environmental strategies' (Buysse and Verbeke, 2002).

Corporate bodies can secure their business even in water-stressed areas through implementation of programmes under water stewardship. As stated, the SDG 6 has several targets which can be aligned with corporate water stewardship. Within the framework of SDG 6, corporations can take into account the following opportunities:

4.8.1 Management improvement

Corporate bodies meaningfully engaged in WSI have dedicated teams for translating plans into action and for midcourse correction in faltering situations. The team consists of directors at the centre who undertake relation-building with shareholders and management. For transparency, the relation-building exercise also takes on board selected employees and suppliers. This team looks after the legal, regulatory, institutional and ethical environment of the basin.

4.8.2   Knowledge asset development

Knowledge assets are 'resources' used to support the storing, sharing and reuse for the purpose of development of the organisation or project development (KAF Knowledge Audit Framework, no date). This is an essential component in building a water strategy because knowledge empowers corporate managers to act more efficiently. There are companies and a consortium of companies who stress developing knowledge assets through collaboration with academic institutions. The knowledge asset can be derived from traditional knowledge or locally sourced knowledge (Nonaka, and Teece, eds. (2001)) or may be built through a process of socialisation. Corporations should also develop systemic knowledge assets in the form of documents, monographs, study reports, academic papers, and so on, which are explicit and codified, and can be evaluated or retrieved. For example, Unilever India Ltd. has published a book entitled *The Story of Water* which is a literature that serves as the water stewardship document of the company (Unilever India Ltd., 2012).

4.8.3   Investment

Capital expenditure on water efficiency is in effect an investment (creating fixed assets like land and water-treatment plants) that ensures rich returns. The simplest and most common example comes from the agricultural sector; by adopting drip irrigation in orchards, a fruit-processing company saves water, and in the process saves millions of dollars. Corporate bodies need to appreciate that water optimisation (in manufacturing processes), in combination with WSI not only broadens the investment horizon but also increases brand value, reputation and, above all, profit. Thus empowered, corporate bodies can then promote research on climate change and natural disaster mitigation, which further boosts company efficiency and visibility through participation of local people and NGOs. Corporate houses with philanthropy funds can create environmental awareness simply by extending WASH facilities to local villages to the point of adoption of a whole ecosystem (e.g. upper catchment recharge zone or a riparian landscape), which can then open up avenues for ecotourism. The possibilities of corporate water stewardship are endless. Buysse and Verbeke (2003) identified six domains in developing corporate strategies for environmental issues which are applicable for water stewardship. The domains where the corporate sector can improve performance through appropriate investment and planning are shown in Table 4.4.

4.8.4   Developing information and database

Through WSI, a company can generate ample data and information from the watershed and the social and natural entities that will be useful in building business strategy and setting up water stewardship goals. In fact, the success of water stewardship is nested in successful water-related data

Table 4.4 Domains of corporate water strategy and investment options.

Domain	Investment options
Technological competencies	Investments in conventional green competencies related to green product and manufacturing technologies
Employee skills	Investments in employee skills, as measured by resource allocation to environmental training and employee participation
Organisational competencies	Investments in organisational competencies, as measured by the involvement of functional areas such as R & D and product design, finance and accounting, purchasing, production, storage and transportation, sales and marketing, and human resources
Routine management	Investments in formal (routine-based) management systems and procedures, at the input process, and output sides. An incentive system to reward environmentally responsible behaviour is an added component for investment in routine management.
Reconfigure the strategic planning process	Efforts to reconfigure the strategic planning process, by explicitly considering environmental issues and allowing the individuals responsible for environmental management to participate in corporate strategic planning.
Knowledge and monitoring	Investing in information, knowledge and monitoring is a major area where the return is indirect but observable in the positive changes in implementation of water conservation methods.
Partnership	Investing in water projects through NGO partners. Corporations can grant project funds based on proposals submitted by competent NGOs intending to implement water resource development, awareness and research projects.

Adapted from Buysse and Verbeke (2002).

generation and information sharing. Major components of water-related data and information generation activities are:

Baseline study

Baseline study is a component of information flow in the decision-support system (SOCOPSE, 2005) on which the strategic model will be built up. To estimate the impact and changes to come in future (say, after five years), a baseline is constructed for the current year. It is generally prepared for a boundary unit of the place of operation. For EIA purposes, the boundary is considered as a land area with a 10 km radius around the industry. For WSI, a watershed is to be taken into account as has been discussed earlier.

Although a business operates within a hydrological boundary (watershed/basin) and withdraws and discharges water within that boundary, the supply chain may be located in another basin in another country. So, the industry is also concerned about the water resource of the country of operation and the country of production of raw materials. For example, sugarcane produced in Belize in Central America is a raw material for the Tate & Lyle Sugar factory in the UK. Both local water resources of Tate & Lyle factory site and the water sustainability of Belize are interlinked symbiotically (Tate & Lyle, no date).

The first strategic step in corporate water management is to know the character of the water resource being developed by the industry, such as

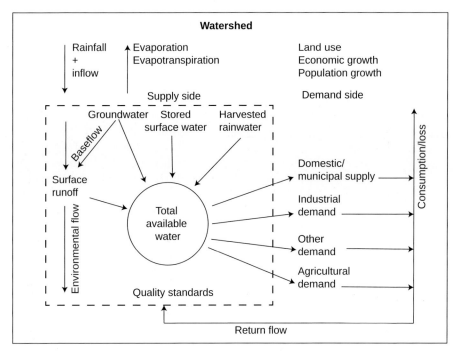

Figure 4.10 Components of baseline and scenario condition.

how much water is abstracted, schedules of withdrawal, treatment facilities and delivery and discharging methods. The company should engage appropriate manpower and technology to generate strategic baseline information in the area of operation. The knowledge may be acquired from accessing historical data and reports, conducting baseline surveys, periodical monitoring and compiling the data into a data management and retrieval system.

The baseline study components within a basin are:

- **Geographical information:** This includes comprehensive land and water data.
- **Supply side component:** Data regarding water availability, sources and infrastructure available for water abstraction.
- **Demand side component:** This includes comprehensive data collection on water demand, withdrawal and use by different sectors. All water conservation measures taken up by different sectors will also come under this head.

In Figure 4.10 the supply-side components and demand-side components of a watershed are elaborated. It may be noted that these components are studied in the baseline year and are to be projected for the scenario year.

Scenario development

There are uncertainties in achieving the goals of corporate water stewardship that are beyond the control of business houses. These uncertainties are on both supply side and demand side. For example, given a particular demand volume, the supply side can be affected by drought or by catastrophic floods damaging the supply line. It needs cautioning in this context that there is no such thing as static demand; it is invariably rising in a growing economy. Long-term observation tends to suggest extreme weather conditions are not random; they fall into a pattern. In addition to the normal weather data on the supply and demand sides, strategic planning must also factor in the projected scenario based on meteorological data on extreme weather and ever-increasing demands. This will help corporate bodies to plan water strategies in advance. Water flow in the baseline year and the scenario year (Merrett, 2007) should be computed considering the variables within a watershed.

Modelling

Resources, infrastructure and demand are the three components of baseline data generation and scenario formulation. Each scenario is linked with population growth, agricultural practice, resource availability, climate change, infrastructure and land use within a certain boundary, that is, geographical, political, administrative, linguistic, cultural and institutional boundaries. The boundary-specific scenario is constructed based on observations made over changes in numerical values of parameters observed in a certain time-frame (say, 10 years). The most important parameters are precipitation, inflow, annual evaporation loss, groundwater availability, baseflow, surface runoff and surface storage on the supply side. Supply-side parameters are influenced by land use change and climatic variability. The demand-side parameters are industrial, agricultural, domestic and ecological demands. When there is any unmet demand in any sector under a specific scenario, quantification of water savings by taking up water conservation measures will give an idea as to what extent the water conservation measure will be effective in meeting the demand for that scenario year. How a scenario modelling works in planning is shown in Table 4.5.

4.8.5 Human resource development

Development has now been reduced into a discourse. Consequently, environmental policy making and the needs of the business community have come to loggerheads on a very particular and volatile question: Development at what cost? A very major component of the answer to that question is the scale of water management, which inevitably gravitates to the availability of the number of scientifically trained personnel, which in turn brings into focus the context of human resource development. Poetic appreciation of nature is, perhaps, genetically embedded in mankind, but the scientific appreciation of nature is in short supply. Science and technology must, therefore, play a central role in water management from the grass-root level upwards. This chapter reviews the possibility of human resource development in different scales of water governance and developing a new

Table 4.5 Impact forecast for a dry scenario year with increased demand

Catchment Condition		Baseline Year (b)		Scenario (s)	Impact
		Resource	Activity	Activity	
Supply side	Rainfall + inflow	P	P_b	$P_s < P_b$	Decremented change in climate activity, an
	Evaporation loss	E	E_b	$E_s < E_b$	increase in demand will impact the
	Groundwater	G	G_b	$G_s < G_b$	ecological flow. Return flow may impact
	Baseflow	B	B_b	$B_s < B_b$	water quality. Water demand may not be
	Surface runoff	R	R_b	$R_s < R_b$	met in any sector. The condition is
	Surface storage	Ss	Ss_b	$Ss_s < Ss_b$	unsustainable. Situation demands
	Total available water		Q_b	$Q_s < Q_b$	immediate measures.
Demand side	Industrial demand	DI	DI_b	$DI_s > DI_b$	
	Agricultural demand	DA	DA_b	$DA_s > DA_b$	
	Domestic demand	DD	DD_b	$DD_s > DD_b$	
	Ecological demand	DE	DE_b	$DE_s < DE_b$	

paradigm of water stewardship. The emerging areas where such human resource development is required are:

- **Strong management skill:** This is developed through management orientation courses and practices in an industrial environment. Senior managers are generally capable of acquiring such skills. The skills also include planning and evaluation of a project or effort. The essential skills are required in the following fields:
- **Water awareness:** This is about an awareness of general water debates (social, environmental, economic), water management (Anonymous and Orr, 2013), the functionality of water institutions, as well as the implications for specific sectors. Awareness is required in the internal water issues of the business. The target for water awareness is inclusive and will engage every stakeholder within the catchment and the governance.
- **Understanding theories and concepts:** Developing a working knowledge of the prevalent concepts and theories related to water stewardship.
- **Water resource data generation, mapping and reporting:** There are specific protocols for collecting data and natural samples. In many countries, there are strict guidelines and reporting formats.
- **Water accounting and disclosure:** Judging capacity in water audit and water footprint studies is a specialised skill. Training programmes are often organised by international and national organisations.
- **Water Act and statutory compliances:** Knowledge and understanding of water-related laws, methods of compliance, processes of applying for licence allocation, and processes of submission of annual returns. Special training programmes are required to impart such knowledge.
- **Water conservation:** Capacity in planning and implementing water conservation projects. This subject is included in engineering and environment science curricula.

- **Wastewater and solid-waste management:** Expert knowledge of wastewater and solid-waste management is also crucial. This also includes knowledge of local geology and hydrogeology. This subject is included in engineering and environment science curricula.

4.9   Water literacy

Water literacy is one of the activities under human resource development. The subject is gaining momentum throughout the world.

4.9.1   Definition and concept

It follows from this that the role of water cannot be overemphasised. The water crisis affecting the present generation is due to uninformed water usage by the preceding generation. If we continue with these practices, that is, uninformed water usage, posterity will be severely affected, which is precisely where corporate water stewardship kicks in so that educated water use, especially in big business, spares the Third Millennium generation from the follies of its parental generation. 'Water Literacy sets standards for water information' (The Alliance for Water Education, no date).

Water literacy is a crucial factor in water stewardship. An alert and responsive corporate manager is expected to know about the natural drainage system or the type of aquifer he would use. At the same time, he must also be concerned with the long- and short-term benefits of both the global as well as local hydrological systems while setting up the water management. Moreover, the manager will not have to depend solely on outsourced expertise while preparing a hydrological report for his project; on the contrary, he will be able to monitor and assess the quality of the work delivered by the consultants. This will lend a cutting edge to arguments defending his case before the regulatory bodies—a faculty respected by regulatory bodies sitting on judgment over corporate water requirements.

Water literacy will help the decision makers or regulatory bodies to make better decisions regarding the laws, regulations or statutes which address case-specific issues and tend to cover the most suitable applications of governance. For example, regulatory authorities in a country or province fix general distance criteria for space between two tube wells (say 500 metres) while giving a permit. This is done because the policy makers are not fully aware of the hydrological variations and hence, simplify the process. This simplification may result in overexploitation of the groundwater and cause interference between the existing and proposed well. A water-literate policy maker will consider location-specific hydrogeological aspects and will take decisions accordingly. Basic knowledge of hydrogeology will help the policy maker to take justified decisions on a case-to-case basis.

Water literacy of a corporate manager or a policy maker will enable him to:

- Understand the hydrologic dynamism of the source water and enable him to plan his water requirement more judiciously.
- Help him to select the discharge area for his effluent.
- Initiate research on source-water protection and water conservation in the value chain of the industry.
- Policy makers will be equipped with the basic knowledge to take an unbiased decision on water policy and water regulations.
- Help in resolving conflicts on water-related issues.
- Develop the proper plan for rainwater harvesting and storm-water management.

The policy dialogue between the corporate manager and the administrative decision maker will be scientifically founded, which is the fundamental aim of educating corporate managers in water resource management.

4.9.2 Water literacy framework

A water literacy framework has been developed in many countries. The Department of Science and Technology, Government of India, has been pursuing water a literacy programme for many years. In developed countries like the United States, Canada and Australia, there are both government and non-government institutions and universities who regularly organise workshops and training programmes on water literacy. It is also a subject under the information literacy framework of Australia and New Zealand. Most of these programmes are targeted for the general public and children. The framework is schematically elaborated in Figure 4.11.

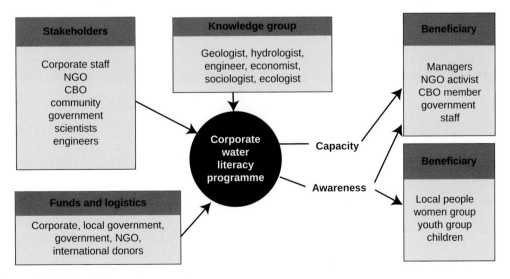

Figure 4.11 Water literacy framework.

4.10   Action programmes under WSI

Several action programmes can be taken up by a company under its water stewardship initiatives. Some of the important programmes are as follows.

4.10.1   Conduct a water resource assessment

The business performs its activities within a river basin, either large or small. The watershed is considered as the geographical unit used for assessment of water resources. The following (Sheng, 1990) watershed challenges are:

- **Data collection and data generation:** Existing hydrological and geological data are required to meet the demands of scenario planning. In many countries, geological, land use, hydrogeological, and meteorological surveys are being done regularly and have a good archive of such resources, including temporal data and maps. These data and maps are required to be updated. But there are several countries where such data are either absent or very difficult to procure. Corporations have to develop proper arrangements to bridge such data gaps, either through lobbying with the government or by using their own machinery to generate baseline data.
- **Assessing physical conditions:** The physical condition of the watershed terrain includes slope stability, fragile geological formations, inaccessibility, protected areas (which require special clearance from the government for data collection), and so forth. Business managers should familiarise themselves with the physical condition of the concerned watershed through the preparation of spatiotemporal maps, either by using their own resources or by outsourcing, whichever is more economical.
- **Resource management:** In a watershed, the two main resources are land and water that need informed management. Problems arise when an uninformed or indifferent management turns a blind eye to deforestation, soil erosion, over-exploitation and so on within the watershed area.

4.10.2   Conduct a water footprint analysis

Companies have to quantify water performance through water footprint study to determine the overall rate of productivity of water. Corporate bodies can develop normative standards for benchmarking water use in direct operations, including the practice followed by suppliers to the company.

4.10.3   Conduct a sustainability analysis

A sustainability analysis involves several subsets. Some of the significant subsets are discussed below.

Assessing water-related social and environmental impacts

The performance of a company in terms of its water withdrawal, consumption, and grey water footprint has actual social and environmental impacts. These impacts are manifest in social and environmental indicators and can be assessed through a set of tools that quantify the impact. The company can also model the projected impact in different business scenarios. Two important tools are available for conducting sustainability analysis, the WBCSD Global Water Tool and the GEMI Water Sustainability Tools.

Business risk assessment

Corporate bodies can be benefitted through water assessment if it is carried out in scales appropriate to the size of the basin and its catchment area. This exercise offers insight into flaws embedded in existing practices and enables quantification of the risk associated with total withdrawal; local water risk is a major threat to business. The major components of water risk are:

- **Inequitable demands and pricing of water:** Water demand management is a stepping stone in poverty alleviation because water also figures prominently in terms of social equity (Tyler, 2008). Tyler also showed through an example from Namibia, that although water pricing has increased water service to rich people, affordability has forced the poor to buy less safe water.
- **Over-appropriation:** Over-appropriation is a degenerating trend found in many countries. It is defined as the withdrawal of water beyond the renewable capacity of the source. Groundwater is believed to be the most over-appropriated resource. Water resource management and water policy should quantify over-appropriation and release water rights accordingly (Sokolova, 2016).

4.10.4   Water accounting and disclosure

Measuring water resources, risks, opportunities and responses is 'water accounting' while 'water disclosure' is placing before investors all relevant data on water and possible risks. Water accounting and disclosure are complementary to each other. Water accounting builds the foundation of the corporate water profile, which then becomes the baseline data for corporate disclosure document.

Disclosure and reporting are vital to a sustainable business strategy. The performance of a company under water stress and its capacity to mitigate water risk are reflected in the disclosure and reporting documents. The main components of a water disclosure document are:

- Water audit report;
- Water footprint study report;
- Water profile of the watershed;
- Water profile of the facility;
- Global water profile of the company; and
- Sustainability assessment

Water disclosure and reporting have been further elaborated in Chapter 9.

4.10.5 *Implement mitigation measures*

Some of the action programmes for mitigation of water-related issues are:

- Water conservation;
- Storm-water management and flood control;
- Retrofitting fixtures to contain water loss;
- Extension of WASH facilities to the community; and
- Watershed management

4.11 Outcome of water stewardship initiatives (WSI)

The outcomes of successful WSI are many. Water stewardship must deliver value for civil society, private and public sector (Alliance for Water Stewardship, no date). The ideal outcome should be business sustainability, well-being of society and most importantly, preservation of ecological balance; a healthy society and green growth are tangible outcomes of WSI. Since the objective of corporate water stewardship is to scale up business through capacity building among local people, engaging the community (Xu, Xi and Spencer, 2016) is vital to WSI. The Alliance for Water Stewardship has developed international standards for water stewardship (Abdel *et al.*, 2014); they have identified four outcomes: good-quality water, good water governance, sustainable water balance, and healthy status of 'Important Water-Related Areas'. Sustainability of WSI depends on local engagement and attainment of the outcomes simultaneously. The outcomes are elaborated in Table 4.6.

4.12 Water stewardship standards

The Alliance for Water Stewardship (AWS) has defined six aspects of stewardship standards and qualified the aspects in two levels: core and advanced. The six aspects of AWS standards are (1) Commit, (2) Gather and process, (3) Plan, (4) Implement, (5) Evaluate, and (6) Communicate and disclose. The standard is based upon a logical sequence (Abdel *et al.*, 2014) of how water stewardship can achieve water security through activities starting from the industrial establishment and extending it to the catchment level. The sequence is shown in Figure 4.12.

AWS has also listed core activities under each aspect and assigned scores against performance under each aspect in their water stewardship standards, which are available for download from their website. An outline of water stewardship activities on which the standards are designed is given in Table 4.7.

Table 4.6 Outcome of WSI.

Outcome	Details	Indicators
Good water governance	Good water governance functions in harmony with the political, social, economic and administrative system and ensures the participation of all sectors for sustainability and equitable growth. The laws are framed without any prejudice in order to promote every stakeholder's participation. The institutional framework is strong, capable of enforcing the law, transparent and prompt. Water governance at the international level and its national counterparts should have the capacity, by law and institutional arrangement, to mitigate water conflict at any level.	• Rational water allocation • Easy process and forms • Timely delivery
Sustainable water balance	Sustainable water balance can be achieved through proactive action so that the water drawn from the catchment or basin is renewed either naturally or by engineered action in collaboration with the government and other stakeholders. Sustainable praxis can be tested through resilience against natural disasters.	• Maintained environmental flow • Rejuvenated springs • No depleting trend in groundwater level
Good water quality status	Abiding by local regulatory standards for water quality is the basic outcome. The corporate body in alliance with the local community is required to maintain ecological balance.	• Aquatic biodiversity population • No eutrophication • Clean water
Healthy status of important water-related areas	Important water-related areas like lakes, waterfalls, streams and water parks, and so on, which are important for both biodiversity and social or religious purposes are well protected. Riparian zones and the groundwater recharge zones are kept free from illegal encroachment and degradation.	
Healthy social status	Successful sustainability practice is automatically reflected in healthy societies, which is the important outcome of WSI.	• Safe drinking water for all. • Health care service for all. • Job opportunities.

Adapted from Abdel *et al.*, (2014) and Xu, Xi and Spencer (2016).

4.13 Global organisations for facilitating water stewardship

It can be appreciated from the title 'Global Water Stewardship Organisations' that start-up companies cannot undertake water stewardship in isolation. They need support from academia and experienced business houses to guide them in delivering water stewardship. Big business houses also seek expert support to deliver WSI. Several institutions have emerged to help the corporate sector in developing water-management capacity. These organisations have developed a consortium of companies with track records in WSI

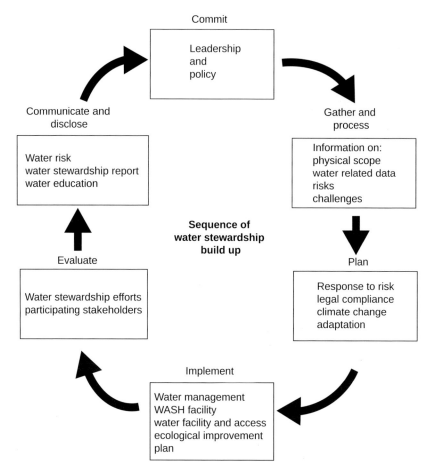

Figure 4.12 The sequence of water stewardship build up. Adapted from Abdel *et al.*, (2014).

as applicable in their respective areas of operation. The UN Global Compact is one such institution that outlines principles of ethically sound business. 'The Global Compact asks companies to embrace, support and enact, within their sphere of influence, a set of core values in the areas of human rights, labour standards, the environment, and anti-corruption' (Hayward *et al.*, 2013). The ethical principles derive from different UN declarations on human rights, labour rights and environment, and are equally relevant to water management. There are international organisations that support corporate bodies in (a) developing a consortium to share and act, (b) guide them in step-by-step performance towards sustainability and (c) help them in all issues related to corporate water stewardship and develop tools to support them. The major international organisations that are mandated to the task of helping corporations in water-related performances are given in Table 4.8.

There are many other agencies, both national and international furthering the cause of WSI by providing professional consultancy at various levels of the government and corporate bodies, and meaningfully engaging stakeholders in crisis mitigation.

Table 4.7 Summary of AWS water stewardship standard (Imane *et al.*, 2013).

Step	AWS core criteria	AWS advanced-level criteria
Commit	1.1 Establish a leadership commitment on water stewardship	1.3 Further the Alliance for Water Stewardship
	1.2 Develop a water stewardship policy	1.4 Commit to other initiatives that advance effective water stewardship
		1.5 Secure a water stewardship commitment from the organization's senior-most executive or the organisation's governance body
		1.6 Prioritize communities' rights to water
Gather and process	2.1 Define the physical scope	2.8 Support and undertake joint water-related data collection
	2.2 Identify stakeholders, their water-related challenges and the site's sphere of influence	2.9 Gather additional, detailed water-related data
		2.10 Review a formal study on future water resources scenarios
	2.3 Gather water-related data for the catchment	2.11 Conduct a detailed, indirect water use evaluation
	2.4 Gather water-related data for the site	2.12 Understand groundwater status or environmental flows and the site's potential contributions
	2.5 Improve the site's understanding of its indirect water use	
	2.6 Understand shared water-related challenges in the catchment	2.13 Complete a voluntary Social Impact Assessment
	2.7 Understand and prioritize the site's water risks and opportunities	
Plan	3.1 Develop a system that promotes and evaluates water-related legal compliance	3.5 Gain stakeholder consensus on the site's water stewardship targets
	3.2 Create a site water stewardship strategy and plan	3.6 Develop a formal plan for climate change adaptation
	3.3 Demonstrate responsiveness and resilience to water-related risks into the site's incident response plan	
	3.4 Notify the relevant (catchment) authority of the site's water stewardship plans	
Implement	4.1 Comply with water-related legal and regulatory requirements	4.9 Achieve best practice results on site water balance
	4.2 Maintain or improve site water balance	4.10 Achieve best practice results on site water quality
	4.3 Maintain or improve site water quality	4.11 Achieve best practice results on Important Water-Related Areas through restoration
	4.4 Maintain or improve the status of the site's Important Water-Related Areas	
	4.5 Participate positively in catchment governance	4.12 Achieve best practice results and strengthen capacity in water governance
	4.6 Maintain or improve indirect water use within the catchment	4.13 Advance regionally specific industrial water-related benchmarking
	4.7 Provide access to safe drinking water, adequate sanitation and hygiene awareness (WASH) for workers on-site	4.14 Re-allocate saved water for social or environmental needs
		4.15 Engage in collective action to address shared water challenges
	4.8 Notify the owners of shared water-related infrastructure of any concerns	4.16 Drive reduced indirect water use throughout the site's supply chain and outsourced water-related service providers
		4.17 Complete implementation of water-related initiatives
		4.18 Provide access to safe drinking water, adequate sanitation and hygiene awareness off-site

(Continued)

Table 4.7 (Continued)

Step	AWS core criteria	AWS advanced-level criteria
Evaluate	5.1 Evaluate the site's water stewardship performance, risks and benefits in the catchment context 5.2 Evaluate water-related emergency incidents and extreme events 5.3 Consult stakeholders on water-related performance 5.4 Update water stewardship and incident response plans	5.5 Conduct executive or governance body-level review of water stewardship efforts 5.6 Conduct a formal stakeholder evaluation
Communicate and disclose	6.1 Disclose water-related internal governance 6.2 Disclose annual site water stewardship performance 6.3 Disclose efforts to address shared water challenges 6.4 Drive transparency in water-related compliance 6.5 Increase awareness of water issues within the site	6.6 Disclose water risks to owners (in alignment with recognized disclosure frameworks) 6.7 Implement a programme for water education 6.8 Discuss site-level water stewardship in the organization's annual report

Printed with permission from The Alliance for Water Stewardship (2014). © 2014 Alliance for Water Stewardship.

Table 4.8 Profiles of International organisations that facilitate water stewardship.

Name of the agency	Short profile	Water management-related engagement
The Water Network	It is a web-based network that provides a neutral platform for multi-stakeholders to share content, best practices, policies and technologies. The portal provides a platform for professional engagement and public education on water issues.	• Network • Publication
Overseas Development Institute	This UK-based international organisation is mainly engaged in pro-poor activities throughout the world. It helps corporations to engage WASH programmes by linking high-quality research with practical policy advice on water issues.	• Research, advocacy • Publication
World Business Council for Sustainable Development	It is a CEO-led organisation of companies working together in the field of sustainability and a leader in providing advocacy services in building strong bonds between the company and stakeholders. 200 member companies under this organisation are committed to sharing best practices in sustainable development and water stewardship. Most of the member companies are located in developing countries.	• Research • Advocacy • Tools
CEO Water Mandate	This UN-based organisation was formed as a special initiative of the UN Secretary-General. CEO Water Mandate works in collaboration with the Pacific Institute and UN Global Compact. The main activities are to (1) conduct research and provide guidance to corporations, (2) help corporations to identify, explore, and solve key water challenges, (3) facilitate meaningful collective action to save river basins.	• Research • Advocacy • Tools
Alliance for Water Stewardship (AWS)	AWS is a global organisation dedicated to promoting the concept of water stewardship through collective action, training and advocacy. In their global-scale endeavour, they engage regional expertise in sustainable water management in industry and other sectors. AWS also provides a voluntary framework for major water users and has developed a water stewardship standard (AWS standard).	• Research • Advocacy • Tools • Standards
Global Environmental Management Initiative (GEMI, 2016)	Global Environmental Management Initiative was established in 2014 as an inter-agency initiative composed of the United Nations Environment Programme (UNEP). GEMI as an organisation of corporate bodies helps its members to develop strategies to assist business to foster global environmental health and safety excellence and economic success. It is also a forum for global corporate environment leaders to share tools and experience. (UN Water, 2016)	• Research • Advocacy • Tools • Standards
Growing Blue	Growing blue is a web-based tool to assimilate data and statistics to assess global water stress by country and region.	• Research • Advocacy • Tools • Standards

(Continued)

Table 4.8 (Continued)

Name of the agency	Short profile	Water management–related engagement
Water Footprint Network	Water Footprint Network provides solutions to water stewardship and balances the economic, social and environmental benefits of water use through resource efficiency and fair allocation by good governance. This is a forum for sharing experience, building community and training practitioners to solve the world's water crises. They are also global leaders in water footprint assessment. (sourced from Water Footprint Network, 2015).	• Research • Advocacy • Tools • Standards
Ceres	Ceres is a not-for-profit organisation of business institutions founded in 1989, advocating for sustainability leadership. It has a network of corporations who get support for sustainable business practices and solutions. One of the core areas of Ceres is water stewardship. More than 130 members including environmental and social non-profit groups and corporations are jointly working for a sustainable world (sourced from Ceres, 2016).	• Advocacy • Tools • Standards
2030 Water Resources Group	The 2030 Water Resources Group secretariat is based in Washington, DC, and works in collaboration with the United Nations, private companies, international funding organisations and non-profit organisations like WWF. The main objective of this organisation is to bring down the gap between the supply and demand of water resources in water-stressed countries. To achieve the goal, they convene on a multi-stakeholder platform to generate an open, trust-based dialogue process on legal requirements in water-sector reforms. They work in collaboration with corporate leaders like Coca-Cola.	• Research • Advocacy • Tools • Data repository
World Resources Institute	Water Resource Institute is a non-profit corporation based in Delaware, USA. The institute carries out scientific and viable 'policy research and analysis on global environment and resource issues'. They are dedicated to bringing about positive environmental changes. In the water sector, they believe that 'critical risks must be measured, mapped, and managed'. They have developed a huge data base and a large pool of information that is freely available for policy makers, corporations and researchers (Source: World Resource Institute).	• Research • Tools • Policy
World Wide Fund for Nature (WWF)	World Wide Fund for Nature is a well-known non-profit organisation having a presence in almost all countries in the world. Corporate water stewardship has been included in their recent mandate, and they are working at both the policy formulation and the implementation levels. WWF works in collaboration with other organisations like CEO Water Mandate and UN-Water. One of the goals of WWF is the protection of freshwater resources, and it recently published a briefing note on the policy and background of water stewardship (WWF, no date).	• Research • Advocacy • Publication • Policy

Table 4.9 Water stewardship tools.

Developed by	Name of the tool	Type	Scope	Primary function
CDP	CDP's 2015 Water Responder Pack (2015)	Guidance, tools	Corporate facility, river basin, value chain	• Reports your water stewardship practices to investors to demonstrate good practices.
CEO Water Mandate	Corporate Water Disclosure Guidelines (2013)	Tools	Corporations	• Determines what water-related topics are relevant to your business. • Communicates and reports water risks, impacts, and practices to key stakeholders.
Ceres	Aqua Gauge (2011)	Tool	Corporations	• Evaluates (qualitatively) the maturity of your water management practices and determines how you might expand and improve those practices. • Demonstrates good practices to investors.
The Water Network (2015)	The Water Network (2015)	Tools	Facility, river basin	• Connects and shares knowledge with water professionals from 185 countries.
World Business Council for Sustainable Development	WASH in the Workplace Self-Assessment Tool (2014)	Tool	Facility, river basin	• Assesses the current status of access to safe WASH at the workplace in a given facility. • Supports decision-making regarding investments and priority actions.
Ecolab USA, Trucost	Water Risk Monetizer (2014)	Tool	Corporate, facility, river basin	• Accesses actionable information to help businesses understand and quantify water-related risks in financial terms.
Alliance for Water Stewardship	International Water Stewardship Standard and Guidance (2014)	Tool	Facility, river basin	• Commits to understand, plan, implement, evaluate, and communicate water stewardship actions at facility-level.
Outdoor Industry Association	Chemicals Management Module (2013)	Tool	Facility/ river basin	• Helps to benchmark, establish, build, maintain and improve your chemical management processes and thus reduces water pollution. • Measures and monitors pollution.
Global Environmental Management Initiative	Local Water Tool (2012)	Tool	Facility, river basin	• Quantifies water-related impacts caused by facility. • Quantifies facility's exposure to water-related risks. • Identifies helpful management responses.
Veolia Water, Growing Blue	Water Impact Index – WIIX (2012)	Tool	Facility, river basin	• Quantifies the impact of companies' water use and wastewater discharge on local water resources in terms of quality, quantity, and scarcity/stress. • Given the technical nature of the tool, use of the WIIX requires some understanding of current conditions and expected changes in quality and quantity use.

(Continued)

Table 4.9 (Continued)

Developed by	Name of the tool	Type	Scope	Primary function
Water Footprint Network	Water Footprint Assessment (2012)	Tool	Corporate, facility, river basin	• Informs 'big picture' strategic planning. • Identifies at what stage within the value chain, the water footprint is in a hotspot. • Builds awareness among public.
Ceres	Aqua Gauge (2011)	Tool	Corporate, facility, river basin	• Evaluates (qualitatively) the maturity of your water management practices and determines how you might expand and improve those practices. • Demonstrates good practices to investors.
CEO Water Mandate, UN Environment Programme	Corporate Water Accounting—An Analysis of Methods and Tools for Measuring Water Use and its Impacts (2010)	Guidance	Corporate, river basin, value chain	• Learns about different early methods for accounting for water-related performance, risks, and impacts.

4.14   Water stewardship tools

International agencies have developed several tools to help corporations in their corporate water stewardship initiatives. Some of such tools are listed in Table 4.9.

4.15   Case studies

Several multinational companies have taken up water stewardship and are performing very well. Some case studies, as available from the websites of respective companies, are mentioned below:

4.15.1   Unilever

Unilever India Ltd, as an agro-based industry, has created a water-conservation potential of 100 billion litres through water-related programmes like revival of natural tanks, integrated watershed development, promotion of appropriate technologies for efficient and sustainable use of rainwater, soil conservation and so on. As a result, 111 tea estates of Assam, West

Bengal and Tamilnadu belonging to this company were declared sustainable in 2014. This company works in partnership with NGOs, local communities, government agencies like NABARD and with state governments in implementing water stewardship programmes and approaches towards governance while implementing water projects in Maharashtra, Gujarat, Puduchery and Dadra Nagar and Haveli in India (Unilever India Ltd, 2010).

4.15.2 BASF

As a chemical manufacturing company, BASF is virtually unrivalled in size. BASF has already achieved its goal set for 2020 regarding the reduction of nitrogen and organic substances emission by 80% and of heavy metals by 60% compared with baseline 2002. BASF is willing to introduce sustainable water management covering 92% of its operational areas. They have already achieved 36.2% of the declared goal by 2015. The assessment of BASF states that 22% of their production sites are located in water-stressed areas from where it abstracts only 1% of their global water requirement.

4.15.3 TOM's of Maine

Tom's Inc. is one of the leading personal care product manufacturers. It uses lots of steam and hot water for sanitising and cleaning. Water stewardship activities, as adopted by Tom's Inc., focus on reducing water use in operations. Tom's uses steam and hot water for sanitising and cleaning. The same water is recycled and reused over and over again. They have an innovative composite pipeline system to deliver raw materials into mixing tanks which do not need frequent cleaning and washing. Tom's has targeted reduced water use per unit of production from 4.5 m^3/MT (in 2011) to 2.5 m^3/MT (Tom's of Maine, 2014).

4.15.4 Mars Inc.

Mars is one of the leading food manufacturers in the world. It is found in seventy-four countries. Mars withdraws about 30.2 million m^3 of water annually (2013 data). Water stewardship activities, as adopted by Mars Inc. (Mars, 2015) are:

- **Wastewater management:** MARS has put in place a water-quality monitoring system in its operations. About 50% of its water from manufacturing plants is directly released to the environment, and the quality of this water is kept the same as natural water.
- **Reduce water withdrawal:** Mars is committed to reducing its water withdrawal by 25% by 2015 with 2007 as baseline, as reported in their website.

- **Reducing water use in operations:** By inculcating RWH (rainwater harvesting) and behavioural changes, MARS has substantially reduced water intake from the grid.
- **Reducing water use in supply chain:** With technical collaboration from different research institutions, MARS is increasing water-use efficiency among farmers in Pakistan, the Mississippi Delta and other members of the supply chain.
- **Augment water resource:** Mars has developed ten large-sized RWH systems. RWH systems installed in UK and Singapore have reduced water intake by 35%.

4.15.5   Nestlé

Nestlé has developed a water stewardship policy that extends stewardship activities in watersheds where suppliers and customers live. Nestlé wishes to work within the regulatory framework of countries of operation and to collaborate with governments in all water stewardship initiatives. The most impacting tasks that they perform are:

- Ensuring the right to water of the local community
- Stimulating innovation in water efficiency
- Reducing water withdrawal through RWH and recycle water.
- Conducting water audit and corporate water reporting

The company has developed an information and experience–sharing platform in their website under their water stewardship programme. Nestlé has funded the Project WET (Water Education for Teachers) Foundation for several decades. This is a global foundation for water education and has reached millions of children. Nestlé is supporting the distribution of education material for WET in fourteen countries where teachers and students equally participate in water awareness programmes (Source: http://www.nestle-waters.com/creating-shared-value/water-stewardship).

4.15.6   Coca-Cola

Coca-Cola is one of the pioneer industries advocating water stewardship. Having a presence in most of the countries across the globe and with a business heavily dependent on water, Coca-Cola has adopted WSI in innovative ways. In India, Coca-Cola has partnered with 'The Energy and Resource Institute' (TERI) to set up an academic department named 'The Coca-Cola Department of Regional Water Studies' (Staff, 2014) to build knowledge and capacity in water-related issues. This department will examine water issues in an interdisciplinary framework, towards a holistic approach to water management.

4.16   Conclusion

Corporate water stewardship may appear elusive to some industries, especially to small and medium farms and start-up companies. But it is not impossible for them too to develop a stewardship framework through developing a water strategy, engaging appropriate human resources and collaborating with stakeholders and other companies. These approaches will safeguard their interest from the detrimental attitude of local people. What the corporations require is to develop a good basic knowledge on water management that covers:

- Water withdrawal and distribution management;
- Water resource assessment methods;
- Methods of water disclosure and statutory compliance;
- Methods of water conservation; and
- Service to the community through CSR.

In the subsequent chapters the tools and methods are elaborated.

Bibliography

Abdel, A., Ahmad, S., Ballestero, M., Bezbaroa, S., Cookey, P., Dourojeanni, A., Galli, C., Langford, J., Mensink, M., Opondo, G., Park, M., Pinero, E., Ruffier, P., Witmer, L., Former, H.X., Jun, M. and Khan, C.R. (2014) *The AWS international water stewardship standard*. Available at https://c402277.ssl.cf1.rackcdn.com/publications/746/files/original/AWS-Standard-v-1-Abbreviated-print_%281%29.pdf?1418140260.

Advameg (2016) *Technology management—strategy, organization, system, examples, advantages, manager, definition, school, model*. Available at http://www.referenceforbusiness.com/management/Str-Ti/Technology-Management.html (accessed: 15 March 2016).

Alliance for Water Stewardship (no date) *Why we need an international water stewardship standard?* Available at http://www.intracen.org/uploadedFiles/intracenorg/Content/Exporters/Sectors/Fair_trade_and_environmental_exports/Climate_change/Nick%20Hepworth.pdf (accessed: 15 March 2016).

The Alliance for Water Stewardship (2014) *The AWS international water stewardship standard*. Available at http://www.allianceforwaterstewardship.org/assets/documents/AWS_Standard_Full_v1.0_English.pdf (accessed: 21 June 2016).

Anonymous and Orr, S. (2013) *5 smart ways to face water stewardship*. Available at https://www.greenbiz.com/blog/2013/08/30/5-ways-promote-corporate-water-stewardship (accessed: 12 March 2016).

Barton, B., Adrio, B., Hampton, D. and Lynn, W. (2011) *A Ceres report: a framework for 21st century water risk management*. Available at http://www.ceres.org/resources/reports/aqua-gauge (accessed: 12 March 2016).

Baue, B. (1998) *Ecosystem depletion presents businesses with risks and opportunities*. Available at http://www.socialfunds.com/news/save.cgi?sfArticleId=2170 (accessed: 8 March 2016).

Buysse, K. and Verbeke, A. (2003) 'Proactive environmental strategies: A stakeholder management perspective', *Strategic Management Journal*, 24(5): 453–470. doi: 10.1002/smj.299.

CEO Water Mandate (2013) *Guide to water-related collective action framewok.* Available at http://ceowatermandate.org/wp-content/uploads/2013/09/guide-to-water-related-ca-web-091213.pdf (accessed: 19 May 2016).

Ceres (2016) *Who we are—Ceres.* Available at http://www.ceres.org/about-us/who-we-are (accessed: 19 March 2016).

Cohen, A. and Davidson, S. (2011) 'The watershed approach: Challenges, antecedents, and the transition from technical tool to governance unit', *Water Alternatives*, 4(1): 1–14.

Davies, J. (2011) *Corporate water strategies: All watersheds are local.* Available at https://www.greenbiz.com/blog/2011/06/12/corporate-water-strategies-all-watersheds-are-local (accessed: 15 March 2016).

Ecouterre (2016) *7 water-saving innovations employed by the garment industry.* Available at http://www.ecouterre.com/7-water-saving-innovations-used-by-the-garment-industry/water-conservation-airdye/?extend=1 (accessed: 6 May 2016).

European Water Partnership (2016) *Benefits from water stewardship.* Available at http://www.ewp.eu/activities/ews/stewardship/benefits/ (accessed: 14 May 2016).

GEMI (2015) *Home.* Available at http://gemi.org/ (accessed: 19 March 2016).

Greenwood, R., Willis, R. and Ross, M.H. (2012) *Guide to water-related collective action acknowledgments project team.* Available at http://pacinst.org/wp-content/uploads/sites/21/2013/02/wrca_full_report3.pdf (accessed: 17 March 2016).

Griffin, C.B. (1999) 'Watershed councils: an emerging form of public participation in natural resource management', *Journal of the American Water Resources Association*, 35(3): 505–518. doi: 10.1111/j.1752-1688.1999.tb03607.x.

Ha, M.L. (2015) *Serving the public interest: Corporate water stewardship and sustainable development.* Available at http://ceowatermandate.org/files/Stockholm/Corporate%20Water%20Stewardship%20and%20the%20SDGs.pdf (accessed: 14 March 2016).

Ha, M.L., Morrison, J., Davis, R., Holzman, B. and Shift, L.L. (2015) *Guidance for companies on respecting the human rights to water and sanitation: Bringing a human rights lens to corporate water stewardship.* Available at http://pacinst.org/wp-content/uploads/sites/21/2015/01/Guidance-on-Business-Respect-for-the-HRWS.pdf (accessed: 18 March 2016).

Hayward, R., Lee, J., Keeble, J. McNamara, R., Hall, C. and Cruse, S. (2013) *The UN Global Compact-Accenture CEO study on sustainability 2013.* Available at https://acnprod.accenture.com/~/media/Accenture/Conversion-Assets/DotCom/Documents/Global/PDF/Strategy_5/Accenture-UN-Global-Compact-Acn-CEO-Study-Sustainability-2013.pdf (accessed: 19 March 2016).

Hindustan Unilever Limited (2010) *HUL launches water conservation project in Gundar Basin with partners* Available at https://www.hul.co.in/news/news-and-features/2010/hul-dhan-foundation-and-nabard-launch-water-conservation-project.html (accessed 26 March 2017)

Hwang, L., Waage, S. and Stewart, E. (2007) *At the crest of a wave: A proactive approach to corporate water strategy.* Available at http://pacinst.org/app/uploads/2013/02/crest_of_a_wave3.pdf (accessed: 27 March 2017).

Imane, A.A., Ahmad, S., Ballestero, M., Bezbaroa, S., Cookey, P., Dourojeanni, A., Galli, C., Langford, J., Mensink, M., Opondo, G., Park, M., Pinero, E., Ruffier, P., Witmer, L., Former, H.X., Jun, M., Chaudry, & and Khan, R. (2013) *The AWS international water stewardship standard; beta version for stakeholder input and field testing.* Available at http://www.allianceforwaterstewardship.org/Beta%20AWS%20Standard%2004_03_2013.pdf (accessed: 14 March 2016).

IUCN (2006) *Business and ecosystems issue brief.* Available at http://www.wri.org/publication/business-and-ecosystems-issue-brief (accessed: 8 March 2016).

Jhunjhunwala, A., Mujumdar, P.P., Pradeep, T., Shankar, H.S., Jain, S.K., Tare, V., Majumdar, A., Rathore, L.S., Jain, R.K. and Kulkarni, M. (2014) *Technology interventions in the water sector 2014; Science Advisory Council to the Prime Minister.* Available at http://www.tenet.res.in/Publications/Presentations/pdfs/SAC-PM_report_on_water.pdf (accessed: 13 October 2016).

KAF Knowledge Audit Framework (no date) *What are knowledge resources? Knowledge audit framework.* Available at https://sites.google.com/site/kafframework/what-are-knowledge-assets#sdfootnote1sym (accessed: 3 October 2016).

Larson, W.M., Freedman, P.L., Passinsky, V., Grubb, E. and Adriaens, P. (2012) *Mitigating corporate water risk: Financial market tools and supply management strategies.* Available at http://limno.com/pdfs/2012_Larson_WatAlt_Art5-3-5%5B1%5D.pdf (accessed: 18 March 2016).

Mars (2015) *Water impact | mars principles in action summary.* Available at http://www.mars.com/global/about-mars/mars-pia/our-operations/water-impact.aspx (accessed: 10 March 2016).

Martin, N. (2011) *Six steps to a successful water strategy.* Available at http://www.thegreentie.org/voices/six-steps-to-a-successful-water-strategy (accessed: 14 March 2016).

Meidinger, E.E., Buffalo, S. and School, L. (2000) *Organizational and legal challenges for ecosystem management (1).* Available at https://www.researchgate.net/profile/Errol_Meidinger/publication/228220074_Organizational_and_Legal_Challenges_for_Ecosystem_Management/links/54788e530cf205d1687f75cd.pdf (accessed: 14 March 2016).

Merrett, S. (2007) *The price of water, studies in water resource economics and management.* 2nd edn. London: IWA Publishing. ISBN 9871 1 84339 177 7

Morrison, J., Schulte, P., Christian-Smith, J., Orr, S., Hepworth, N. and Pegram, G. (2010) *Guide to responsible business engagement with water policy drafting team.* Available at https://ceowatermandate.org/files/Guide_Responsible_Business_Engagement_Water_Policy.pdf (accessed: 14 March 2016).

Nonaka, I. and Teece, D.J. (eds.) (2001) *Managing industrial knowledge: Creation, transfer and utilization.* London: SAGE Publications.

OECD, and World Bank Goup. (2014) Global value chains: challenges, opportunities, and implications for policy. Available at: http://www.oecd.org/tad/gvc_report_g20_july_2014.pdf (accessed: 13 March 2016).

Orr, S. (2014) *The similarity between circular economy and water stewardship.* Available at http://www.theguardian.com/sustainable-business/2014/sep/05/the-similarity-between-circular-economy-and-water-stewardship (accessed: 25 March 2016).

Pedley, S., Pond, K. and Joyce, E. (2011) *Interventions for water provision.* Available at http://www.who.int/water_sanitation_health/publications/2011/ch7.pdf (accessed: 12 March 2016).

Rabinowitz, P. (2015) *Section 8. Identifying and analyzing Stakeholders and their interests.* Available at http://ctb.ku.edu/en/table-of-contents/participation/encouraging-involvement/identify-stakeholders/main (accessed: 21 May 2016).

Rozza, J.P., Richter, B.D., Larson, W.M., Redder, T., Vigerstol, K. and Bowen, P. (2013) 'Corporate water stewardship: Achieving a sustainable balance', *Journal of management and sustainability*, 3(4). doi: 10.5539/jms.v3n4p41.

Schulte, P. and Morrison, J. (2014) *Private sector water policy engagement.* Available at http://ceowatermandate.org/files/private-sector-water-policy-engagement.pdf (accessed: 14 March 2016).

Sheng, T.C. (1990) *Watershed management field manual. GUIDE 13/6.* Available at http://www.fao.org/docrep/006/t0165e/t0165e00.htm#cont (accessed: 13 March 2016).

Simon, P. (2011) *How do you create your partnerships?* Available at https://www.petersimoons.com/2011/09/how-do-you-create-your-partnerships/ (accessed: 12 January 2016).

SOCOPSE (2005) *3: Definition of a baseline scenario.* Available at http://www.socopse.se/decisionsupportsystem/steps/3definitionofabaselinescenario.4.3d9ff17111f6fef70e9800052780.html (accessed: 22 March 2016).

Sokolova, D. (2016) *Groundwater management plan to focus on over appropriation.* Available at http://pvtimes.com/news/groundwater-management-plan-focus-over-appropriation.html (accessed: 22 March 2016).

Staff, J. (2014) *TERI university and coke launch water studies program.* Available at http://www.coca-colacompany.com/stories/teri-university-and-coke-launch-water-studies-program/ (accessed: 9 May 2016).

Tate & Lyle (no date) *Our story—Tate & Lyle.* Available at http://www.tasteandsmile.com/our-story (accessed: 17 March 2016).

Tom's of Maine (2014) *Tom's of Maine is making a difference with water use and conservation. Explore the #GoodnessReport.* Available at http://www.tomsofmaine.com/goodness-report/water (accessed: 10 March 2016).

Tyler, S. (2008) *Water demand management research series by the regional water demand initiative in the Middle East and North Africa water demand management, poverty & equity.* Available at http://www.crdi.ca/EN/Documents/wdm-poverty-and-equity.pdf (accessed: 22 March 2016).

United Nations (2015) *The millennium development goals report 2015 United Nations.* Available at http://www.un.org/millenniumgoals/2015_MDG_Report/pdf/MDG%202015%20rev%20(July%201).pdf (accessed: 14 March 2016).

UN Global Compact (2015) *Understanding integrity & integrity risks.* Available at http://ceowatermandate.org/integrity/understanding-integrity/ (accessed: 18 March 2016).

UN Water (2015) *UN-water: Water and gender.* Available at http://www.unwater.org/topics/water-and-gender/en/ (accessed: 9 May 2016).

UN Water (2016) *UN-water: GEMI background and objectives.* Available at http://www.unwater.org/gemi/gemi-background-and-objectives/en/ (accessed: 19 March 2016).

UNESCO (2015) *Water for a sustainable world: The United Nations world water development report 2015.* 2 vols. Paris: United Nations Educational Scientific and Cultural.

Unilever India Ltd. (2012) *Story of water.* Available at https://www.hul.co.in/Images/story-of-water_tcm1255-451516_en.pdf (accessed: 10 May 2016).

University of Cambridge (2013) *Sustainable water stewardship.* Available at http://www.cisl.cam.ac.uk/business-action/natural-resource-security/natural-capital-leaders-platform/pdfs/sustainable-water-stewardship-innovation-through-c.pdf (accessed: 18 March 2016).

The value chain map 22 (no date) Available at http://www.ser.nl/nl/~/media/894adb278220446bb91e4e0ae41ec753.ashx (accessed: 18 March 2016).

Vox Global (2012) *Making the pitch: selling sustainability from inside corporate America 2012 report of sustainability leaders full report.* Available at http://voxglobal.com/wp-content/uploads/VOX-Global-2012-Sustainability-Leaders-Survey-Full-Report.pdf (accessed: 20 May 2016).

Water and sanitary sewer easement information (2010) Available at http://www.hcwsa.com/system/media_files/attachments/153/original/HCWSA_EasementFAQ.pdf?1293715083 (accessed: 14 March 2016).

Water Footprint Network (2015) *Introductory leaflet new.* Available at http://waterfootprint.org/media/downloads/Introductory_leaflet_new.pdf (accessed: 19 March 2016).

Water Integrity Network (2015) *What is corruption in the water sector—WIN*. Available at http://www.waterintegritynetwork.net/2015/03/11/what-is-corruption-in-the-water-sector/ (accessed: 15 May 2016).

WWF (no date) *Water stewardship*. Available at http://wwf.panda.org/what_we_do/how_we_work/conservation/freshwater/water_management/ (accessed: 14 March 2016).

Xu, Z., Xi, M. and Spencer, M. (2016) *Developing a global water stewardship system*. Available at http://chinawaterrisk.org/opinions/developing-a-global-water-stewardship-system/ (accessed: 14 March 2016).

Zhu, K. and Weyant, J.P. (2003) 'Strategic decisions of new technology adoption under asymmetric information: A Game-Theoretic model', *Decision Sciences*, 34(4).

5 Water Governance Framework and Water Acts

5.1 Introduction

In the transition of human society from the time of hunter gatherers to modern society, water management and water governance were motivated by a number of drivers. The drivers are individual and community stakes, concern for environment, emerging uncertainties and political interests. Development of science and technology and flow of information to the common people catalysed the demand for a systematic and controlled management of water.

Though not properly manifested, some sort of water governance in the form of ownership and monopolisation of water for a particular sect or clan prevailed in the past. In Africa, tribal groups used to be the owners of natural springs. There were many local ethnic customs regarding sharing the water with other tribes.

The caste system of India played a big role in water allocation and on access to water. Lower castes were deprived from access to freshwater wells owned by the local landlords. There were caste-based divisions in surface water resources also. Upper-caste people had better managed ponds while lower-caste people had to depend on water of lesser quality. Though ethical value of water had been appreciated in Indian society, water justice was absent.

From the beginning of democratic reforms, society has been constantly searching for solutions to resolve water crises by imposing social values in water allocation and governing it for a controlled distribution. In almost all countries, some sort of water governance has emerged, either for equitable management or to monopolise the resource for some vested interest. In Central and South African countries, the modern water framework directives either assimilated or overruled the prevailing tribal systems of water

Industrial Water Resource Management: Challenges and Opportunities for Corporate Water Stewardship, First Edition. Pradip K. Sengupta.
© 2018 John Wiley & Sons Ltd. Published 2018 by John Wiley & Sons Ltd.

management. In South Asian countries, the prevailing water governance changed from autocratic or sectarian to democratic governance through several reforms.

The beginning of 1970s witnessed the emergence of a revolutionary movement that added a new dimension to science: namely, environmental sciences. This was, of course, a bloodless revolution focused on formulating sustainable development plans, in which scientists, the governments and their agencies, and the civil society came face to face recurrently in international and national conventions. Water was marked as being essential to human dignity in these conventions. The United Nations Water Conference (1977) held at Mar del Plata in Argentina, recognised water as a fundamental right. This conference adopted a number of resolutions on assessment of water resources and community water supply and recommended plans of action. Subsequently, in two other international conventions, the Convention on the Elimination of Discrimination against Women (1979) and the Convention on the Rights of the Child (1989), the rights of women and children to safe drinking water were recognised (UN, 1996). The resolutions of these two conventions built up the background of water governance, nationally and internationally.

Efforts to develop a model for good water governance led to the realisation that one single universally applicable model covering all scenarios is not feasible; each country must formulate its own water governance policies to ensure feasibility. One thing, however, is common to all policies. Governance must bind itself seamlessly with the social, cultural, economic and environmental particularities of a country (UNDP, 2013). Some recent models suggest that good governance should address water management from the perspective of sustainability, ethics and inclusive decision making. A suggestive framework is given in Figure 5.1.

Several models on water governance have been studied by Rogers (2002) in the countries of Latin America. He has evaluated water governance in federal systems where political interests function in three different ways. His 'Bureaucratic politics and process model' evaluates a condition in which the political players or the elected legislature act through the bureaucratic system and the politicians do not become involved. In the 'Congressional behaviour model', the politicians are interested in the development of their constituencies and frame laws accordingly to gain political benefits. In the 'Interest group model', Rogers focussed on the existence of several interest groups or stakeholder groups who try to influence the political system.

5.2   What is water governance?

Water governance is defined as an administrative cum political instrument to (1) regulate withdrawal, use and disposal of water without causing any serious adverse effect on the ecosystem and (2) maintain a

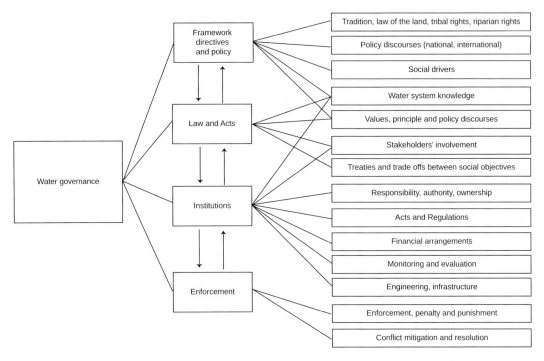

Figure 5.1 A multidimensional framework for water governance. Adapted from Dai (2015).

balance of interest among stakeholders in the system. A water governance framework oversees how freshwater from different sources is exploited, managed and cared for by different users such as the manufacturing sector, agriculture and society at large. The water governance machinery prioritises water allocation so that essential sectors do not suffer from water uncertainty.

Water governance is driven by a guiding concept which is also termed as 'Water Framework Directives' as is in vogue in the European Union. According to Hirsch *et al.* (2006), the framework or guiding concept behind water governance is driven by an integrated water-resource management (IWRM) and other similar paradigms. A water framework may be defined as 'a framework for the protection of all waters including rivers, lakes, estuaries, coastal waters and groundwater, and their dependent wildlife/ habitats under one piece of environmental legislation' (EPA Ireland, 2016). Water framework directives or laws are instrumental in developing water governance in a country or a group of countries (e.g. the European Union). Many countries including both developing and developed have either water policy or water framework directives for efficient water governance. In India, the first draft on water policy was released in 1972 which was modified several times and was later replaced by the National Water Framework Act 2012 based on the national water policy of 2012. The domain of water policy covers water management in river basins, flood management, water transport, withdrawal uses, water supply, pollution, ecosystem management and sanitation.

Water policy also covers government level activities that include:

- Formulation and implementation of interrelated policy instruments for the management and development of water resource systems. It covers both directives (top down) and implementation (bottom up) activities.
- Transfer, development and implications of governance principles and policies for water resources management and development.
- Formulation of structures, policies, goals and provisioning fund for organisations.
- Preparation of government policies to deal with international water policies.
- Research, education, and capacity development.

5.3   Water laws

Laws are based upon principles. Water governance is backed by responsible stakeholders, pressure groups like NGOs, administrative departments of governments and user groups like farmers and private sectors. Politicians also often take initiatives, but they often try to harvest political benefits and to manipulate decisions in their favour. Framing the water law in a country is a very complex matter as it has to deal with local customs, values and prevailing rights. The fundamental basis of water laws are:

1 The constitution of a country should have a provision for the law and if such provision exists, the law should be consistent with constitution.
2 Water law will actively promote the values of water and uphold human rights.
3 Law should demolish the concept of ownership of water and establish the right to water.
4 All water, wherever it occurs, is a resource common to all, and the use of it should be controlled by the government through appropriate authority.
5 The relation between ownership of land and right to water should be well defined by the law.
6 The law should recognise the uniqueness of all water sources and their interdependence and connection with the elements of the water cycle.
7 The law should provide scope for scientific data generation, sharing data with the public and formation of water research institutions.

5.4   Tasks of water governance

The Royal Geographical Society (2012) in a document identified the basic tasks related to water governance. The most prominent tasks are:

- **Catering to human needs:** Human needs are considered prime and diverse. Apart from the daily water need for drinking, cooking, washing, and toilet, there are other needs like irrigation, landscaping, recreational and religious uses, and manufacturing goods. To cater to these needs, people have to regulate certain things like flow of water (by quantity and direction) in the streams, water reserve in the aquifer and surface waterbodies, and the amount of flow finally required for the environment. Regulation of groundwater withdrawal is a major component of water governance in many countries. A good example of such laws is the Groundwater Regulation Acts of different states of India.

- **Maintaining water quality:** Regulation is required to maintain water quality so that discharges from industries and irrigation overflow do not contaminate surface water bodies, groundwater and streams or the marine environment. Many countries have developed pollution control acts and laws to protect the quality of effluent water. The US-EPA is a very good example of institutional governance on the quality of water in the USA.

- **Regulating changes in land use:** Industry and agriculture play crucial roles in distribution and quality of water. Regulation is required in this sector to avoid contamination and conflict. In many countries, land use boards, urban development authorities, municipal authorities and land development boards are formed, who often act as regulatory bodies for water resources in the urban context or work collaterally with other regulatory bodies.

- **Regulate environmental demands and impacts:** The environment provides us with a number of ecosystem services. To ensure proper delivery of such services, a regulatory framework is required that will assess the health of the ecosystem, guide human behaviour towards sustainability, impose conditions on industrial and urban operations, and enforce laws to keep the ecosystem functioning. There are several acts in many countries like wetland protection acts, coastal zone regulation acts and so forth, that look after the functioning of the ecosystem.

5.5   Challenges in water governance

Challenges and tasks of water governance vary according to the dimensions of its jurisdiction and the scale of operation; in a densely populated and plural society as in India, national-level water governance becomes a huge task with multifarious objectives. Water governance, like any other governance, evolves from the collective action of stakeholders; in the context of water, it goes without saying that the largest stakeholder is the general public; water governance plans that disregard the needs of people are doomed to failure. Basic challenges in water governance are shown in Table 5.1.

Not all countries face the same water challenges, and there may be issues which are yet to be listed. International organisations like UNDP, WHO and FAO are continually in the process of negotiating cross-cultural communication (Hearns, Paisley and Henshaw, 2013), framing international water and energy laws, international ocean law and laws to govern environmental impacts.

Table 5.1 Challenges and measures of water governance.

Challenges	Measures
Resource distribution and use	Frame guidelines and methods on water allocation for every sector in accordance with local and national priorities.
Water pricing	Regulate pricing of water so that water is affordable to everyone.
Water allocation	Formulate allocation plans on a common platform including government agencies, scientists, corporations and civil society representing people, especially marginal people.
Management at river basin–level	This deals with the natural boundary of an ecosystem unit. At this scale, a 'river basin authority' is the agency formed to regulate and manage water at river basin–level.
Maintain ecosystem services	Determine the carrying capacity of an ecosystem to ensure sustainability, that is, maintain environmental flow.
Control pollution and maintain water quality	Controlling water pollution by industry, agriculture and urban centres through sector-specific guidelines.
Land-use management	Regulating land-use changes due to urbanisation and neo-colonial activities to ensure balanced sharing of freshwater resources.
Managing public and stakeholders' demands and aspirations	Motivate and educate the public and stakeholders and engage them in resource assessment and monitoring.
Behavioural change that impacts water use	Recover cost of water from the users and motivate them to pay for water.
An international perspective and virtual water trade	Regulate virtual water trade to reduce internal stress on water resources.
Adapting to future pressures	Long-term scenario analysis for sustainability modelling.
Groundwater depletion	Prepare guidelines to ensure aquifer recharging.

Source: Royal Geographic Society (2012)

5.6   Legal framework

Water governance is guided by laws framed under the constitution of the country. Under such framework, authorities and appraisal committees are formed in each country (member country, in the case of the European Union) that control the whole process of allocation, permits, sanctions, notifications, guidelines, and so on. In a federal system, it may either be a national or federal law or a state law. There are other authorities in the lower levels of administrative hierarchy who can also frame laws, acts and rules for water governance for the areas under their jurisdiction.

In a federal system, the provincial or state government makes laws (in conformity with the national framework) for controlling water withdrawal, discharge or recharge. In some cases, before a national framework is formed or a state law is enacted, local municipalities regulate extraction of groundwater or protection of waterbodies/urban wetlands, which may be superseded later by new enactments or state laws. At the national level or provincial level, there may be separate authorities to look after domains like groundwater, surface water, water pollution, wetlands, hills and so on. Each of the authorities may have their own rules, often cross-cutting, but binding upon industries. Industries need to satisfy statutes framed by several authorities.

The policy structure of water governance is not universal; like human society it is diverse and driven by locally developed concepts, ethics and geo-scientific basics. The boundaries of jurisdiction of one water governance system may be different from another. In a federal structure like the USA or India, water governance depends on whether the state or the national government is empowered by the constitution to look after matters related to water. In general:

1 The concept or requirement of water governance is included in the national Constitution.
2 The federal or state government through its respective departments initiates the instrument for water governance.
3 Government departments are empowered to frame legal documents.
4 Empowered institutions are formed to guide and enforce the regulation or advise the government to make necessary amendments from time to time.

In certain cases, the court of justice under the provisions of the national constitution may instruct the government (national or federal) to initiate a water governance structure and frame laws accordingly. In general, the composition of the water governance instrument is in the form of committees who are empowered to frame legal documents, statutes and procedures. There may be separate committees or separate departments involved to look after different aspects of water, such as pollution, water allocation, industrial water, wetlands, surface water, groundwater, river basins and so on. In such cases, several authorities are formed. In a federal structure, the national-level governing body generally releases the water framework and prepares guidelines, scientific literature, quality standards, and so on. The state or provincial government frames the laws, acts and rules and implements the act through appropriate institutions. In a statutory body of water governance, the representatives from stakeholder groups generally participate as members. In Figure 5.2 the functional domains of water governance is illustrated.

5.7   Institutional framework

The institutional framework of water governance takes into consideration three main functional domains: (1) socioeconomic aspects (ownership, tenure and allocation), (2) qualitative and quantitative aspects (water pollution and effluent discharge, monitoring) and (3) ecological aspects (ecosystem health and environmental flow).

Water supply to the public by the government as a service is different from managing water as a resource (Asian Development Bank, 2014). These two aspects function differently and demand separate institutional

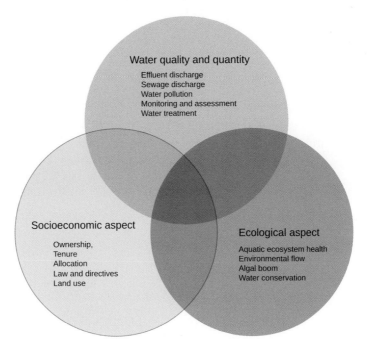

Figure 5.2 Functional domains of water governance.

arrangements. The water resource governance regulates and controls the water service sector, which is a separate entity. The institutional framework of efficient water governance should develop personal effectiveness, leadership and integrity in the total hierarchical system and integrate many organisations (SSWM, 2003) and actors together.

The institutions formed for water governance will perform duties under one or all of the functional domains, but it is not uniform. In a water-rich country, the system will act differently from a water-stressed country. The system will also function differently in a federal government and in a unitarian state. The main objective of governance being to regulate water sourcing, supply, delivery, uses and outflow within a country or a river basin; the involvement and evolvement of institutions are demand driven and often driven by political interests.

A general overview of the water governance of different countries reveals that in almost all countries the governance works for:

1 Framing and enforcing laws and regulatory norms for optimal use of water resources.
2 Managing water through long-term plans.
3 Forming institutions like water board and river basin authorities.
4 Coordinating different sectors of water use like drinking, agriculture, health, energy, and so on.
5 Capacity building in institutions and individuals engaged in water management.
6 Maintaining integrity and transparency.
7 Delegate power to local authorities.

8 Generate required temporal and spatial data related to water resource estimation.

9 Formulate guidelines on water withdrawal, discharge and licensing systems.

10 Water pricing.

Common political institutions, administrative departments and enforcing institutions that get involved in the water governance in a country are ministries and departments dealing with environment, agriculture, land and water.

Water governance bodies are categorised as (1) political institutions (ministries) (2) government departments (3) authorities (boards, agencies and commissions) and (4) R & D institutions.

5.7.1   Ministries

The functions and power of a ministry to frame laws on water depends on the constitution and water policy or framework of a country. In a federal structure, both national and provincial-level ministries hold the power to frame laws and acts. In large federal countries like the United States, Canada, India and Australia, water governance is the responsibility of the state. Political institutions at the state level hold the power to regulate water. In general the following ministries of a country can assume the authority to frame laws on water:

- Ministry of the Environment
- Ministry of Water Resources
- Ministry of Agriculture
- Ministry of Forest and Natural Resources
- Ministry of Public Works, Transportation, and Communication
- Ministry of Planning and Urban Development
- Ministry of the Economy and Finance
- Ministry of Irrigation and Waterways
- Ministry of Rural Development

The water resource ministry may be a separate ministry, or it may be merged with another ministry. For example, in Kenya, the Ministry of Water Resources was merged with the Ministry of Environment and Natural Resources in 2001.[1]

5.7.2   Government Departments

Government departments act under their respective ministries and are authorised to implement laws. There may be several departments under one ministry looking after different aspects. For example, in South Africa the water resource department is a subset under the Ministry of Environment.

In India, the National Ministry of Water Resources, Rural Development and Ganga Rejuvenation is a single ministry, under which there are several water resource departments and authorities.

5.7.3 Authorities

The Water Resources Authority (WRA) is responsible for the management, protection and controlled allocation of water. There are separate authorities to look after different aspects of water governance, such as surface water resources, groundwater, water quality and so on. The WRA maintains a hydrological database and provides data, information, and technical assistance to government and non-government institutions. In the United States, the EPA is the authority that looks after water quality and functions through its state counterparts, while the United States Geological Survey (USGS) deals with groundwater mainly. In India, there are authorities like pollution control boards, river commissions, and groundwater boards at both the national and the state level. In Bangladesh, water governance functions through (1) the Water Resources Planning Organisation (WARPO), (2) the Bangladesh Water Development Board, (3) Bangladesh Haor and Wetland Development Board and (4) Joint River Commission of Bangladesh.

5.7.4 Institutions

The problem of freshwater supply has assumed such proportions that today every country has R & D institutions. They maintain the national hydrological database to provide data, information and technical assistance to government, non-government organisations and corporations. In the United States, the State Water Resources Research Institute (WRRI) Program, authorised by section 104 of the Water Resources Research Act of 1984, is a federal-state partnership which (1) plans, facilitates, and conducts research to aid in the resolution of state and regional water problems, (2) promotes technology transfer and the dissemination and application of research results, (3) imparts training to scientists and engineers through participation in R & D programmes and (4) provides for competitive grants to be awarded under the Water Resources Research Act. Similar institutions are established in almost all countries which have water governance supported by water acts.

5.8 Principles of water governance

Water governance is guided by principles that form the basis for enacting laws, rules and regulations. Some ethical principles for good governance are given in Table 5.2.

Table 5.2 Principles of good water governance.

Principle	Description
Participatory	The different stakeholders involved need to be identified and included in policy and decision making. Inclusive processes build confidence in the resulting policies, and in the institutions. Two-way communication using engaging language creates trust and a sense of democracy.
Transparent and accountable	Information flows freely and steps taken in policy development are visible to all. This helps ensure legitimacy by being seen to be fair to all the parties. It implies the need to be seen to be ethical and equitable, for the roles and responsibilities of both institutions and stakeholders to be clear, and for the rule of law to apply.
Integrative	A holistic approach is taken to the primary influences within the water system, be they landscape components such as land use or river groundwater connections, different community worldviews or diverse scientific interpretations. Integration recognises linkages within the management system; in turn, policies and action must be coherent and aligned—this requires political leadership and consistent approaches amongst institutions.
Efficient	Governance should not impede effective action. Transaction costs are minimised, including financial and time costs of decision making and compliance, administrative costs, complexity, and ease of understanding of how the system operates.
Adaptive	The system incorporates collaborative learning, is responsive to changing pressures and values, and anticipates and manages threats, opportunities and risks. It recognises that the system is complex and constantly in flux.
Competent	Decisions must be based on sound evidence. Competence requires development of capability at all levels: skills, leadership, experience, resources, knowledge, social learning, plans and systems to enable sustainable water management.

Fenemor *et al.* (2011). Printed with permission from the Editor, Policy Quarterly. Copyright: Policy Quarterly.

5.9 Spatial scale of water governance

Understanding the spatial scale (Moss and Newig, 2010), that is, understanding the size and location of the domain taken up for water governance is a vital prerequisite for planning governance. Any domain, an individual user, such as a farmland owner, or a community or an industry, has a specific location and size defined by natural, political or administrative boundaries. For instance, delineating the administrative boundaries for water governance can range from rural community to national scale. The scale of water governance defined by natural boundaries ranges from small watersheds at the lowest level to large transnational river basins at the highest level; in between there are administrative units such as districts, provinces, nations and unions of nations. Political boundaries are determined by electoral constituencies. Water governance that works at a local scale may take adaptive or regulatory measures related to local water resources, groundwater or local self-management of water resources (Olsson *et al.*, 2007); but larger issues like water flow regulation or withdrawal of water from international rivers, effects of climate change, virtual water trade are managed by a governance that functions in international scale.

5.10 Hierarchical governance

Like boundaries, there are three parallel hierarchical systems in water governance: natural, administrative and political. The natural boundary of a river basin or a watercourse has a natural hierarchical system in the river catchment. A natural hierarchy is the hierarchy of the river basin in which water in large river basins is aggregated as a bottom-up system because water from smaller basins flows into larger basins. This basin hierarchy is in reverse order when it is institutionally administered by a river basin authority. See Figure 5.3.

The administrative hierarchy of water governance follows the administrative boundaries from the national level to the local level, and power flows from top to bottom through different stages of administration. In some federal systems, the authority of water governance is vested solely in the provincial governments; in that case, the national authority only performs the task of framing guidelines, notifying over-exploited areas, and so on, and in some cases, in the absence of any provincial law in a particular federal state, implements the water laws directly. For instance, in India there is a national groundwater authority and state groundwater authorities in eleven out of twenty-nine states so far; national authority functions in those states where state law is absent.

Power sharing by political parties in water resource governance exists in very rare instances like the People's Republic of China, where the political authority at all levels shares power with the administrative counterpart.

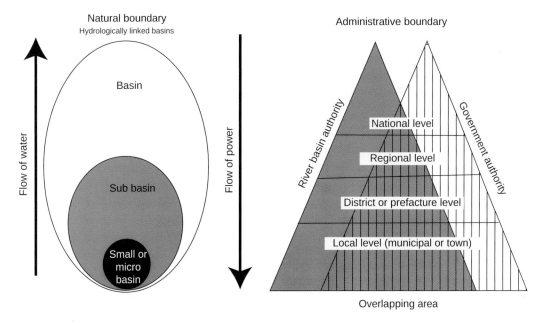

Figure 5.3 Difference between natural and administrative boundaries.

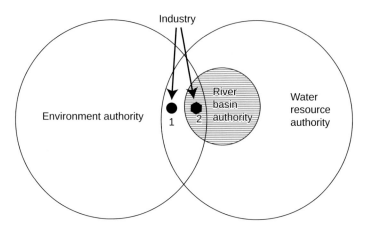

Figure 5.4 Cross-cutting issues in water governance.

5.11   Cross-cutting authority of governance

In any country, government departments and agencies, local authorities (Boyle, 2006) are structured in such a way that the mandate and the issues are balanced. But, water being everybody's business, several departments may claim authority on water resources. For example, the state groundwater board is empowered to issue permits for groundwater withdrawal, but if the site is located on a large river basin for which a separate authority exists, the second authority will also have its own mandate to act on that issue. The environment department also has its role to play. In such cases, coordinated action is required to resolve horizontal cross-cutting issues. For example, in a location there are three authorities for water governance: (1) water resource department (authority), (2) environment department and (3) a river basin authority functioning under the water resource department. If an industry is located under the jurisdiction of the river basin authority (Industry 2 as in Figure 5.4), then it will have to comply with the norms of all three authorities. If it falls outside the jurisdiction of the river basin authority, it will have to satisfy two departments only. In the event that there are local authorities like industrial development boards, land reform department and urban development authorities, industries will have to follow their norms as well.

5.12   Stakeholders engagement in water governance

While discussing available literature on stakeholders' participation in decision making, Abelson (2006) emphasises that involving ordinary citizens in the policy process makes decision-making 'more collaborative' and effective.

This indicates a trend of governance to shift from the traditional top-down system to a bottom-up system.

In water governance, a formal or informal participation of stakeholders exists (OECD, 2015), but in good governance, meaningful inclusive engagement is essential. In any geopolitical entity (country, province, village or locality) or in a river basin, stakeholders come from diverse sections of the society, and their inclusion in the governance requires specific policies, goals and visions. Stakeholders, as defined by OECD, are the persons or groups who have a stake in a water-related topic, have the ability to contribute in the policy making or 'may be directly or indirectly affected by water policy'. The policy makers have to engage these stakeholders such that their engagement yields the best result. The engagement mechanism involves several processes, which are:

- **Identification of various stakeholders' groups and individuals:** It had long been advocated for engagement of experts, policy-makers and citizens in environmental governance who can provide valuable inputs in the functioning of water governance. Indeed, in almost all water governance frameworks science plays a valuable role where 'scientists inform policy-makers and policy-makers turn to science for knowledge and technical assistance' (Bäckstrand, no date). The United Nations plays a positive role in engaging scientists and technicians in the global assessment of climate change, ozone depletion, water assessment and so on (Biermann, 2002), which supports policy-makers at the national or international level in framing water laws and regulations. Stakeholders may represent various groups like civil society, NGOs, the private and public sector, governments, farmers' groups, and so. Individuals who perform as activists and can contribute to the cause may also be included.
- **Assess capacity and types of involvement in water sector:** Representatives from stakeholders' groups and individuals possess various types of skill and capacity which can be important inputs. Schwartz *et al.* (2009), while discussing the water policy of Indonesia, mentioned some essential skills required in the delivery of water governance, which are (1) awareness of challenges in the arena of IWRM, (2) concept of river basin, (3) technical and scientific skills in hydrology, (4) mental set-up and outlook for collaborative work, (5) an understanding of policies and (vi) capability to develop policy statements. A single individual or an organisation may not have all the above-mentioned capacities, but they must be ready to engage in collaborative work with others.
- **Engagement policy:** For delivering service in water governance, an engagement policy is required to give appropriate importance to individual stakeholders. Engagement depends on (1) competence of the individual or organisation in the technical, managerial aspects of governance and (2) capacity to deliver continuous learning and innovation inputs to the system. An engagement framework is shown in Figure 5.5.

5.13 Functions and functionaries of the water governance

The OECD defined the functions of the water governance as actions to manage *too much*, *too little* and *too polluted* water in a sustainable, integrated and inclusive way, at an acceptable cost, and in a reasonable

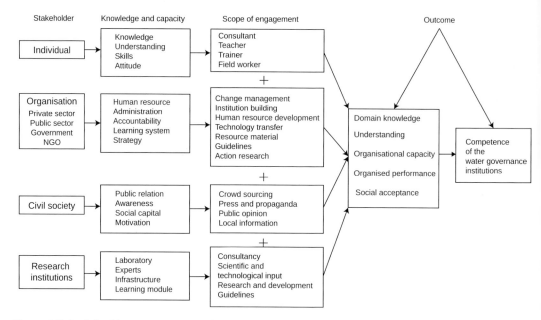

Figure 5.5 Stakeholders' engagement framework. Adapted from Alberts and Kaspersma (2009).

time-frame (OECD, 2015).Though it has been said in a very simple way that the success of good governance lies in solving water challenges using an appropriate approach that involves society and government working together, there are no simple or uniform functions of governance for all sorts of situations. Good water governance has control over 'activity domains of what people do with water' (Wiek and Larson, 2012). The actors and in water governance and their functions are briefed in Table 5.3.

5.14 Role of civil society organisations (CSO)

The Organisation for Economic Co-operation and Development (OECD) describes CSOs as 'the multitude of associations around which society voluntarily organises itself and which represent a wide range of interests and ties. These can include Community Based Organisations (CBOs), indigenous peoples' organisations and non-government organisations' (Bouchane and Coulby, 2011).

Throughout the developing world, there are significant social, economic, ecological and capacity obstacles to meeting the Millenium Development Goal (MDG) for water and sanitation. Overcoming them will require more effective water governance, improved water management, enhanced capacity at all levels, and greater empowerment of the poor (UNDP, 2004).

NGOs can play an important role in the development of democratic and inclusive public policy making in water governance, but the capacity of NGOs to bring about change greatly depends on the socio-political context and on the networks they are able to forge with grassroots organisations, state agencies, funders and other third-sector actors (Hoogesteger, 2015).

Table 5.3 Actors in water governance and their functions.

Actors and partners	Functions
National constitution	Has the provision for framing the water laws through parliament or legislative assembly or congress.
Service providers	Water governance personnel may come from civil society, political parties, the private sector, union government or local government. They provide, as a part of the institutional structure, services like licensing, helping people or organisations in tendering applications for licences or tenure, legal assistance, technical and scientific research, documentation, awareness generation, mobilisation, and so on.
Government	Government is the principal regulatory and functioning authority. It enjoys the power to allocate water rights, environmental management related to water, quality of water, land use planning, protection of coastal zone, water-bodies, and so on. It also has the power to fix tax rates, fines, pollution charges, etc. (Global Water Partnership, 2013) and earns revenues by providing these services. The principal authority through its agencies enforces the law, sets up advisory committees and arranges for hearing on appeals and grievances.
Financial institutions	Financial institutions are also stakeholders in water governance. Before providing funds for an industrial project, they must satisfy themselves that the availability of water is assured and legal requirements are complied with.
Private sector	The private sector is listed among users, but often emerges as one of the parties in water governance through business consortiums, chambers of commerce, and so on. Representatives from private-sector organisations may provide services to water governance or regulatory bodies through arranging conventions, public meetings, organising awareness camps, and so on.
Local authority	A local authority provides a number of regulatory instruments like permits, tenures, application procedures, and so on and a number of economic instruments like fees, special taxes, surcharges, fines and penalties (IWMI, no date). A local authority generally enjoys only a limited power of execution, but it may forward cases to higher authority if the regulatory demand exceeds its power.
Civil society	• Advocates on behalf of nature and environmental protection; • Develops and tests new models and tools in water management; • Increases awareness of the need for sustainable water management; • Mobilises local communities to get involved; • Generates awareness among the general public; and • Carries out activities related to 'right to information'.
Educational/research institutions	Conduct research on water resource, water demand and water resource, modelling and guide governing bodies by providing research findings.
Elected representatives	Elected representatives in the parliament or legislative assembly debate on the proposed legislation. They also propose amendments and raise questions in parliament or the legislative assembly on the functioning of water governance. At the local-government level, they are often empowered to provide oversight of the implementation of legislation.

Adapted from Global Water Partnership (2013).

5.15 Water governance framework of different countries (case studies)

Water policies and water governance frameworks are not uniform in all countries. But most countries try to maintain a global standard of institutions to govern their own water resources and cooperate with neighbouring countries when they share water. The most important examples of water governance framework are the Water Framework Directives of the European Union and water policies of large federal states. Some of the case studies are as follows:

5.15.1 European union water framework directives

The Water Framework Directive of the European Union is a directive to the member states to achieve good qualitative and quantitative water status of all water bodies including surface water, groundwater and seawater. It prescribes steps to reach the common goal rather than adopting a traditional limit value approach. The Directive was adopted in the European Parliament and the Council on 23 October 2000. The European Water Framework Directive (WFD) is a legally binding document that requires member states (MS) to implement water management measures to achieve good overall quality of European bodies of water by 2015 (Steyaert and Ollivier, 2007).

Objective

The main objective of the Directive is to improve the ecological health of surface water resources of member states by (1) improving management, (2) reducing water pollution and (3) improving water quality by standardising methodology. The Directive aims for 'good status' for all ground and surface waters (rivers, lakes, transitional waters, and coastal waters) in the EU (Water framework directive, 2016).

Articles

The Directive has 26 Articles that are summarised in Table 5.4 as sourced from Directive 2000/60/EC of the European Parliament and of the Council (2000).

Each member country has adopted its water acts in accordance with the WFD and implemented methods of improving the quality and quantity of ground and surface water resources. One important aspect of the Water Framework Directive is the introduction of *River Basin Districts*. River basin districts are defined as the area of land and sea, made up of one or more neighbouring river basins together with their associated groundwater and coastal water. If a river crosses national borders, representatives from several Member States have to cooperate and work together for the management of the basin according to river basin management plans with specific objectives and goals.

Table 5.4 Brief description of Articles in the European Union Water Framework Directives.

Article	Heading	Salient features
Article 1	Purposes	To establish a framework for protection of inland surface waters, transitional waters, coastal waters and groundwater and prevent further deterioration and promote sustainable practice among users.
Article 2	Definitions	Standardisation of terminology so that all member countries use the same definition. This Article introduced the term 'River basin district' under Article 3(1).
Article 3	Coordination of administrative arrangements within river basin districts	This Article deals with river basin districts and defines the administrative set-up and boundaries. This Directive also covers the responsibilities of member states in maintaining uniformity in managing water resource of river basin districts.
Article 4	Environmental objectives	This article contains directives for protecting (a) surface waters and (b) groundwater by the Member States, who shall implement the measures necessary to prevent pollution and deterioration of the status of all water bodies.
Article 5	Characteristics of the river basin district	This article directs the review of environmental activity in river basin districts to ensure analysis of economic impact of human activities.
Article 6	Register of protected areas	Member states will maintain a register of protected areas.
Article 7	Waters used for the abstraction of drinking water	This Article directs the member states to monitor all bodies of water which are used for human consumption and establish safeguard zones around such bodies of water.
Article 8	Monitoring	Monitoring of surface water status, groundwater status and protected areas
Article 9	Recovery of costs for water services	Member States shall take account of the recovery of the costs of water services, including environmental and resource costs, having regard to economic analysis.
Article 10	The combined approach for point and diffuse sources	This Article proposes that Member States shall undertake establishment and implementation of: (a) emission control based on best available techniques; (b) impose emission limit values and, in case of diffused impacts (c) frame controls and inculcate best environmental practices.
Article 11	Programme of measures	Each member state will take up programmes of measures, taking into account the results of analyses required under Article 5, in order to achieve the objectives established under Article 4.
Article 12	Unresolved issues	Issues which cannot be dealt with at Member State level should be taken up by the Commission
Article 13	River basin management plans	This article explains how a Member State will manage river basin districts falling entirely within its territory or an international river basin. River basin management plans shall be reviewed and updated at the latest 15 years after the date of entry into force of this Directive and every six years thereafter.
Article 14	Public information and consultation	This Article directs an involvement of stakeholders in water management and protection of water resources.
Article 15	Reporting	Contains instructions and guidelines for reporting (a) river basin districts falling within the national territory and (b) international river basin districts.
Articles 16 and 17	Strategies against pollution of water	This Article guides the European Parliament, and the Council shall adopt specific measures against pollution of water by individual pollutants or groups of pollutants for both surface and groundwater resources.

(Continued)

Table 5.4 (Continued)

Article	Heading	Salient features
Article 18	Commission report	The Commission shall publish a report on the implementation of this Directive at the latest 12 years after the date of entry into force of this Directive and every six years thereafter, and shall submit it to the European Parliament and to the Council. This report will contain (a) review of progress of implementation of Directives, (b) review of status of water resources, (c) survey of river basin management plans and (d) review of the responses to the reports and recommendations made to the Commission.
Article 19	Plans for future Community measures	The Commission shall review the measures that have impact on water legislation and review the Directive at the latest 19 years after the date of its entry into force.
Article 20	Technical adaptations to the Directive	Enforcement of guidelines on scientific and technical progress.
Article 21	Regulatory committee	The Commission shall be assisted by a committee. The Committee shall adopt its rules of procedure.
Article 22	Repeals and transitional provisions	A list of previous directives which are repealed after enforcement of the WFD.
Article 23	Penalties	Member states will determine penalties applicable to breaches of the national water laws framed under the WFD.
Article 24	Implementation	Member States shall bring into force the laws, regulations and administrative provisions necessary to comply with this Directive.
Articles 25 and 26	Entry into force and Addressees	The date of enforcement is the date of publication of the Directives in the Official Journal of the European Communities. The Directive is addressed to the member states.

Data source: European Union (2000).

5.15.2   Water governance in Australia

Australian water governance dates back to 1915 with the signing of the Murray-Darling Basin agreement with states under a federal government structure. It has been in force for a long time and has given rise to a number of agreements and institutions provided under the Australian Constitution that determined the extent of state and Commonwealth influence over river management (Kildea and Williams, 2010). In 2010, the Australian environmental water management report was released that is the basis of the Australian water framework.

The Australian water framework (Government of Australia, National Water Commission, 2012) released in 2012 aimed at improving the management of environmental water in Australia. This framework addresses four major criteria and has 27 subsections. In summary, the key provisions of environmental water governance in Australia are:

- Water management through water managers having the necessary authority to perform activities related to environmental water protection;
- Giving recognition, entitlement and security to environmental water; and
- Environmental water accounting and annual reporting of the operation of environmental water rules.

Murray-Darling Basin Authority

The *Water Act 2007* establishes an independent Murray-Darling Basin Authority (MDBA) with the functions and powers, including enforcement powers, needed to ensure that Basin water resources are managed in an integrated and sustainable way. The MDBA will oversee water planning considering the Basin as a whole, rather than state by state, for the first time.

In December 2008 the Water Amendment Act amended the Water Act 2007. This resulted in the functions of the Murray-Darling Basin Commission being transferred to the Authority, creating a single body responsible for overseeing water resource planning in the Murray-Darling Basin. Key functions of the MDBA include:

- Preparing a Basin Plan for adoption by the Minister, including;
- Setting sustainable limits on water that can be taken from surface and groundwater systems across the Basin;
- Advising the Minister on the accreditation of state water resource plans;
- Developing a water rights information service which facilitates water trading across the Murray-Darling Basin;
- Measuring and monitoring water resources in the Basin;
- Gathering information and undertaking research engaging the community in the management of the Basin's resources; and
- The MDBA provides independent advice to the Commonwealth Minister for Climate Change and Water.

Source: Australian government (2009) CC BY.

Water Act

Australia has implemented its Water Rules 2007 with some amendments which broadly covers the provisions for:

- The Murray-Darling Basin Authority. See Box 5.1;
- Management of basin water resources;
- Water charge rules and water market rules;
- Regulated water charges; and
- Water information.

Licensing methods

Water rights in Western Australia are governed by the *Rights in Water and Irrigation Act 1914*. Under the Act, one may need a water licence for taking or interfering with surface water, overland flow water or groundwater.

Types of licences

The department issues several types of water licences and permits.

5C Licence: A 5C licence allows the licence holder to 'take' water from a watercourse, wetland or underground source.

26D Licence: A 26D licence allows construction or alteration of wells. This licence is needed before licensees commence, construct, enlarge, deepen or alter any *non-artesian* well in a proclaimed groundwater area or commence, construct, enlarge, deepen or alter an *artesian* well.

Permits

When granted, permits authorised interference or obstruction of the bed and banks of a watercourse or wetland. The Act provides for several types of permits:

- Section 11 permit—taking of water in a proclaimed area where access is via road or crown reserve;
- Section 17 permit—taking of water in a proclaimed area;
- Section 21A permit—taking of water[2] in an un-proclaimed area where access is via road or Crown reserve.

5.15.3   *Water governance in Brazil*

Brazil has a federal government, and the member states have power to make water-related acts. A water policy and management system has been established in Brazil as demanded by the 1988 Constitution. Prior to that the states moved ahead of the federal government and passed their own legislation putting pressure on the federal government. Rapid economic growth with increasing demands for water and water services, globalisation issues and the experiences of other countries with concessions dictated some of the actions by the provincial governments. The National Water Resources Policy was spelled out in the new water legislation to achieve: (1) sustainability, to ensure that future generations have adequate availability of water of suitable quality; (2) integrated management, to ensure the integration among users in order to guarantee continuing development; and (3) safety, to prevent and protect against critical events, due either to natural causes or inappropriate uses.

The new legislative acts are:

- National Water Act (1997); and
- Legislation to establish Agencia Nacional de Agua (ANA) (2000).

Under these Acts the following new Institutions have been formed:

- Agencia Nacional de Agua (ANA), who oversee the implementation of the water law;
- National Water Resources Council;
- State Water Resources Councils;
- River Basin Committees composed of representatives of the federal government, states or the federal districts, the municipalities in which they are located, entirely or in part, the water users in their areas of action and civil water-resources agencies that have a demonstrated record of action in the basin; and
- Water Agencies that shall serve as the executive secretariats of the River Basin Committees.

Under Article 19 of the National Water Resource Policy, the fees charged for the use of water are intended to (1) reinforce the idea that water is an economic good so that the user has a sense of the value of water, (2) to encourage rationalisation of water use and (3) to raise revenue for financing the programs and interventions provided for in the Water Resources Plans (*Brazil national water resources policy*, 1997). The salient features under National Water Resource Act are IWRM plans, decentralisation of governance, participation of stakeholders and public, water pricing, permit systems, water resources assessment and shadow pricing of bulk water.

5.15.4 Water governance in Canada

More than 84% of Canadian people are concerned about their water security, because both quality and quantity of Canadian water are under threat. Canada does not have a consistent, national approach to its management and is handicapped by fragmented water policy, turf wars and one of the lowest levels of environmental performance in the developed world. Major environmental issues in Canada, as sourced from Lagacé (2011) are (1) sharing of river basins with the United States, (2) strained relationships across jurisdictions of water authorities and (3) lack of transparency in knowledge transfer.

Actors in water governance

The Canadian constitution provides the premises for dividing responsibilities of water governance between the federal and the state governments. The water governance framework in the federal system is managed and influenced by various sectors of stakeholders (FLOW, 2009). Key water policy actors in Canada are:

- **Federal government:** The federal government is mandated to (1) create institutions with specific goals of water governance (e.g. the Inland Waters Directorate and the Canada Centre for Inland Waters) and (2) pass domestic legislation (e.g. *Canada Water Act, 1970*) dovetailed with international agreements (e.g. the *Great Lakes Water Quality Agreement,* 1972). Recognising the enormous scale of water development, the Canadian Government played an active role in framing the Canada Water Act in 1970 to implement joint plans for management of water resources. Comprehensive river basin planning proceeded in several regions under this act (Pearse and Quinn, 1996). Canada also introduced the Federal Water Policy of 1987, which is comprehensive and has application to specific issues. It addressed strategies of realistic pricing, science leadership, integrated planning with policy framework directives for twenty-five areas of federal concern. But gradually the federal government reduced resource flow to water management, and most of the national-level water institutions like Inland Waters Directorate and Freshwater Science Programme stopped functioning.
- **Provincial government:** Water quality, regulating drinking water systems and water allocation are specific responsibilities of the provincial governments in Canada, and many of them have developed their own policy

frameworks. Some of these frameworks are the Alberta Water for Life Strategy (2003), the Quebec National Water Strategy (2002) and the British Columbia Living Water Sustainability Act 2014 (Queen's Printer, 2016). Ontario has implemented the Clean Water Act (2006). Manitoba has created the only stand-alone water ministry in Canada (FLOW, 2009).

- **Local government:** Protection of local water sources often falls to local governments and watershed agencies—with their influence over important areas of water management including land-use planning, water and wastewater infrastructure and municipal water supplies. Local actors are particularly well suited to understand activities occurring within their own watersheds and to develop locally tailored solutions.

  Over time, there has been an increasing trend to rely more on local action to manage water resources. This movement, however, has occurred outside the national framework and without effective legislative and scientific support. As a result, local actors are significantly under-resourced, leaving Canada's water resources vulnerable to existing and emerging threats.

- **Non-government organisations:** Non-governmental organisations, like citizen's groups, environmental organisations, watershed agencies and irrigation districts, demonstrate leadership and commitment to water management in Canada. These organisations support the government through activities like mass awareness, water-quality monitoring, enforcement and mapping. Industries have also taken an interest in water. For instance, the Royal Bank of Canada has invested 40 million USD in its Blue Water Project (Royal Bank of Canada, 2016) that funds projects on sustainable water use through different organisations.

- **Intergovernmental agencies:** There are also intergovernmental agencies in Canada that act as institutions of water policy management and action. The roles of these agencies are vital in resolving intergovernmental water issues related to fisheries, wildlife, energy and agriculture. The following intergovernmental agencies are active in Canada:

International Joint Commission (IJC);
Canadian Council of Ministers of the Environment (CCME);
The Council of the Federation; and
Western Water Stewardship Council.

Water Acts

The main federal Acts related to water are:

- Canada Water Act 1970;
- Canadian Environmental Assessment Act 1992;
- Canadian Environmental Protection Act 1999;
- International Boundary Waters Treaty Act;
- Canada National Parks Act;
- Fisheries Act; and
- International River Improvements Act, 1985.

Water rights and permit system

The Canadian water rights and permit system varies from province to province. There are four approaches to water rights. These are: prior allocation, public authority, riparian rights, and civil code. Aboriginal water rights play an important role in each province' (Watergovernance Canada, 2008).

- **Prior allocation:** Prior allocation means that the person who has applied first for allocation will get priority in acquiring the right, provided that water must be used for socially accepted purposes defined by regulation.
- **Public authority:** In the North (Yukon, Northwest Territories, Nunavut), a Public Authority at the provincial level makes decisions about water use that are then implemented by local water boards. All uses of water require a permit, with the exception of domestic and emergency uses.
- **Riparian rights:** Riparian rights are granted to residents living close to water-bodies and granted only for domestic water use. This right is non-transferable and does not apply to groundwater.
- **Civil code:** Civil code management is another form of water governance practiced in Quebec, and water rights are granted on the basis of water use and by different authorities. For example, when water is used for power, the permit is issued by the Ministry of Natural Resources; in the case of irrigation water, the Ministry of Agriculture and Fisheries is the permitting body. Drinking water and water supply permits fall under the purview of the Ministry of Municipal Affairs.
- **Aboriginal water right:** Aboriginal water right is a right to water for the indigenous people of America who resided in Canada prior to colonisation. The rights of the aboriginal people and their water laws work in tandem with the current water laws.

5.15.5 Water governance in China

The People's Republic China is under a federal system of governance, and water governance is guided by this system with multiple and cross-cutting authorities. The Ministry of Water Resources of the People's Republic of China shares its authority with the Ministry of Housing and Urban-Rural Development, the Ministry of Land and Resources, the Ministry of Environmental Protection and the Ministry of Agriculture. The Table 5.5, adapted from Circle of blue (no date) shows the authoritative shares in water governance in China. The National People's Congress is the law-enforcing supreme authority. Yang and Griffiths (2010) studied China's legal framework supporting water management, which starts at the highest level with decrees issued by the State Council and followed by provincial regulations and rules issued by provincial people's congress. Local regulations and rules are issued by the prefectural people's congress. The laws that support water governance are (1) the Water Law of the Peoples' Republic of China, (2) the Environmental Protection Law of the People's Republic of China, (3) the Flood Prevention Law of the People's Republic of China, (4) the Law of the People's Republic of China on Water and Soil Conservation (FAOLEX, 2010), (5) the Law of the People's Republic of China on the Prevention and Control of Water Pollution (revised in 1996 and revised again in 2008), (6) the Law of the People's Republic of China on Evaluation of Environmental Effects and (7) Law of the People's Republic of China on Promotion of Cleaner Production. As a consequence there are several ministries involved in water governance.

Table 5.5 Ministries involved in water governance in China and their functions.

Ministries	Function
Ministry of Water Resources (MoWR)	This ministry runs most of the water-related aspects and plans for all major water development, water management and conservation projects on river basins. Among the specific duties are IWRM, water resource protection planning, monitoring water quantity and quality in rivers and lakes, issuing water resource extraction permits and proposing water pricing policy. There are a few subsidiary authorities under this ministry like Water Resource Bureau (WRB) and the National River Commission (NRC). Under the WRB there is an irrigation department providing agricultural water services.
Ministry of Land and Resources	This is the main department dealing with geological and geographical entities in China. Groundwater falls under the jurisdiction of this department, and it shares power with the Ministry of Housing and Urban-Rural Development, the Ministry of Environmental Protection and the Ministry of Water Resources.
Ministry of Environment Protection	This ministry is responsible for environmental protection and water quality; it enforces environmental laws and shares power with the Ministry of Land and Resources and the Ministry of Water Resources. MEP is the authority to release guidelines on water quality and standards for water pollution from different sectors. This ministry also organises water pollution–monitoring programmes, implementing water pollution laws, regulations/standards and supervises/enforces all environment related matters and issues.
Ministry of Agriculture	This department is responsible for management and use of water in the agricultural sector. It develops and implements policies regarding irrigation in consultation with the Ministry of Water Resources. The Agriculture Ministry also manages and regulates non-point pollution in agricultural fields. It also formulates and implements guidelines. It mainly shares power with the Ministry of Water Resources.
Ministry of Construction	This ministry has the power to regulate use of groundwater for urban, industrial and domestic water deliveries. It primarily shares responsibilities with the Ministry of Land and Resources and the Ministry of Water Resources.
Ministry of Finance	This ministry looks after the economic aspects of water use and pollution by regulating pollution levy proceeds, decides wastewater treatment charges and water price policy.

Adapted from Circle of Blue (no date) and (Levie, 2006).

The bureaucratic framework is hierarchical. It starts with State Council at the top and successively passes through provincial government, prefecture government, country or township government, and finally to the village government at the bottom of the ladder. The provincial water resource bureau is the state level authority under MoWR, and at each bureaucratic level, there is a water resource bureau up to the town level. China is governed through a dual leadership system run by administrators and political instruments. 'Offices and agencies of the two bureaucracies with the same bureaucratic rank (horizontal bureaucracy) cannot issue binding orders to each other' (Water bureaucracy, http://www.circleofblue.org/Waternews_MultiMedia/BYU/CPC_YellowRiver/bureaucracy.html#).

Wang (no date) has described the problems of China's water management system as having (1) 'fragmented approach and absence of coordination', (2) inadequate legal framework and (3) inadequate participation of stakeholders.

Dai (2015) comments that China tends more to apply political instruments, rather than a sound administrative framework.

China is on the point of major reforms in the water framework and is heading towards water demand management, conservation, prevention of pollution, ordered development, efficiency and comprehensive management.

5.15.6 Water governance in India

India has a federal government structure, and the country is divided into a number of states and union territories. According to the constitution of India, states are responsible for protection of water resources. There are several federal acts that are meant to protect the environment and have been in operation since 1897. Surface water belongs to the state, and the right to extract groundwater is attached to land ownership. Some of the essential National Acts are:

- 1897: The Indian Fisheries Act;
- 1956: The River Boards Act;
- 1970: The Merchant Shipping Act;
- 1974: The Water (Prevention and Control of Pollution) Act;
- 1977: The Water (Prevention and Control of Pollution) Cess Act;
- 1978: The Water (Prevention and Control of Pollution) Cess Rules; and
- 1991: The Coastal Regulation Zone Notification.

Ministries

The ministries that deal with water-related laws are:
- **Ministry of Environment:** Deals with water pollution and water-based ecology. There is a Central Pollution Control Board under this ministry. In every state, there are state-level pollution control boards.
- **Ministry of Water Resources, Rural Development and Ganga Rejuvenation:** This ministry deals with all other aspects of surface and groundwater, including water allocation and guidelines on water. The institutions created for water governance are shown in Table 5.6

State Departments

Most of the states in India have state departments dealing with water; water governance is generally entrusted to the State Department of Environment and Forest, State Groundwater/Water Resource Departments and State Irrigation Departments. These state departments are headed by ministers, and each department has the power to frame regulatory laws. Water laws of any state are derived from the problems specific to the state. The licensing authorities at the state level are:

- State Irrigation Department;
- State Groundwater Board/State Groundwater Authority; and
- State Pollution Control Board.

Table 5.6 Institutions created by Government of India for water governance.[1]

Licensing authorities	River boards and commissions	R & D Institutes
Central Water Commission and Central Ground Water Board	• Upper Yamuna River Board • Tungabhadra Board • Brahmaputra Board • Betwa River Board • Bansagar Control Board • Sardar Sarovar Construction Advisory Committee	• National Water Development Agency • National Institute of Hydrology • Central Water and Power Research Station • Central Soil and Material Research Station

[1] The Geological Survey of India, a premiere institute in India, is also involved in R & D activities related to groundwater. This organisation is under the Ministry of Mines.

Though India has a well constituted administrative system down to the village level (called Panchayets) the Panchayets are not the water licensing authorities; the district headquarters (of which the village Panchayet is a constituent part) is the licensing authority, and the service is delivered through empowered government institutions.

Local authorities

Municipalities of some cities also regulate water abstraction and water discharge into environment in some metropolitan cities in India. After promulgation of state laws, municipal control over water is restricted to protection of urban wetlands, sewage discharge and industrial effluent discharge.

Water Framework draft bill

In May 2016, the Ministry of Water Resource has tabled a National Water Framework Bill 2016, which is under public review[3]. The rationale behind such bill is described as follow:

> Under the Indian Constitution water is primarily a State subject, but it is an increasingly important national concern in the context of: (a) the right to water being a fundamental right; (b) the emergence of a water crisis because of the mounting pressure on finite resources; (c) the inter-use, intra-State and inter-State conflicts that this is leading to, and the need for a national consensus on water-sharing principles, and on the arrangements for minimising conflicts and settling them quickly without resort to adjudication to the extent possible; (d) the threat to this vital resource by the massive generation of waste by various uses of water and the severe pollution and contamination caused by it; (e) the long-term environmental, ecological and social implications of efforts to augment the availability of water for human use; (f) the equity of distribution based on types of use and users, area specific issues, different economic sectors, location of countries or states, i.e., upstream or downstream, that share river water; and particularly to safeguard future generations against water crisis; (g) the international dimension of some of India's rivers; and (h) the emerging concerns about the impact of climate change on water and the need for appropriate responses at local, national, regional, and global levels (Draft national water framework bill, 2016).

Industrial water under the draft bill

In this draft bill, the industry-specific water rights are categorically outlined with the stricture that the industries are to take concrete steps to reduce water footprint. All companies using large volumes of water must disclose their water footprint and include, in their annual report, information such as water utilisation per unit of production, effluent discharge details, rainwater harvesting, water reuse details and freshwater consumption. In water-short areas, industries can withdraw water to the extent of the makeup water requirement only and must recycle and reuse water. There is also provision for penalties for misuse and abuse of water by the industries.

5.15.7 Water governance in Indonesia

Water in Indonesia is regulated by the Laws on Water Resources 2004. The basic framework of the laws is that the state guarantees everyone's right to obtain water for their minimum daily basic needs in order to achieve a healthy, clean and productive life. Water resources shall be controlled by the state and used for the welfare of people. Management of water resources shall be carried out by the government and/or regional governments by continuing to recognise the traditional right of the local traditional community and any similar rights, to the extent that it does not contradict the national interest as well as the relevant laws and regulations (LEAD, 2008).

For the business community, the law provides a licensing system to extract water from natural resources. Regulations for obtaining water permits come under Regulation No. 37/PRT/M/2015 on Utilization Permits of Water and/or Water Resources (Regulation No. 37/2015) issued by the Minister of Public Work and Public Housing on 24 July 2015.

The salient features of the permit system are given in the Box 5.2.

5.15.8 Water governance in Namibia

Namibia has large areas of private lands and some tribal lands whose water sources are privately managed. As the most arid country in sub-Saharan Africa, Namibia faces several challenges in meeting the water needs of its people. The two lead organisations in the water sector are the Namibia Water Corporation (NamWater), which is responsible for bulk water supply to municipalities, and industries, and the Department of Water Affairs and Forestry (DWAF), responsible for water resources management and rural water supply.

The Republic of Namibia had an old Water Act 54 of 1956 which had been amended several times. Recently Namibia adopted its new water act, the Water Resources Management Act, 2013. This Act contains provisions for managing all water resources of Namibia including watercourses, aquifer, sea and meteoric water. The Act has 134 sections divided in 22 parts based on the water policy of the country which broadly recognises:

Permits for Water Extraction in Indonesia.

Authority for Granting the Utilization Permits of Water Resources

There are 2 (two) types of utilization permits of water resources, namely: (i) the surface water on rivers, lakes, swamp and other surface water resources and (ii) in-land seawaters.

The government institutions that have the authority to grant the above mentioned permits are:

a   Minister of Public Works and Public Housing ("Minister"), as licensor to acquire and utilise water resources in the cross-provincial river area, cross-national river area, and national strategic river area.
b   Governor, as licensor to acquire and utilise water resources in the river area of cross-regency/city; or
c   Regent/mayor, as licensor to acquire and utilise water resources in the river area within a regency/city.

Application of Water Resources Utilization

The application to obtain the water-resources utilization permit for the surface of rivers, lakes, swamps, and other surface water resources and/or in-land sea is submitted to the Minister of Public Work and Public Housing cq., Director General of Water Resources through the Licensing Service Unit with a copy to the management of water resources in the cross-national river area, cross-provincial river area, and national strategic river area.

The requirements for the utilization permits of water resources are granted by considering the technical recommendations from the management of water resources in the cross-national river area, cross-provincial river area, and national strategic river area, which containing technical considerations and suggestions. The results of the technical recommendations are then verified by the licensing verification team assigned by the Director General of Water Resources. The licensing verification team performs examination of:

- technical recommendations and permit suggestions;
- conformance between the permit application and technical recommendation; and
- technical feasibility of the permit.

Granting of Utilization Permits

Within a period of 30 (thirty) days from the receipt of application from the applicant, the Minister of Public Work and Public Housing cq. Director General of Water Resources shall give an approval or a rejection towards the application. In case of the rejection, the Minister of Public Work and Public Housing cq. Director of General Water Resources shall inform the reason of rejection in writing to the applicant, and the applicant cannot resubmit the application by using the same data.

Extension of Utilization Permits

The utilization permits of water resources can be extended by submitting the application of extension permit in writing, at the latest 3 (three) months before

BOX 5.2

the expiration of permit period. If within 3 (three) months prior to the expiration of permit period the application of extension has not been submitted, then the permit cannot be extended but it is possible to apply for a new permit. Within 30 (thirty) days from the receipt of application, the Minister of Public Works and Public Housing cq. Director General of Water Resources shall give the approval or rejection towards the application.

Revocation of Utilization Permits

The utilization permits of water resources can be revoked if:
The permit holder does not carry out the obligations contained in the utilization permits of water resources;
The permit holder misuses the utilization permits of water resources or
The validity period of the utilization permits of water resources expired.

Prohibition

The holder of utilization permit of water resources is prohibited to lease and/or assigns part or all of the utilization permits of water resources to other parties. (Leks, 2015)

Printed with permission from Eddy Leks (Leks & Co.).

- Equitable access to a sufficient quantity of safe water within a reasonable distance from their place of abode with emphasis on human rights;
- Harmonisation of human water needs with the water requirements of environmental ecosystems in terms of quantity and quality;
- Promotion of sustainable development based on a combination of water resources and human resource development;
- Transparent public information service on water resources;
- Recognition of the economic value of water in water allocation;
- Promotion of water awareness and participation of stakeholders; and
- Prevention of water pollution and implementation of the principle that a person disposing of effluent or waste has a duty of care to prevent pollution and the polluter is liable to pay all costs to clean up any intentional or accidental spill of pollutants.

The policy also recognises the requirement of management and control over international water and formation of water basin management frameworks.

The act in its first three parts describes the duties, obligation and power of the Minister and formation of a Water Advisory Council. The act requires the Minister to prepare an Integrated Water Resources Management Plan, the scope of which is defined by Section 32. The matter of water abstraction and use is included in Part 9, which empowers the Minister to reasonably conserve aquatic and wetland ecosystems while meeting the requirements of domestic water use. The act provides for licences to discharge effluent or construct or operate wastewater treatment facilities or waste-disposal sites. For the protection of water resources and the environment from pollution, the Minister may invoke prescribed standards and other measures listed in Section 76. The salient points of the act pertinent to industrial interests are:

Licence to abstract and use water (Part 11)

The application for abstracting water should be in a form approved by the Minister accompanied by the requisite fee and the required information and documents. If the location of the proposed abstraction site is under the jurisdiction of a water basin committee, it will require recommendation from that authority also. The Minister will also examine the impact of the proposed withdrawal on the interest of the ethnic groups and international obligations of Namibia relating to internationally shared water.

Groundwater Licence (Part 12)

A licence to construct a borehole to withdraw water will be given in response to an application in prescribed form along with supporting documents and prescribed fees. The Minister will examine recommendations of the relevant basin management committee. The committee recommendations are made in relation to hydro-geological conditions, expected groundwater potential and the conformity of the proposed abstraction to efficient water management practices. The proposed abstraction will not have any effect on any customary rights and practices related to the water resource concerned or the needs of any community dependent on that water resource.

Water pollution control (Part 13)

Under the act no individual, community, industry or water service provider can discharge wastewater to the environment. Section 72 stipulates provisions to apply for a licence to discharge effluent or to construct or operate a wastewater treatment facility or a waste-disposal site. This application must be in a prescribed form accompanied by relevant documents and a licence fee. The Minister will examine the effect of waste disposal on water resources and owners and occupiers of land in the area where effluent is proposed to be discharged or the proposed wastewater treatment facility or waste-disposal site is to be located. An environmental impact analysis must be done.

5.15.9   Water governance in South Africa

South Africa is the 30th driest country worldwide. As South Africa traditionally had several local tribal administrations, the Union of South Africa, formed in 1910, got its first Irrigation and Conservation of Water Act in 1912 'to codify all the water laws of the Union' (Thompson *et al.* 2001). The first South African water law under the republic came in force under Water Act No. 54 of 1956, which was later replaced by National Water Act 1998 based on the National Water Policy (1997). The polity of South African

Government entitles every person to lead a dignified and healthy life. This act also recognises people's participation in water management and water as a resource for livelihood and economic benefits. In South Africa, water and environment are not looked upon as separate entities; so there is a single water and environment ministry in the federal system eliminating unnecessary bureaucratic formalities. The main areas covered under this law are (1) basic legislation, (2) water conservation zone, (3) authorisation/permit, (4) water charges, (5) freshwater quality/freshwater pollution, (6) irrigation, (7) water rights, (8) inspection, (9) institution, (10) subsidy/incentive, (11) groundwater, (12) rainwater, (13) well sinking/boreholes, (14) effluent wastewater/discharge, (15) servitude, (16) waterworks (17) water supply (18) hydropower generation, (19) industrial water use, (20) recreational water use, (21) free water use, (22) water shortage/drought and (23) court/tribunal (FAO, no date).

South Africa (SA) now has a National Water Resource Strategy (Department of Water Affairs and Forestry, SA, 2004) that describes the modalities by which water will be protected, used, developed, conserved, managed and controlled under different policies and laws. In the document, the SA government provides guidelines for (1) institutional arrangement, (2) financing the water sector, (3) monitoring and information management, (4) research and innovation and (5) enhancing water sector skills and capacity.

The policies or strategies behind South Africa's water governance are the National Water Resources Strategy (NWRS), the National Disaster Management Framework and the National Strategic Framework for Water Services. The Acts that govern the management framework are the National Water Act (1997), the National Environmental Management Act (1998), the Environmental Conservation Act (1969), and the National Environmental Biodiversity Act (2004).

Administrative framework

The administrative structure is the National Government, which holds full legislative and administrative powers. The national government delegates power to provincial governments. The local authorities also enjoy power to regulate water withdrawal and discharge of effluent and wastewater.

- **National government:** The Minister of Water Affairs is responsible for managing and administering water resources as the public trustee, ensuring that water is allocated equitably, and that environmental values are promoted. General water management functions are delegated to the Department of Water (DWA). The DWA is responsible for implementing the two major legal instruments relating to water: the Water Services Act No. 108 of 1997 and the National Water Act (NWA) No. 36 of 1998. The NWA is to be implemented through the NWRS.
- **Provincial government:** South Africa has nine provinces, each with its own provincial government with legislative and executive authority (South African Environmental Law, 2016). Executive powers accorded to provincial executives include:
  - Implementing provincial legislation.
  - Implementing all national legislations within the functional areas listed in Schedules 4 and 5 of the Constitution.

- ◦ Developing and implementing provincial policy.
- ◦ Co-ordinating the functions of provincial administration and its departments.
- ◦ Preparing and initiating provincial legislation (South African Environmental Law 2016).
- **Local authority:** The local authority constituted as municipalities deliver services and execute powers related to specified water and sanitation services. In post-colonial South Africa, the existing legal framework is still criticised as being complex and needing reforms. Licensing is delivered after the proposal is channelled through an initial assessment and full assessment, depending on the nature of withdrawal. The process of full assessment will also examine the requirement of an EIA of the project. If EIA is not required, the project is passed.
- **River basin authorities:** The river basin authorities formed under the acts and international treaties work with a very complex system that covers multiple international administrative jurisdictions. Particularly with regard to the Orange-Senqu Basin, the institutional structure is highly complex because it involves five countries (ORASECOM, 2009) and is governed by the Orange-Senqu River Commission (ORASECOM) Agreement.

5.16   Conclusion

A well-constituted water framework is always a strength in the water governance of a country. It honours human rights, water ethics and the value of water. The water framework guides the water governance, promotes the formation of laws and institutions and paves ways for multi-stakeholders' participation. According to OECD, water is a fragmented sector and is sensitive to and dependent on multi-level governance and connects across sectors, countries and geographical boundaries. It becomes more complex when hydrological boundaries and political perimeters do not coincide. Water governance also involves a 'plethora of public, private and non-profit stakeholders in the decision-making, policy and project cycles' (OECD, 2015). It is dependent on capital-intensive infrastructures that are critical to the ecosystem, energy, human habitat, economic growth and regional development. Water governance needs to be free from human rights violations, monopolisation and corrupt practices. Water allocation, a major task of water governance, must be transparent, pro-people and proactive for the betterment of the society.

Notes

1. Further information is available at http://www.water.go.ke/?page_id=6.
2. 'Taking of water' means taking of water from a source in an area which is either proclaimed or not proclaimed by the government. For further reading, see https://

www.environment.sa.gov.au/managing-natural-resources/water-use/water-planning/water-licences-and-permits.
3. DRAFT OF 16 MAY 2016 is available at http://wrmin.nic.in/writereaddata/Water_Framework_May_2016.pdf (accessed 11 March 2017).

Bibliography

Abelson, J. (2006) *Assessing the impacts of public participation: concepts, evidence, and policy implications.* Available at http://www.cprn.org/documents/42669_fr.pdf (accessed 1 March 2016).

Alberts, G. and Kaspersma, J. (2009) 'Progress and challenges in knowledge & capacity development', in Blokland, M., Alaerts, G., Kaspersma, J. and Hare, M. (eds.), *Capacity development for improved water management UNESCO-IHE editors Pre-Print copy.* Delft: UNSCO-IHE., 3–30.

Asian Development Bank. (2014) *Improving water governance in the Asia-Pacific region: Why it matters.* Available at http://www.adb.org/features/improving-water-governance-asia-pacific-region-why-it-matters (accessed: 6 March 2016).

Australian Government (2009) *Key features of the water act 2007.* Available at https://www.environment.gov.au/water/australian-government-water-leadership/water-legislation/key-features-water-act-2007 (accessed: 8 July 2016).

Bäckstrand, K. (no date) *Civic science for sustainability: Reframing the role of experts, policy makers and citizens in environmental governance.* Available at http://citeseerx.ist.psu.edu/viewdoc/download?doi=10.1.1.473.851&rep=rep1&type=pdf (accessed: 23 March 2016).

Biermann, F. (2002) 'Institutions for scientific advice: Global environmental assessments and their influence in developing countries', *Global Governance*, 8(2): 195–219. doi: 10.2307/27800338.

Bouchane, K. and Coulby, H. (2011) *Learning from experience.* Available at http://www.wateraid.org/~/media/Publications/Learning-from-experience-Rights-and-governance-advocacy-in-the-water-and-sanitation-sector.pdf (accessed: 16 July 2016).

Boyle, R. (2006) *Committee for public management research the management of cross-cutting issues.* Available at http://www.ipa.ie/pdf/DiscussionPaper_8.pdf (accessed: 13 March 2016).

Brazil national water resources policy (1997) Available at http://www.gwp.org/Global/ToolBox/About/IWRM/America/Brasil%20National%20Water%20Resources%20Policy.pdf (accessed: 7 August 2016).

Circle of Blue (no date) *China's water governance.* Available at http://www.circleofblue.org/Waternews_MultiMedia/BYU/CPC_YellowRiver/index.html# (accessed: 8 July 2016).

Dai, L. (2015) *Transition in Chinese water resource management—a governance analysis.* Poster presentation, University of Utrecht, The Netherlands

Department of Water Affairs and Forestry, SA (2004) *National water resource strategy.* Available at http://www.orangesenqurak.org/UserFiles/File/National%20Water%20Departments/DWEA-DWAF/NATIONAL%20WATER%20RESOURCE%20STRATEGY.pdf (accessed: 4 October 2016).

EPA Ireland (2016) *Water framework directive.* Available at http://www.epa.ie/water/watmg/wfd/(accessed: 10 June 2016).

European Union (2000). Directive 2000/60/ec of the European Parliament and of the Council. Available at http://eur-lex.europa.eu/resource.html?uri=cellar:5c835afb-2ec6-4577-bdf8-756d3d694eeb.0004.02/DOC_1&format=pdf (accessed 10 September 2016).

FAO (no date) *Statutes of the republic of south africa-water.* Available at http://faolex.fao.org/docs/pdf/saf1272.pdf (accessed 24 January 2016).

FAOLEX (2010) *Water and soil conservation law of the People's Republic of China.* Available at http://faolex.fao.org/cgi-bin/faolex.exe?rec_id=107231&database=faolex&search_type=link&table=result&lang=eng&format_name=@ERALL (accessed: 4 October 2016).

Fenemor, A., Neilan, D., Allen, W. and Russell, S. (2011) 'Coordination of policies and governance regime requirements in Dutch freshwater management improving water governance in New Zealand: Stakeholder views of catchment management processes and plans single worthwhile policy, seeking legitimacy and implementa', *Policy quarterly,* 7(4): 10–19.

FLOW (2009) *Water governance in Canada.* Available at http://www.flowcanada.org/policy/governance (accessed: 16 June 2016).

Global Water Partnership (2013) *Regulatory bodies and enforcement agencies (B1.05)—creating an organisational framework (b1)—institutional roles (b)—tools—toolbox—root—global water partnership.* Available at http://www.gwp.org/en/ToolBox/TOOLS/Institutional-Roles/Creating-an-Organisational-Framework/Regulatory-bodies-and-enforcement-agencies/(accessed: 23 March 2016).

Government of Australia, National Water Commission (2012) *National water commission framework criteria.* Available at http://www.nwc.gov.au/__data/assets/pdf_file/0020/22169/Australian-environmental-water-management-framework-criteria.pdf (accessed: 8 July 2016).

Hearns, G.S., Paisley, R.K. and Henshaw, T.W. (2013) *Institutional architecture and the good governance of international Transboundary waters.* Available at http://www.internationalwatersgovernance.com/uploads/1/3/5/2/13524076/institutional_arch_paper_06_august_2013_1.pdf (accessed: 6 March 2016).

Hirsch, P., Mørck, K., With, J., Boer, B., Carrard, N., Fitzgerald, S. and Lyster, R. (2006) *National interests and transboundary water governance in the Mekong in collaboration with Danish international development assistance.* Available at http://sydney.edu.au/mekong/documents/mekwatgov_mainreport.pdf (accessed: 4 October 2016).

Hoogesteger, J. (2015) 'NGOs and the democratization of Ecuadorian water governance: Insights from the Multi-Stakeholder platform el Foro de los Recursos Hídricos', *VOLUNTAS: International journal of voluntary and nonprofit organizations,* 27(1): 166–186. doi: 10.1007/s11266-015-9559-1.

IWMI (no date) *B1.10 local authorities.* Available at http://www.cawater-info.net/bk/iwrm/b1-10_e.htm (accessed: 23 March 2016).

Kildea, P. and Williams, G. (2011) *The constitution and the management of water in Australia's rivers.* Available at https://sydney.edu.au/law/slr/slr_32/slr32_4/Kildea_and_Williams.pdf (accessed: 4 October 2016).

Lagacé, É. (2011) *Shared water one framework.* Available at http://poliswaterproject.org/sites/default/files/SharedWaterOneFramework_BriefingNote.pdf (accessed: 26 March 2017).

LEAD (2008) 'Law on Water Resources—Indonesia', *2/1 Law, environment and development journal (2006),* p.118, available at http://www.lead-journal.org/content/06118.pdf (accessed: 19 September 2016).

Leks, E. (2015) 'Utilization permits of water resources', *Leks & Co lawyers.* Available at http://www.lekslawyer.com/utilization-permits-of-water-resources/(accessed: 19 September 2016).

Levie, C. (2006) *Water quality Management—policy and institutional considerations environment and social development East Asia and Pacific region, The World Bank.* Available at http://siteresources.worldbank.org/INTEAPREGTOPENVIRONMENT/Resources/China_WPM_final_lo_res.pdf (accessed: 8 July 2016).

Moss, T. and Newig, J. (2010) 'Multilevel water governance and problems of scale: Setting the stage for a broader debate', *Environmental Management*, 46(1): 1–6. doi: 10.1007/s00267-010-9531-1.

OECD (2015) *OECD principles on Water Governance*. Available at https://www.oecd.org/gov/regional-policy/OECD-Principles-on-Water-Governance-brochure.pdf (accessed: 15 April 2016).

Olsson, P., Folke, C., Galaz, V., Hahn, T. and Schultz, L. (2007) 'Ecology and society: Enhancing the fit through Adaptive Co-management: Creating and maintaining bridging functions for matching scales in the Kristianstads Vattenrike Biosphere reserve, Sweden', *Ecology and Society*, 12(1).

ORASECOM (2009) *Agreement between the governments of the Republic of Botswana, the Kingdom of Lesotho, the Republic of Namibia*. Available at http://www.orangesenqurak.com/UserFiles/File/ORASECOM/ORASECOM%20Agreement%202000.pdf (accessed: 4 October 2016).

Pearse, P.H. and Quinn, F. (1996) 'Recent developments in federal water policy: one step forward, two steps back', *Canadian water resources journal*, 21(4): 329–340. doi: 10.4296/cwrj2104329.

Queen's Printer (2016) Available at http://www.bclaws.ca/civix/document/id/complete/statreg/14015 (accessed: 16 June 2016).

Royal Geographical Society (with IBG) (2012) *Waterpolicy in theUK: the challenges. RGS-IBG Policy Briefing*. Available at https://www.rgs.org/NR/rdonlyres/4D9A57E4-A053-47DC-9A76-BDBEF0EA0F5C/0/RGSIBGPolicyDocumentWater_732pp.pdf (accessed: 6 March 2016).

Rogers, P. (2002) *Water governance in Latin America and the Caribbean*. Available at http://idbdocs.iadb.org/wsdocs/getdocument.aspx?docnum=1441619 (accessed: 14 October 2016).

Royal Bank of Canada (2016) *RBC blue water project*. Available at http://www.rbc.com/community-sustainability/environment/rbc-blue-water/index.html (accessed: 12 August 2016).

Schwartz, K., Nursyirwan, I., Nes, A.V. and Luijandijk, I. (2009) 'Capacity building in Indonesian water resource sector', in Blokland, M., Alaerts, G., Kaspersma, J., and Hare, M. (eds.) *Capacity development for improved water management UNESCO-IHE editors Pre-Print copy*. Delft: UNSCO-IHE., pp. 141–158.

SSWM (2003) Building an institutional framework. Available at http://www.sswm.info/category/implementation-tools/water-distribution/software/creating-enabling-environment/building-ins (accessed 6 March 2016).

Steyaert, P. and Ollivier, G. (2007) 'The European water framework directive: How ecological assumptions frame technical and social change', *Ecology and society*, 12(1).

Thompson, H., Stimie, C.M., Richters, E., Perret, S. and Area, S. (2001) *Policies, legislation and organizations related to water in South Africa, with special reference to the Olifants river basin* Available at http://www.iwmi.cgiar.org/Publications/Working_Papers/working/WOR18.pdf (accessed: 4 October 2016).

UNDP (2004) *Water governance for poverty reduction key issues and the UNDP response to Millenium Development Goals United Nations Development Programme*. Available at http://www.undp.org/content/dam/aplaws/publication/en/publications/environment-energy/www-ee-library/water-governance/water-governance-for-poverty-reduction/UNDP_Water%20Governance%20for%20Poverty%20Reduction.pdf (accessed: 16 July 2016).

UNDP (2013) *Concepts and approaches for effective water governance in the Arab region*. Available at http://www.ae.undp.org/content/dam/rbas/doc/Energy%20and%20Environment/Arab_Water_Gov_Report/Arab_Water_Report_AWR_Chapter%203.pdf (accessed: 13 March 2016).

United Nations (1996) *Convention on the rights of the child.* Available at http://www.ohchr.org/en/professionalinterest/pages/crc.aspx (accessed: 6 March 2016).

Water framework directive (2016) in *Wikipedia.* Available at https://en.wikipedia.org/wiki/Water_Framework_Directive (accessed: 7 August 2016).

Watergovernance Canada (2008) *Water rights across Canada.* Available at http://watergovernance.ca/wp-content/uploads/2010/04/FS_Water_Rights.pdf (accessed: 12 August 2016).

Wang, Y., (no date) *China's Water Issues: Transition, Governance and Innovation.* Available at http://admin.cita-aragon.es/pub/documentos/documentos_WangYi_286d0ba6.pdf (accessed 8 July 2016).

Wiek, A. and Larson, K.L. (2012) 'Water, people, and Sustainability—A systems framework for analyzing and assessing water governance regimes', *Water resources management*, 26(11): 3153–3171.doi: 10.1007/s11269-012-0065-6.

Yang, X. and Griffiths, I.M. (2010) 'A comparison of the legal frameworks supporting water management in Europe and china', *Water science & technology*, 61(3), p. 745.doi: 10.2166/wst.2010.899.

6

Water Quality Standards and Water Pollution

6.1 Water quality-standards

6.1.1 Introduction

Water is a unique material because of its polarity (Chemwiki, 2013). Oxygen has a slightly negative charge, while two hydrogen atoms have a slightly positive charge. It is unique because its density is higher in the liquid phase than that in the solid phase. The other unique properties of water are high specific heat, high latent heat of vaporisation, and high surface tension. Due to this uniqueness, water has become the most useful material for mankind. These properties, both chemical and physical, make water more susceptible to pollution. While it is true that water supports life, it also hosts and becomes a site for the menacing growth of microorganisms and vectors. The chemical bond of water ion can associate or dissociate almost all chemical compounds in different quantities. Solid substances that do not dissolve in water can remain suspended in water in the form of minute particles. The unique qualities of water are:

1 **Temperature:** Water remains in the liquid form between 0°C and 100°C under normal pressure.
2 **Dissolved chemicals (dissolved salts/organic or inorganic compounds):** The colour, odour, toxicity and utility depend on the dissolved chemicals.
3 **Suspended particles:** Solid particles of very minute size (usually not less than 0.45 micron) can remain suspended in water. Some physical properties and acceptability of water depend on a number of suspended particles.
4 **Microorganisms:** The presence of microorganisms is a quality indicator of water. Its potability depends on this property (Chemwiki, 2013).

Industrial Water Resource Management: Challenges and Opportunities for Corporate Water Stewardship, First Edition. Pradip K. Sengupta.
© 2018 John Wiley & Sons Ltd. Published 2018 by John Wiley & Sons Ltd.

5 **Nutrients:** The presence of nutrients is very common in natural surface water and in water discharged after human use. High level of nutrients is considered a pollutant.

Pure water is not available in nature. Even if it is free from suspended particles or microorganisms, it is rarely free from dissolved solids and gases. The presence of some dissolved chemicals (salts) in limited quantity in drinking water is also essential for human health. Pure water is not good for drinking.

The quality of water is a prime concern for industries, especially for water supply, food and beverage industries that need potable water for their products. Other industries use water for various purposes like heating, cooling, cleaning, manufacturing and fire fighting. Drinking and sanitary uses of water are also concerns for all industries because they are liable to extend water, sanitation and hygiene (WASH) facilities for their employees. Industries also deliver water to industrial colonies and townships.

Irrespective of the quality of water that industries take in, they are always responsible for the quality of the discharge they contribute to the environment. Water quality is also a major concern in agro-based industries, because pesticides and fertilisers are applied in the farms belonging to the supply chain of the food and beverage industry. Every industry has the responsibility of supplying safe drinking water for the staff or employees living within the industrial campus. Water supply industries, beverage and medicine manufacturers need purified water that meets the potable standard or sometimes in the ultrapure state. They need to maintain drinking water quality standards for their raw materials and as well as for their products.

6.1.2   Quality parameters for drinking water

There are many sources of water quality criteria and standards. Each country has its own standards and guidelines. These guidelines do not vary widely from country to country. However, many countries have adapted guidelines of other countries or organisations to frame their own standards. The United Nations (UN Water, 2015) has published the *Compendium of Water Quality Regulatory Frameworks: Which Water for Which Use?* which is a very useful guideline and a comparative study of water quality frameworks of different countries.

Water quality not only influences the water strategy of the business but also impacts capital and non-capital investment in the water sector. The quality parameters may be classified as follows:

- *Organic or inorganic*, according to its biological properties;
- *Physical or chemical*, based on its physical and chemical properties;
- *Potable or non-potable*, depending on its use for human consumption; and
- *Hazardous or non-hazardous*, based on its effect on human health and ecology.

For example, fluoride found in the groundwater of South Africa (Ncube and Schutte, 2005) and India is an inorganic non-industrial chemical contaminant and a severe health hazard.

The presence of heavy metals in water is highly dangerous as they are essentially toxic and may cause chronic biological damage. Heavy metals are undesirable in potable water. Industrial wastes generally contain heavy metals in high concentrations and, if unattended, may cause serious environmental damage. The following elements are heavy metals: antimony, cobalt, nickel, tin, arsenic, copper, selenium, titanium, beryllium, lead, silver, uranium, cadmium, mercury, tellurium, vanadium, chromium, molybdenum, thallium and zinc.

Some essential parameters for drinking water quality and standards are described briefly in Table 6.1 to Table 6.5.

6.1.3 Microbiological contaminants

There are a large number of microorganisms which live in water either permanently or for a certain period during their life cycle. Disease-causing microorganisms are commonly classified as coliform or staphylococci bacteria. Industries do not generally discharge these organisms to the environment, but lack of proper WASH facility in the industrial environment may cause health hazards.

6.1.4 Physical parameters

Physical parameters may or may not relate to industrial pollution. These parameters relate to potability and suitability for industrial uses. Physical parameters also quantify the aesthetic value of water. Change in physical parameters due to anthropogenic activities may adversely affect aquatic ecosystems. See Table 6.3.

6.1.5 Organic chemical pollutants

Organic chemicals are mainly industrial products used as pesticides and herbicides in agriculture and in textile industries. These are harmful chemicals with long-term effects on human health. In Table 6.4, a list of such chemicals, their sources and effects on human health is given.

6.1.6 Parameters indicative of environmental pollution

There are some parameters of water quality which are measured as pollution indicators and not as pollutants; by testing these parameters, it may be determined whether the water is linked with a potential source of pollution or not. See Table 6.5.

Table 6.1 Source and potential health hazards of inorganic contaminants.

Contaminant	Sources	Potential health and other effects
Aluminium (Al)	Aluminium occurs in nature as aluminium minerals like silicates and oxides. Aluminium is released in water from mines and aluminium industries.	Can precipitate out of water after treatment, causing increased turbidity or discoloured water.
Antimony (Sb)	Enters environment from natural weathering, industrial production, municipal waste disposal and manufacturing of flame retardants, ceramics, glass, batteries, fireworks, and explosives.	Decreases longevity, alters levels of glucose and cholesterol in blood.
Arsenic(As)	Arsenic is a very common element in the earth's crust (Sengupta, 2014). It is present as arsenic minerals in rocks and as arsenic ions in water. Arsenic enters into hydrological environment through natural processes. Arsenic compounds present in the aquifer are released from sand grains and contaminate groundwater. It is used in chemical and metallurgical industries and often released into environment through effluent. Chatterjee, Das and Chakraborti (1993) reported that industrial contamination of arsenic had been detected in groundwater of Kolkata.	Arsenic is highly toxic and may cause liver, kidney and other problems if consumed for a long time. It is carcinogenic also.
Barium (Ba)	Barium is a hazardous contaminant (Purewater Occassional, 2013). Barium minerals (e.g., in baryte) is released into water from mines. In normal surface water, barium level is low because it reacts with the sulphate present in water and forms insoluble barium sulphate.	Causes cardiac, gastrointestinal and neuromuscular disorders.
Beryllium (Be)	Beryllium occurs naturally in soils, groundwater, and surface water. Beryllium is used in electrical, nuclear power and space industries and enters the environment from mining wastes and mineral-processing plants. It also enters as natural flow from rocks, coal, and petroleum.	Causes acute and chronic toxicity; can cause damage to lungs and bones. Possible carcinogen.
Cadmium (Cd)	Cadmium in water generally comes from industrial discharges (e.g. electroplating, paint-making, manufacture of plastics) and landfill leachates. Cadmium is used in manufacturing pigments, batteries, metal-plating and plastics. WHO (2011) reports that fertilisers produced from phosphate ores constitute a major source of diffused cadmium pollution.	Highly toxic to both terrestrial and aquatic life US-EPA (2000). It replaces zinc biochemically in the body and causes high blood pressure, liver, kidney damage and anaemia.
Chloride (Cl)	Chloride is associated with sodium chloride, that is, common salt. Occurs naturally in seawater, saline aquifers, lakes and estuaries.	Affects taste of water. Deteriorates plumbing, water heaters and municipal water-works equipment.
Chromium (Cr)	Enters environment from old mining operations runoff and leaches into groundwater. Industries release chromium through fossil-fuel combustion, cement-plant emissions, mineral leaching and waste incineration. It is used in tanneries, metal plating and as a cooling-tower water additive.	Chromium VI is more toxic than Chromium III and causes liver and kidney damage, internal haemorrhaging, respiratory damage, dermatitis and ulcers on the skin.

Table 6.1 (Continued)

Contaminant	Sources	Potential health and other effects
Copper (Cu)	Enters into water sources naturally from leaching of copper ores. Copper mines, metal alloy and metal plating industries release copper into environment.	Copper is an essential trace element in human biological system, though it can cause stomach and intestinal distress, liver and kidney damage, and anaemia if consumed in high doses. Its presence in water causes bad taste and stains on plumbing fixtures.
Cyanide (Cn)	Often used in electroplating, steel processing, plastics, synthetic fabrics and fertiliser production. Released into water due to improper waste disposal.	Poisoning results in damage to spleen, brain, and liver.
Total dissolved solids (TDS)	TDS is a parameter to indicate the total quantity of dissolved salts in water.	High TDS concentration imparts bad taste. Deposits scales on water heating elements and utensils.
Fluoride (F)	Fluoride occurs naturally and is found in groundwater in India, China, South Africa, Ghana and many other places in the world.	Naturally occurring fluoride in groundwater above permissible limit is a health hazard. Causes bone and teeth decay. Fluoride deficiency is also a health issue; it is added in minute quantities to drinking water in many countries in America, Africa, Europe and Australia.
Iron (Fe)	Iron is the most common contaminant present in natural water. Generally iron content is high in groundwater and comparatively low in surface water. Iron is present in water as soluble ferrous hydroxide, but in contact with air it oxidises into insoluble ferric oxide and appears as suspended solid.	Imparts a bad taste. Creates brownish colour to laundered clothing and plumbing fixtures. Water contaminated with iron will show muddy appearance due to presence of insoluble ferric oxide. Water containing high levels of iron is not suitable for industries.
Lead (Pb)	Enters into environment from industries and lead, zinc and silver mines.	It affects red blood cell chemistry, delays normal physical and mental development in babies and young children. Can cause increase in blood pressure in adults.
Manganese (Mn)	Released into water from manganese mines. It is widely used in ferrous and nonferrous metal industries and hence, a common industrial contaminant.	It affects taste of water, and causes dark brown or black stains on plumbing fixtures. It is relatively non-toxic to animals but toxic to plants.
Mercury (Hg)	Occurs both as inorganic salts and as organic mercury compounds. Enters the environment from industrial waste, mining, pesticides, coal, electrical equipment (batteries, lamps, and switches), smelting, and fossil-fuel combustion. Mercury discharged into sea by industries undergoes/initiates the process of biomagnification in the sea fish and sea shells which cause mercury poisoning.	Causes acute and chronic toxicity. Targets the kidneys and can cause nerve disorder. Widespread mercury poisoning from fish consumption became a concern after the tragedy caused at Mina Mata, Japan, popularly known as 'minamata disease'.
Nickel (Ni)	Principal sources are minerals and industrial wastes. Used in electroplating, stainless steel and alloy products and refining.	Very limited health/sanitary significance as far as humans are concerned. Toxic to plant life and fish.

(Continued)

Table 6.1 (Continued)

Contaminant	Sources	Potential health and other effects
Nitrate (NO^{-3})	Occurs naturally in mineral deposits, soils, seawater, freshwater systems, atmosphere and biosphere. Enters the environment from fertiliser, feedlots, and sewage.	Toxicity results from the body's natural breakdown of nitrate to nitrite. Causes 'bluebaby disease', or methemoglobinemia, which threatens oxygen-carrying capacity of the blood.
Nitrite (NO^{-2})	Enters environment from fertiliser, sewage, and human or farm-animal waste.	Nitrite causes toxicity manifested by 'bluebaby disease' or methemoglobinemia, which threatens the oxygen-carrying capacity of the blood.
Selenium (Se)	Comes from weathering of rocks/soils, but major environmental sources are man-made.	Although an essential biological requirement for both man and animals, selenium in more than very small amounts is toxic, causing a variety of illnesses. There have been conflicting reports as to whether or not the element is carcinogenic
Silver (Ag)	Enters environment from mining, processing, product fabrication (notably solder) and ornament industry.	Can cause argyria, a blue-grey coloration of the skin, mucous membranes, eyes and organs in humans and animals with chronic exposure but not considered particularly toxic to humans, as it is found only in very low concentrations.
Sodium (Na)	Derived geologically from leaching of surface and underground deposits of salt and decomposition of various minerals. Human activities contribute through de-icing and washing products.	Can be a health risk for those on a low-sodium diet.
Sulphate (SO_4)	Elevated concentrations may result from saltwater intrusion, mineral dissolution, and domestic or industrial waste.	Forms hard scales on boilers and heat exchangers; can change the taste of water, and has a laxative effect in high doses.
Thallium (Tl)	Enters environment from minerals, but more often from discharges in electronics, pharmaceuticals, glass and alloy industries.	Causes a wide variety of effects including nausea, vomiting, pain and, ultimately, death.
Zinc (Zn)	Found naturally in water, most frequently in areas where it is mined. Enters environment from industrial sludge, metal plating and galvanized pipes.	Aids in the healing of wounds. Causes no ill health effects except in very high doses. Imparts an undesirable taste to water. Toxic to plants at high levels.
Uranium (U)	Rare natural occurrence; equally rare in effluents. Uranium may contaminate water by leaching from nuclear power plants affected by accident or natural disaster; WHO (2004) reports 'Concentrations in excess of 20 µg/litre have been reported in groundwater from parts of New Mexico, USA'.	Uranium is a radioactive mineral and highly hazardous to health. Widespread contamination in Pacific Ocean occurred due to breakdown of Fukushima nuclear power plant in Japan in 2011(Burr, 2015).

Source: USGS (2016a), US-EPA (2000), WHO (2004), Burr (2015), EPA Ireland (2001).

Table 6.2 Source and potential health hazards of microbiological contaminants.

Contaminant	Sources	Potential health and other effects
Coliform bacteria	One of the major pathogenic pollutants in water. It originates in humans and other warm-blooded animals and goes into water through faeces. Non-faecal bacteria, if present in water, serve as indicators of presence of other forms of bacteria that may cause disease. The most common indicator bacteria are *E. coli*, which is a subgroup of coliform bacteria in water, which is generally measured to find presence of pathogenic bacteria in water.	Bacteria cause polio, cholera, typhoid, dysentery and infectious hepatitis, to name a few; list of bacterial infection is virtually endless.
Pathogenic Staphylococci	It is a kind of pathogenic micro-organisms found in sewage-contaminated waters and bath water discharge (e.g. swimming pools).	Causes a wide variety of diseases notably skin disease and food poisoning.

Source: EPA Ireland (2011).

Table 6.3 Source and potential health hazards of physical parameters.

Contaminant	Sources	Potential health and other effects
Turbidity	Caused by the presence of suspended particles such as clay, silt and organic and inorganic matter. Dissolved iron hydroxide can cause turbidity when oxidised to insoluble ferric oxide. It is measured in terms of the amount of light transmitted through the water sample.	Objectionable mainly for aesthetic reasons.
Total suspended solids (TSS)	When water with high turbidity is to be processed for use, then another parameter needs to be measured, namely TSS.	Impacts aquatic environment by restricting penetration of sunlight.
Colour	Can be caused by decomposed organic matter and mine waste.	Mine waste may be toxic to human and aquatic life.
pH	By definition pH of water is the negative logarithm of the hydrogen ion concentration (EPA Ireland 2001). The range goes from 0 to 14, 7 being neutral. Natural water generally has pH from 4.5 to 10. If pH of water is above 7 it is called alkaline. Best suitable pH range for human consumption, aquatic life and industrial use is from 6.5 to 8.5. The pH of water determines the solubility (amount that can be dissolved in the water) and biological availability (amount that can be utilised by aquatic life) of chemical constituents such as nutrients (phosphorus, nitrogen and carbon) and heavy metals (lead, copper, cadmium etc.). (USGS, 2016b)	High pH causes a bitter taste; water pipes and water-using appliances become encrusted. Low-pH also imparts bad taste and corrodes metals.

(Continued)

Table 6.3 (Continued)

Contaminant	Sources	Potential health and other effects
Odour	Odour is an indicative of organic waste or decomposed biota.	Malodourous water is unfit for consumption.
Taste	Some organic salts impart taste without causing malodourousness.	Bad taste makes water unfit for consumption.
Salinity	Relates to tidal waters or water which has a hydraulic interface with seawater. Seawater has salinity as high as 35000 mg/L.	High salinity makes water unfit for consumption, creates salt encrustation and corrosion.
Alkalinity	Alkalinity is the measure of the presence of bicarbonate salt in water. Alkalinity is a measure of the buffering capacity of water, or the capacity of bases to neutralise acids.	Not a health hazard but unsuitable for industries for corrosiveness.
Electrical conductivity	Electrical conductivity is a major physicochemical property of water. It is a measure of the capacity of water to transmit electric current through it. Fresh water assumes conductivity due to concentration of ions in water which comes from dissolved salts and inorganic materials (Fondriest, 2015). The commonly used units for measuring electrical conductivity of water is microSiemens/cm.	It is not a definite indicator of water pollution but an indicator of suitability of water for domestic, industrial or agricultural purposes.
Hardness	Hardness is caused due to the presence of certain dissolved salts. The compounds responsible for developing hardness in water are generally salts of calcium and magnesium. Hardness is measured by the amount of calcium carbonate expressed as mg/l of $CaCO_3$.	Decreases the lather formation of soap and increases scale formation in hot-water heaters and low-pressure boilers at high levels. 'General guidelines for classification of waters are: 0 to 60 mg/L (milligrams per litre) as calcium carbonate is classified as soft; 61 to 120 mg/l as moderately hard; 121 to 180 mg/l as hard; and more than 180 mg/l as very hard' (USGS, 2016c).

Source: EPA Ireland (2001), USGS (2016b), USGS (2016c) and (Fondriest, 2015).

6.1.7 Guidelines for standard quality parameters

Different countries have different guidelines for drinking water quality. Numbers of parameters for which the guidelines are prescribed also differ from country to country. Water quality standard guidelines of India, the United States, Canada, Australia and India are compared against WHO and EU specifications in Table 6.6.

6.1.8 Water quality requirements of industries

Industrial water quality depends on the type of industries (Ford *et al.*, 2008). Different countries have different guidelines for water-quality requirements for industrial and other uses; for example, India has five

Table 6.4 Source and potential health hazards of organic contaminants.

Contaminants	Sources of contaminants	Potential health effects from long-term exposure
Acrylamide	Added to water during sewage/wastewater treatment	Nervous system or blood problems; increased risk of cancer
Alachlor	Runoff from herbicides used on row crops	Eye, liver, kidney or spleen problems; anaemia; increased risk of cancer
Atrazine	Runoff from herbicide used on row crops	Cardiovascular system or reproductive problems
Benzene	Discharge from factories; leaching from gas storage tanks and landfills	Anaemia; decrease in blood platelets; increased risk of cancer
Benzo(a)pyrene (PAHs)	Leaching from linings of water storage tanks and distribution lines	Reproductive difficulties; increased risk of cancer
Carbofuran	Leaching of soil fumigant used on rice and alfalfa	Problems with blood, nervous system, or reproductive system
Carbon tetrachloride	Discharge from chemical plants and other industrial activities	Liver problems; increased risk of cancer
Chlordane	Residue of banned termiticide	Liver or nervous system problems; increased risk of cancer
Chlorobenzene	Discharge from chemical and agricultural chemical factories	Liver or kidney problems
2,4-D	Runoff from herbicide used on row crops	Kidney, liver or adrenal gland problems
Dalapon	Runoff from herbicide used on rights of way	Minor kidney changes
1,2-Dibromo-3-chloropropane (DBCP)	Runoff/leaching from soil fumigant used on soybeans, cotton, pineapples and orchards	Reproductive difficulties; increased risk of cancer
o-Dichlorobenzene	Discharge from industrial chemical factories	Liver, kidney, or circulatory system problems
p-Dichlorobenzene	Discharge from industrial chemical factories	Anaemia; liver, kidney or spleen damage; changes in blood
1,2-Dichloroethane	Discharge from industrial chemical factories	Increased risk of cancer
1,1-Dichloroethylene	Discharge from industrial chemical factories	Liver problems
cis-1,2-Dichloroethylene	Discharge from industrial chemical factories	Liver problems
trans-1,2-Dichloroethylene	Discharge from industrial chemical factories	Liver problems
Dichloromethane	Discharge from drug and chemical factories	Liver problems; increased risk of cancer
1,2-Dichloropropane	Discharge from industrial chemical factories	Increased risk of cancer
Di(2-ethylhexyl) adipate	Discharge from chemical factories	Weight loss, liver problems, or possible reproductive difficulties.
Di(2-ethylhexyl) phthalate	Discharge from rubber and chemical factories	Reproductive difficulties; liver problems; increased risk of cancer
Dinoseb	Runoff from herbicide used on soybeans and vegetables	Reproductive difficulties

(Continued)

Table 6.4 (Continued)

Contaminants	Sources of contaminants	Potential health effects from long-term exposure
Dioxin (2,3,7,8-TCDD)	Emissions from waste incineration and other combustion; discharge from chemical factories	Reproductive difficulties; increased risk of cancer
Diquat	Runoff from herbicide use	Cataracts
Endothall	Runoff from herbicide use	Stomach and intestinal problems
Endrin	Residue of banned insecticides	Liver problems
Epichlorohydrin	Discharge from industrial chemical factories; an impurity of some water-treatment chemicals	Increased cancer risk, and over a long period of time, stomach problems
Ethylbenzene	Discharge from petroleum refineries	Liver or kidneys problems
Ethylene dibromide	Discharge from petroleum refineries	Problems with liver, stomach, reproductive system, or kidneys; increased risk of cancer
Glyphosate	Runoff from herbicide use	Kidney problems; reproductive difficulties
Heptachlor	Residue of banned termiticide	Liver damage; increased risk of cancer
Heptachlor epoxide	Breakdown of heptachlor	Liver damage; increased risk of cancer
Hexachlorobenzene	Discharge from metal refineries and agricultural chemical factories	Liver or kidney problems; reproductive difficulties; increased risk of cancer
Hexachlorocyclopentadiene	Discharge from chemical factories	Kidney or stomach problems
Lindane	Runoff/leaching from insecticide used on cattle, lumber, gardens	Liver or kidney problems
Methoxychlor	Runoff/leaching from insecticide used on fruits, vegetables, alfalfa, livestock	Reproductive difficulties
Oxamyl (Vydate)	Runoff/leaching from insecticide used on apples, potatoes and tomatoes	Slight nervous system effects
Polychlorinated biphenyls (PCBs)	Runoff from landfills; discharge of waste chemicals	Skin changes; thymus gland problems; immune deficiencies; reproductive or nervous system difficulties; increased risk of cancer
Pentachlorophenol	Discharge from wood-preserving factories	Liver or kidney problems; increased cancer risk
Picloram	Herbicide runoff	Liver problems
Simazine	Herbicide runoff	Problems with blood
Styrene	Discharge from rubber and plastic factories; leaching from landfills	Liver, kidney or circulatory system problems
Tetrachloroethylene	Discharge from factories and dry cleaners	Liver problems; increased risk of cancer
Toluene	Discharge from petroleum factories	Nervous system, kidney, or liver problems
Toxaphene	Runoff/leaching from insecticide used on cotton and cattle	Kidney, liver or thyroid problems; increased risk of cancer
2,4,5-TP (Silvex)	Residue of banned herbicide	Liver problems
1,2,4-Trichlorobenzene	Discharge from textile-finishing factories	Changes in adrenal glands
1,1,1-Trichloroethane	Discharge from metal-degreasing sites and other factories	Liver, nervous system or circulatory problems
1,1,2-Trichloroethane	Discharge from industrial chemical factories	Liver, kidney or immune system problems
Trichloroethylene	Discharge from metal-degreasing sites and other factories	Liver problems; increased risk of cancer
Vinyl chloride	Leaching from PVC pipes; discharge from plastic factories	Increased risk of cancer
Xylenes (total)	Discharge from petroleum factories; discharge from chemical factories	Nervous system damage

Table 6.5 Sources and potential hazards of some indicating parameters.

Contaminant	Sources	Potential health and other effects
Biochemical oxygen demand (BOD)	BOD measures the amount of oxygen required or consumed by microbiological organisms while stabilising decomposable organic matter (Sawyer, McCarty and Parkin, 2002). It is an indicator of the health of a body of water and has an application in designing water treatment for plants.	Has no direct health implications, but an important indicator of overall water quality.
Chemical oxygen demand (COD)	This parameter 'is a measure of the capacity of water to consume oxygen during the decomposition of organic matter and the oxidation of inorganic chemicals'. (Gary, 2016).	COD test is done for quantifying oxidisable pollutants in surface water.
Dissolved oxygen (DO)	Dissolved oxygen is vital to aquatic life.	Low DO indicates pollution.

Source: UN Water (2015), EPA Ireland (2011), Sawyer, McCarty and Parkin (1994), Gary (2016).

water-quality classes from A to E and China has classes from I to V as shown in Table 6.7. Category E in the Indian standard and Category IV in the Chinese standard are meant for industrial uses.

Modern water-treatment technology makes it possible to adjust the quality of any raw water to meet the needs of any industrial use (Murali Krishna, 2008). The cost of purification for very low-quality water may be high, which may not be a deterrent, depending on the industry concerned.

Guidelines for industrial water intake are not so restrictive except for the chemical (including pharmaceutical), food and beverage industries. Any single industry may have several different production units that require water of various qualities. In a cement industry, along with a captive power plant, for example, there may be several production units like cement-grinding units, boilers and cooling towers. Water requirements will be for (1) boiler water, (2) cooling water, (3) water for drinking, (4) water for sanitation and housekeeping, (5) water for green belt development and fire fighting. A single water-treatment plant generally satisfies all these uses, but sometimes more than one treatment plant may be necessary. For some purposes like fire fighting and green belt development, treatment may not be required at all. Raw water quality requirement for different industries is given in Table 6.8. (Although the Table lists specifications for Indian industries it is in general agreement followed globally.)

Raw water as mentioned above is not usable unless it is properly treated to achieve the required quality of specific use. In Table 6.9 the water quality requirement (maximum) for point use in different types of industries is given.

6.1.9 Water quality of effluent

In almost all countries, there are standard guidelines on the quality of water discharged by various industries; for example, US-EPA has released effluent

Table 6.6 Drinking water quality guidelines of different countries and international organisations. Values are in mg/l if not otherwise stated.

Parameter	WHO	EU	Australia	US-EPA	Canada	India
				Guideline value		
1,1-Dichloroethylene	0.03	–	–	–	0.014	–
1,2-Dichlorobenzene	1	0.0001	1.5	–	0.2	–
1,2-Dichloroethane	0.03	0.003	–	0	0.005	–
1,4-Dichlorobenzene	0.3	0.0001	0.06	–	0.005	–
2,3,4,6-tetrachlorophenol	–	0.0001	0.04	–	0.1	–
2,4,6-trichlorophenol	0.2	0.0001	–	–	0.005	–
2,4-D	0.03	0.0001	0.03	0.07	–	–
2,4-Dichlorophenol	–	0.0001	–	–	0.9	0.0003
Aldicarb	0.01	0.0001	0.004	–	0.009	–
Aldrin/Dieldrin	0.00003	0.00003	0.0003	–	0.0007	0.00003
Alkalinity	–	–	–	–	–	200
Aluminium (Al)	–	–	–	–	0.03	–
Ammonia	–	–	–	–	0.5	–
Antimony (Sb)	0.02	0.005	0.003	0.006	0.006	–
Arsenic (As)	0.01	0.01	0.01	0	0.01	0.05
Atrazine	0.002	0.0001	0.2	0.003	0.005	0.002
Azinphos-methyl	–	0.0001	0.03	–	0.02	–
Barium (Ba)	0.7	–	2	2	1	0.7
Bendiocarb	–	0.0001	–	–	–	–
Benzene (C$_6$H$_6$)	0.01	0.001	0.001	–	0.005	–
Benzo[a]pyrene	0.0007	0.00001	–	–	0.00001	–
Boron (B)	0.5	1	4	–	5	0.5
Bromate	0.01	0.01	0.01	0	0.01	–
Bromoxynil	–	0.0001	–	–	0.005	–
Cadmium (Cd)	0.003	0.005	0.002	0.005	0.005	0.003
Calcium (Ca)	–	–	–	–	–	75
Carbaryl	–	0.0001	0.03	–	0.09	–
Carbofuran	0.007	0.0001	0.01	0.04	0.09	–
Carbon tetrachloride	0.004	0.0001	0.003	0	0.005	–
Chloramines-total	–	–	–	4	–	–

Chloride (Cl)	—	—	—	—	—	0.3
Chlorpyrifos	0.03	0.0001	—	—	0.09	0.003
Chromium (Cr)	0.05	0.05	0.05	0.1	0.05	0.05
Coliforms, faecal and total	nil	nil	nil	nil	nil	nil
Colour	—	—	—	—	—	5 unit
Copper (Cu)	—	—	2	—	—	0.05
Cyanazine	0.0006	0.0001	—	—	—	—
Cyanide (CN)	0.07	0.05	—	0.2	0.2	0.05
Cyanobacterial toxins	—	—	—	—	0.0015	—
Diazinon	—	0.0001	0.004	—	0.02	—
Dicamba	—	0.0001	0.1	—	0.12	—
Dichloromethane	0.02	—	0.004	0.005	0.05	—
Diclofop-methyl	—	0.0001	—	—	0.009	—
Dimethoate	0.006	0.0001	0.007	—	0.02	—
Dinoseb	—	0.0001	—	0.007	0.01	—
Diquat	—	0.0001	0.007	0.02	0.07	—
Diuron	—	0.0001	0.02	—	0.15	—
Epichlorohydrin (ECH)	—	—	0.0005	0	—	—
Ethylbenzene	0.3	—	0.3	0.7	—	—
Ethylene di bromate	—	—	—	—	—	nil
Faecal streptococci and enterococci	nil	nil	nil	nil	nil	nil
Fluoride (F)	1.5	1.5	1.5	4	1.5	1
Giardia	—	—	1	0	—	—
Glyphosate	—	0.0001	Not required	0.7	0.28	—
Hardness	—	—	—	—	—	200
Iodine	—	—	0.5	—	—	—
Iron (Fe)	0.2	0.2	—	—	—	0.3
Lead (Pb)	0.01	0.01	0.01	0	0.01	0.01
Magnesium (Mg)	—	—	—	—	—	30
Malathion	—	0.0001	—	—	0.19	—
Manganese (Mn)	—	—	0.5	—	—	0.1
Mercury (Hg)	0.001	0.001	0.001	0.002	0.001	0.0019
Methoxychlor	0.02	0.0001	0.3	0.04	0.9	0.001
Methylene Blue	—	—	—	—	—	—

(Continued)

Table 6.6 (Continued)

Parameter	Guideline value					
	WHO	EU	Australia	US-EPA	Canada	India
Metolachlor	0.01	0.0001	0.3	0.04	0.05	–
Metribuzin	–	0.0001	0.07	–	0.08	–
Molybdenum (Mb)	–	–	0.05	–	–	0.07
Monochlorobenzene	–	–	–	–	0.08	–
Nickel (Ni)	11	11	0.02	10	–	0.02
Nitrate (measured as Nitrogen)	–	–	50	–	10	45
Nitrilotriacetic acid	0.2	–0.2	0.2	0.4	–	–
Nitrite (measured as nitrogen)	–	–	0.2	–	–	–
Oxygen, dissolved (DO)	–	–	–	–	–	–
Paraquat	–	0.0001	0.02	–	0.01	–
Parathion	–	0.0001	0.02	–	0.05	–
Pathogenic Staphylococci	nil	–	–	–	–	–
Pentachlorophenol	0.009	0.0001	0.01	0	0.06	–
pH (unitless)	–	–	6.5–8.5	–	–	6.5–8.5
Phorate	–	0.0001	–	–	0.002	0.002
Phosphates (PO_4)	–	–	–	–	–	–
Picloram	–	0.0001	0.3	0.5	0.19	–
Selenium	0.01	–	0.01	0.05	0.01	0.01
Silver (Ag)	–	–	0.1	–	–	0.1
Simazine	0.002	0.0001	0.02	0.004	0.01	–
Sodium	300–600	–	–	–	–	–
Solids, total dissolved	–	–	–	–	–	500
Sulphate (SO_4)	–	–	250	–	200	–
Terbufos	–	0.0001	0.0009	–	0.001	–
Tetrachloroethylene	0.04	0.01	0.05	0.005	0.03	–

Thallium (Tl)	–	–	–	–	–	–
TDS	300–600	–	–	–	500	–
Toluene	0.7	–	0.8	1	–	–
Trichloroethylene	0.07	0.01	C	0.005	0.005	–
Trifluralin	0.02	0.0001	0.09	–	0.045	–
Trihalomethanes	–	0.1	0.25	0.08	0.1	–
Turbidity in NTU	0.015	–	0.017	–	–	1
Uranium	0.0003	0.0005	–	0	0.02	–
Vinyl chloride	0.5	–	0.0003	0	–	–
Xylenes-total	0.5	–	–	10	0.09	–
Zinc	–	–	–	5.0	–	5

Source: US-EPA (2007), Australian Government (2015), Bureau of Indian Standards (2013), Government of Canada and Safe Environments Directorate (2014), WHO (2009), European Union (2014).

Table 6.7 Categorisation of water quality for designated uses in India and China.

India		China	
Class	Criteria	Class	Criteria
A	Drinking water source without conventional treatment but after disinfection.	I	Mainly applicable to the water from sources, and the national nature reserves.
B	Outdoor bathing (organised)	II	Mainly applicable to first class of protected areas for centralised sources of drinking water, the protected areas for rare fish, and the spawning fields of fish and shrimps.
C	Drinking water source after conventional treatment and disinfection	III	Mainly applicable to second class of protected areas for centralised sources of drinking water, protected areas for the common fish and swimming areas
D	Propagation of wildlife and Fisheries	IV	Mainly applicable for industrial use and entertainment which is not directly touched by human bodies.
E	Irrigation, industrial cooling, controlled waste disposal	V	Mainly applicable to the bodies of water for agricultural use and landscape requirement.
Below-E	Not meeting any of the A, B, C, D and E criteria	Nil	

Source: http://cpcb.nic.in/Water_Quality_Criteria.php, the National Standards of the People's Republic of China (Ministry of Health of China, 2007).

guidelines for a number of industries, which are available on the US-EPA website. A comparative table of quality guidelines of seven major countries regarding industrial effluent is given in Table 6.10.

6.2 Industrial water pollution

6.2.1 Definition

There are many ways of defining water pollution. Michael Hogan (2014) defines it as 'the contamination of natural bodies of water by chemical, physical, radioactive or pathogenic microbial substances'. The scale of water pollution at present can be registered from the fact that it is found in *all* aquatic environments, that is, surface water, groundwater, wetlands and the oceans; and, in proportion with the scale of pollution, one can also see its effects in *all* ecosystems.

Industrial water pollution is caused in many ways; disposal of wastewater, drained effluents, waste dumped by industries in landfill sites near a water source (both surface and groundwater) and discharge from mines. The visible cause of industrial water pollution is not merely the toxic emissions, but what it actually reflects is the indifference of industries, society, government, elected representatives and civic bodies.

Table 6.8 Maximum concentration of constituents in raw waters for various industrial operations (mg/l).

Characteristics	Boiler water	Cooling water	Textile plants	Pulp and paper	Chemical industry	Petrochemicals
Silica	150	50	–	80	–	85
Aluminium	3	3	–	–	–	–
Iron	80	14	0.3	2.6	10	15
Manganese	10	2.5	1.0	–	2	–
Calcium	–	500	–	–	250	220
Magnesium	–	–	–	–	100	85
Ammonia	–	–	–	–	–	40
Bicarbonate	600	600	–	–	600	480
Sulphate	1400	680	–	–	850	900
Chloride	19000	600	–	200	500	1600
Nitrate	–	30	–	–	–	8
Dissolved solids	35000	1000	150	1080	2500	3500
Suspended solids	15000	5000	1000	–	10	5000
Hardness	5000	500	120	475	1000	900
Alkalinity	500	500	–	–	500	500
Colour (units)	1200	–	–	300	500	25

Table 6.9 Maximum concentration of constituents at point of use in industrial operation (mg/l).

Characteristics	Boiler water	Cooling water	Textile plants	Pulp and paper	Chemical industry	Petrochemicals
Silica	0.7	50	25	50	–	–
Aluminium	0.01	–	–	–	–	–
Iron	0.05	–	0.3	0.2	0.3	0.3
Manganese	0.01	–	0.05	0.1	0.2	0.05
Calcium	–	200	–	20	150	–
Magnesium	–	–	–	–	–	–
Ammonia	0.1	–	–	–	–	–
Bicarbonate	48	600	–	–	–	–
Sulphate	–	680	100	–	250	500
Chloride	–	600	100	200	2250	500
Dissolved solids	200	1000	200	300	–	–
Suspended solids	0.5	5000	5	–	–	–
Hardness	0.07	850	50	100	40	–
Alkalinity	40	500	200	75	–	85
Colour (units)	–	–	5	25	5	5
Turbidity (units)	–	–	5	40	0	–
M P N	–	–	–	–	1	1

Table 6.10 Maximum concentration of constituents at point of discharge in industrial operation (mg/l).

	Uganda	South Africa	India	China	Brazil	Mexico	Canada
1,1,1, -trichloroethane	3.0	–	–	–	–	–	–
1,1,2.- dichloroethyelene	0.2	–	–	–	–	–	–
1,1, 2,- Trichloroethne	1.06	–	–	–	–	–	–
1,2- Dichloroethane	0.04	–	–	–	–	–	–
1,3- dichloropropene	0.2	–	–	–	–	–	–
Aluminium	0.5	–	–	–	–	–	–
Ammonia Nitrogen	10	–	50	15	–	–	–
Aniline	–	–	–	1.0	–	–	–
Arsenic	0.2	0.1	–	–	1.5	0.75	1
Barium	10	–	–	–	–	–	–
Benzene	0.2	–	–	0.1	–	–	–
BOD5	50	–	30	20	–	200	300
Boron	5	5	2.0	–	–	–	–
Cadmium	0.1	0.05	2.0	–	1.5	0.75	0.7
Calcium	100	–	–	–	–	–	–
Carbon tetrachloride	–	–	–	0.03	–	–	–
Chloride	500	–	1000	–	–	–	–
Chlorine (total residual)	–	–	1.0	–	–	–	–
Chlorine	1.0	nil	–	–	–	–	–
Chlorobenzene	–	–	–	0.2	–	–	–
Chromium (total)	1.0	–	2.0	–	–	–	4
Chromium (VI)	0.05	0.05	0.1	–	0.5	0.75	–
Cobalt	–	–	–	–	–	–	–
COD	100	30	250	100	–	–	–
Coifform Organisms counts/100	10,000	nil	–	–	–	–	–
Color TCU	300	–	–	50	–	–	–
Copper	1.0	0.02	3.0	0.5	1.5	15	2
Cyanide	0.1	–	0.2	0.5	0.2	1.5	–
Detergents	10	–	–	–	–	–	213

(*Continued*)

Table 6.10 (Continued)

	Uganda	South Africa	India	China	Brazil	Mexico	Canada
Dichloromethane	0.2	–	–	–	–	–	–
DO	–	75	–	–	–	–	–
Ethylobenzene	–	–	–	0.4	–	–	–
Fluoride	–	1.0	2.0	10	10	–	10
Petroleum hydrocarbon	–	–	–	5.0	–	–	–
Iron	10	0.3	–	–	15	–	–
Lead	0.1	0.1	0.1	–	1.5	1.5	1
Magnesium	100	–	–	–	–	–	–
Manganese	1.0	–	–	2.0	–	–	–
Mercury	0.01	0.02	0.01	0.1	1.5	0.015	0.01
Methylebenzene	–	–	–	–	–	–	–
Nickel	1.0	1.0	3.0	0.1	2	6	2
Nitrate - N	2.0	10	–	–	–	–	–
Nitrogen total	10	–	–	–	–	–	–
Nitrobenzene	–	–	–	2.0	–	–	–
Oil and grease	10	–	10	–	150	75	150
Paraxylene	–	–	–	4.0	–	–	–
Pesticide	–	–	nil	Nil	–	–	–
pH (no unit)	6–8	5.5–9	5.5–9	6–9	6–10	5.5–10	–
Phenols	0.2	0.01	1.0	0.5	5	–	–
Phosphate (total)	10	–	–	–	–	–	–
Phosphate (soluble)	5.0	–	5.0	0.5	–	–	–
Phosphorus	–	–	–	–	–	20	10
Selenium	1.0	0.05	0.05	–	–	–	–
Silver	0.5	–	–	–	–	–	–
Sulphate	500	–	1000	–	–	–	–
Sulphide	1.0	–	2.0	1.0	1	–	–
TDS	1200	–	–	–	–	–	–

Temperature °C	20–35	35	40	—	40	<40	—
Tetra Cholera ethylene	0.1	—	—	0.1	—	—	—
Tetrachloromethananc	0.02	—	—	—	—	—	—
Tin	5	—	—	—	—	—	—
Total Suspended Solids	100	90	—	70	200	200	350
Tricholoroethylene	0.3	—	—	—	—	—	—
Turbidity NTU	300	—	—	—	—	—	—
Zinc	5.0	0.3	5.0	2.0	5	9	2

Sourced from the National Environment (standards for discharge of effluent into water or on land) Regulations, (2008), General and special standards (2009), Wastewater discharge standards in Latin America (2016), China Water risk (1996), Central Pollution Control Board (2011), US-EPA (2016).

The main challenge to water stewardship is control and management of emissions and effluents.

6.2.2   Direct reasons of water pollution

1   **Lack of awareness:** Lack of awareness and inclination of industries to protect the environment is one of the main causes of dumping the waste into natural systems.

2   **Tendency to reduce cost in waste management:** Some industries cut costs in solid-waste and effluent management to increase profit. This callous indifference towards nature comes from the profiteering tendency of industries, especially small establishments.

3   **Unplanned growth:** In old industrial hubs, many small industries had grown up within a small area, long before public awakening to environmental pollution. In such areas of unplanned growth, during earlier development, there is not enough space available for the construction of effluent treatment plants (ETP).

4   **Technological drawbacks:** Technologically backward industries sometimes neither upgrade their old and energy-intensive production machineries, nor their old and inefficient ETP installations, thus contributing heavily to environmental pollution.

5   **Lack of maintenance:** Due to lack of proper maintenance of water-treatment plants and ETP, pipelines and conduits result in increased discharge pollutants into the environment.

6   **Policy and enforcement:** Some countries lack properly defined water policies, while others have well defined water policies but no enforcement machinery; both problems amount to the same thing—water pollution.

7   **Improper corporate policy:** The corporate policy itself may not include a waste-disposal policy and therefore management may be uninformed about waste disposal. As a rule, absence of corporate policy and management for implementation is the defining part of the industrial pollution scenario.

6.2.3   Indirect reasons of pollution

The title of this sub-section is a misnomer in the sense that the pollution caused is direct but it results from ignorance and causes beyond control.

1   **Lack of scientific knowledge:** Environmental pollution as an unintended consequence is generally negative and results from systems being more complex and interconnected than the industry realises (Dictionary of Sustainable Management, 2016). Industries need to educate themselves about:
   • Discharging into the recharging zone of an aquifer system;
   • Dumping waste in the riparian zone; and
   • Landfills in an unconfined groundwater system

2   **Natural disaster and system breakdown:** Natural disasters like earthquakes or tsunamis can cause widespread water pollution through destruction of plants and factories.

The industrial pollution scenario under various conditions is shown in Figure 6.1

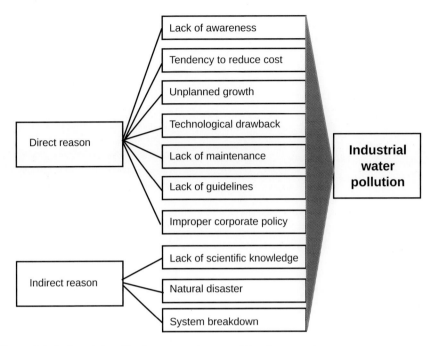

Figure 6.1 Industrial pollution scenarios resulting from various reasons.

6.2.4 Indicators of industrial water pollution

Nature by itself, in spite of being an enormously complex multi-component system, remains in balance till disturbed. Water pollution causes imbalance in the ecosystem. There are many ways in which an environmental indicator can be defined, and there are several indicators to assess how an ecosystem is functioning. I have followed the definition by the Department of Environment and Heritage Protection, Queensland Government (2016) which states, 'Environmental indicators [are] … physical, chemical, biological or socio-economic measures that best represent the key elements of a complex ecosystem or environmental issue. An indicator is embedded in a well-developed interpretative framework and has meaning beyond the measure it represents. For an indicator to be effective it must provide a true measure of a component of the ecosystem. Selection of effective indicators is best achieved by developing conceptual models of the ecosystem and using these to pinpoint indicators that provide the required information.' Listed below are the chemical indicators of water pollution. See Table 6.11.

6.2.5 Socio economic indicator of water pollution

The effects of water pollution can be studied in abstraction, but their effects are real and manifest in loss of sanitation, loss of septic facilities, lack of potable water, health degradation, epidemics, loss of ecological services and above all loss of livelihood. These effects often result in migration of

Table 6.11 Water pollution parameters and indicators.

Category	Indicator	Parameter	Indications
I	Temperature (T) and dissolved oxygen, and oxygen demand	DO, COD, BOD	Rise in temperature lowers DO level in water. Discharge from cooling towers increases water temperature in the environment and lowers the DO level, affecting the metabolic process of aquatic life. Normal DO level in natural water is 8 mg/l at 30°C. High DO indicates good health of the ecosystem. Industrial waste discharge increases COD and BOD and decreases DO of water required for ecosystem.
II	Conventional variables (pH, total dissolved solids, conductivity and suspended sediment)	pH, TDS, conductivity, SS	These are physico-chemical indicators. Elevated concentrations of these indicators are clear pointers to pollution. Industrial waste and effluent discharge increase the concentration of both TDS and SS. pH may either increase or decrease beyond the permissible limit. (6.5– 8.5)
III	Nutrients	N, P	Compounds of nitrogen and phosphorus are main indicators nutrients in water and are found as soluble and as well as suspended particles. Industrial wastes from fertilisers and, petrochemical plants may increase nutrients that result in eutrophication.
IV	Metals	As, Fe, Cr, Hg, Al, etc.	Metals are released into water both naturally and from anthropogenic activities. The elevated concentration of metals in water may indicate anthropogenic water pollution if the body of water is linked with industrial or urban a wastewater disposal system.
V	Hydrocarbons	Polycyclic aromatic hydrocarbons (PAHs) and naphthenic acids	These two hydrocarbons are common indicators of pollution and are toxic if present in high concentrations. The sources of these pollutants are petrochemical industries, oil sand mines and chemical and photographic industries.
VI	Industrial chemicals	PCBs and dioxins/ furans	Industrial chemicals include carbon-based and synthetic chemicals used in agro-industries, pharmaceutical, chemical and other industries. Some of these industrial products are potentially toxic, non-biodegradable (PCBs) and clear indicators of industrial pollution.

Adapted from Queensland Government (Department of Environment and Heritage Protection), 2016 (under CC BY 3.0 licence) and RAMP (no date).

people away from polluted areas. Shanthi and Gajendran (2009) have shown that industrial water pollution caused ecological plunder accompanied by the loss of livelihood for the fishermen in Ennore creek, India. Likewise, Guimarães *et al.* (2012) linked deteriorating water quality to reduced tourism around the Guadiana River in Spain.

6.2.6 *Biological indicators of water pollution*

Biological indicators are direct measurements of population and the health of fauna and flora in a body of water. Commonly used biological indicators are macro-invertebrates, fish diversity, benthic algal growth and benthic

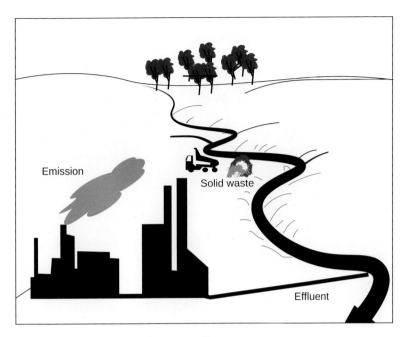

Figure 6.2 Sources of industrial water pollution.

oxygen demand. For estuaries, biological indicators are less developed. The only commonly used biological indicator in estuaries is chlorophyll-a, which is a measure of phytoplankton population density; in coastal embayments, sea-grass condition and the condition of fringing coral reefs are used as indicators. Factors other than water quality can also be an indicator of degraded aquatic ecosystems, such as habitat degradation and changes to natural flow patterns. Therefore, it is vital to monitor biological indicators to assess water pollution. (Adapted from the Queensland Government/ Department of Environment and Heritage Protection, 2016).

6.2.7   Industrial sources of pollution

Waste generation is a phenomenon that exists within the industrial process itself. Industrial waste exists in all its three forms, that is solid waste, effluent water and emitted gases, each of which is a major cause of water pollution. See Figure 6.2.

6.2.8   Water pollution from industrial emission

Industries emit particulate matters that go into the atmosphere and come down with rain. Volatile organic compounds (VOC) and Hg are also atmospheric pollutants emitted by industries. Oil refineries and industries that burn fossil fuels emit carbon dioxide and oxides of sulphur and nitrogen.

These gases react with atmospheric water turning it acidic, which comes down as acid rain creating widespread corrosion, and lowering the pH of bodies of water; two strongly acid-forming gases are sulphur dioxide and oxides of nitrogen (Science Daily, 2006). Sulphur dioxide in its gas phase reacts with hydroxyl ion and forms $HOSO_2$,

$$SO_2 + OH^+ \rightarrow HOSO_2$$

which is followed by:

$$HOSO_2^+ + O_2 \rightarrow HO_2 \cdot + SO_3$$

Sulphur trioxide in the presence of water forms sulphuric acid.

$$SO_3(g) + H_2O(l) \rightarrow H(aq)$$

Nitrogen dioxide reacts with OH to form nitric acid

$$NO_2 + OH^+ \rightarrow HNO_3$$

H_2SO_4 and HNO_3 are the two most common chemicals in acid rain. What makes atmospheric pollution an international issue is the fact, that the emitted gases may travel far from the source and can cause acid rain in neighbouring countries. See Figure 6.3.

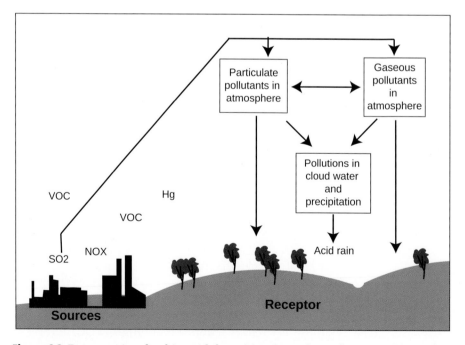

Figure 6.3 Processes involved in acid deposition (note that only SO_2 and NO_x play a significant role in acid rain). Adapted from Wikimedia Commons.

6.2.9   *Water pollution from industrial effluent*

Industrial effluents contain many toxic substances like heavy metals, oil, grease, toxic chemicals, gases, pesticides, and so on. When these pollutants come in contact with the natural ecosystem through bodies of water, such as the sea, streams or groundwater, the natural water becomes contaminated. When the level of contamination rises above the permissible limit, it renders the ecosystem unsuitable for aquatic biodiversity or for human consumption. Cause and effect of water pollution from different industries is shown in Table 6.12.

Table 6.12 Direct water pollution from industrial liquid wastes (effluent).

Sector	Cause	Pollutants
Iron and steel	Fugitive emission from handling, crushing, loading and unloading, and smelting. Major polluting processes are pig iron and steel manufacturing.	High organic carbon, suspended solids, dissolved solids; cyanide fluoride, COD, zinc, lead, chromium, cadmium, zinc, fluoride, and oil and grease.
Textile and leather	Polyvinyl chloride to size fabrics, chlorine bleach, benzidine and toluidine as dyeing agents, formaldehyde, lead and mercury (Rogers, 2016); Chromium for tanning.	BOD, solids, sulphates, chromium.
Pulp and paper	Pulping and bleaching are major sources of pollutants. Paper machine water flows, fibre and liquor spill.	Volatile organic compounds (VOCs) such as terpenes, alcohols, phenols, methanol, acetone, chloroform, methyl ethyl ketone; detergents and surfactants; dyes and pigments; acids; and alkaline solutions (Frost and Sullivan, 2006).
Petrochemicals and refineries	Process wastewater, tankage wastewater, cooling tower blow-down, ballast water tank flow.	Total suspended solids, heavy metals, NH_3, H_2S, trace organics, BOD5, COD., oil, phenolic compounds
Chemical and pharmaceutical	Chemicals for washing and cleaning of floors and chemical by-products.	Pesticides, VOC, arsenic, cadmium, cyanide, mercury, chromium and lead, organic chemicals (Bahadori, 2004).
Nonferrous metals	Handling and re-use of scraps, furnace emissions, primary gas cleaning, effluent plant, coke oven by-products.	Oils and grease, cyanides, suspended solids.
Microelectronics	Soldering.	COD and organic chemicals.
Mining	Acid mine-drainage, heavy-metal contamination and leaching, processing chemicals, erosion and sedimentation (Buccini, 2004).	From common metals like iron, aluminium to cadmium, selenium, etc.
Food and beverage	Grey water from washing of raw food, cooking, additives, colouring, preservatives and decomposed food wastes.	BOD, total suspended solids (TSS), excessive nutrient loading (nitrogen and phosphorus compounds), pathogenic organisms, and residual chlorine and pesticide levels.
Infrastructure	Construction and demolition of infrastructure.	Tar, suspended solids, oil.
Thermal energy	Fossil-fuel burning.	Fly ash, heavy metals, thermal pollution, and acid rain.

6.2.10   Water pollution from solid-waste disposal

Industrial solid waste is defined as the waste in solid form, generated from industrial activities including manufacturing processes, providing service and establishment maintenance. Wastes are classified as hazardous or non-hazardous depending on their effects on environment and human health. Wastes like construction material, metal scrap and so on physically affect the drainage system and land use pattern without causing chemical pollution, but cause hazard by choking drainage.

Government of UK (2015) issued directives against hazardous waste disposal (produced or handled in the business) ensuring that it causes no harm or damage to the environment or people. As per this directive, industries must now comply with the statutory requirements, depending on whether the business is a waste producer (produce or store waste) or holder (collect and transport waste), carrier or consignee (receive waste, e.g. for recycling or disposal). In Table 6.13, a brief overview of different types of wastes generated by industries and their toxic effects are given. As per these guidelines waste is generally considered hazardous if it (or the material or substances it contains) is harmful to humans or the environment.

6.2.11   Impacts of mining on water quality

Mining activities include (1) mining and excavation, (2) mineral processing, (3) disposal of mining wastes, (4) mine dewatering, (5) post-mining flooding and (6) uncontrolled discharge of polluted waters (Younger and Wolkersdorfer, no date). Water pollution from mining is caused by:

- Siltation of sediments loosened by excavation;
- Mobilising waters of poor quality; and
- Aggravated erosion in open cast mines.

Different industries pollute water in different ways, and the contaminants released are industry-specific. The process of waste disposal of two major industries is discussed below.

6.2.12   Water pollution potentiality in petrochemical and power industry

Petrochemical industries generate wastewater from washing of raw crude, product washing, cooling water and other operations. The pollution stream in a petrochemical industry is shown in Table 6.14.

Thermal power plants account for more than half of all the toxic pollutants: notably, mercury, arsenic, lead, and selenium, which are discharged into surface bodies of water of the United States (Godwin, 2016). The pollution stream in a thermal power plant is shown in Figure 6.4.

Table 6.13 Different types of waste generated by the industries and their toxic potentialities.

Solid waste	Industries	Effect on water sources
Asbestos	Infrastructure, construction industry, automobile industry, asbestos-sheet factories, asbestos mines.	Not a water pollutant directly.
Empty chemical containers	Chemical, pharmaceutical, fertilisers.	Hazardous.
PCB contaminated wastes	Plastic factory.	Polychlorinated biphenyls (PCBs) consist of 209 chlorinated hydrocarbons, banned since 1979.
Foundry wastes	Metal fabrication industries.	Non-recyclable wastes sometimes contain heavy metals and may cause water pollution.
Fly ash	Thermal power plant.	Heavy metals.
Paint residue, filters and dust	Paint manufacturing industry, painting service industry.	Oil and pigment, change colour of water, formation of film on water.
Sludge	Sludges are by-products of a variety of wastewater treatment processes or commercial operations, such as the production of paper or lime, or milling of lumber.	May be hazardous if containing toxic materials.
Tyre and rubber goods	Commonest solid waste in the sense that all industries use automobiles, especially transport and aviation industries.	Non-pollutant but hazardous
Spent carbon filters	Water industry.	Non pollutant but hazardous.
Contaminated soil	Originating from various industries like petrochemicals, automobile, infrastructure, and so on.	Pollution by benzene, toluene, ethyl benzene, xylene, (BTEX), and total petroleum hydrocarbons (TPH), gasoline.
Medical waste	Health care industry products, cytotoxic and cytostatic medicines, infectious clinical waste; yellow bag.	Hazardous waste; contaminates water with pathogen.
Machining wastes	Wastes in the form of chips and expended cutting tools (Agrawal and Khare, 2012).	Not essentially toxic, but often may contain heavy metals that may leach into groundwater.
Electrical component wastes	Electronic waste contains cathode-ray tube (CRT), circuit boards, and plastics etc. Very common solid waste from industry and commercial establishments.	Arsenic, barium, beryllium, cadmium, CFCs, chromium, dioxins, lead, mercury, polychlorinated biphenyls (PCBs), Polyvinyl chloride (PVC), selenium (EMPA, no date)
Industrial non-recyclable plastics	Any industry, plastic industry, water industry.	Hazardous waste causes widespread pollution if disposed in watercourses.

Adapted from Agrawal and Khare (2012) and EMPA (2009).

6.2.13 Groundwater pollution from industrial effluents and leachates

Industries are the greatest source of groundwater contamination (Welsch and Lieber, 1955). Groundwater is contaminated by unattended and unmanaged solid-waste dumps and effluent water discharged directly into aquifers or

Table 6.14 Major wastewater streams in a petrochemical industry.

Wastewater	Description
Desalter water	Water produced from washing the raw crude prior to topping operations.
Sour water	Wastewater from steam stripping and fractionating operations that comes into contact with the crude being processed.
Other process water	Wastewater from product washing, catalyst regeneration and dehydrogenation reactions.
Spent caustic	Formed in extraction of acidic compounds from product streams.
Tank bottoms	Bottom sediment, and water settles to the bottom of tanks used to store raw crude. The bottoms are periodically removed.
Cooling tower	Once-through cooling tower water and cooling tower blow-down to prevent build-up of dissolved solids in closed-loop cooling systems.
Condensate blow-down	Blow-down from boilers and steam generators to control build-up of dissolved solids.
Source water-treatment system	Source water must be treated prior to use in the refinery. Waste streams may include water from sludge dewatering if lime softening is used; ion-exchange regeneration water; or reverse osmosis wastewater.
Storm water	Process area and non-process area runoff from storm events.
Ballast water	Waste water from product tankers.

Source: US-EPA.

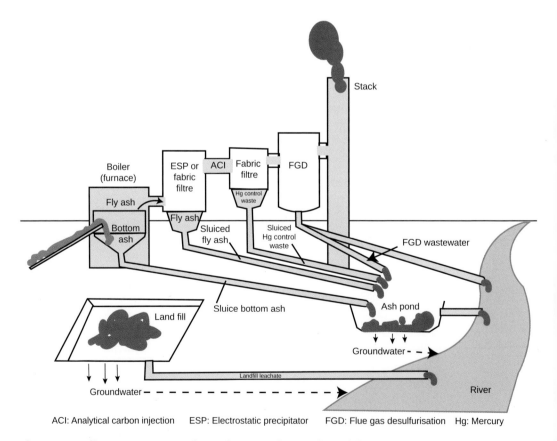

ACI: Analytical carbon injection   ESP: Electrostatic precipitator   FGD: Flue gas desulfurisation   Hg: Mercury

Figure 6.4 Pollution stream in a thermal power plant. Adapted from US-EPA.

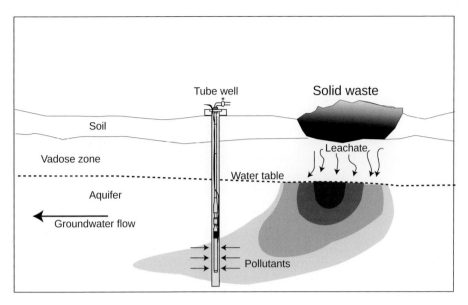

Figure 6.5 Groundwater contamination due to leaching from landfill. Adapted from Waterloo Hydrogeologic (2015).

in the recharge zone of aquifers. Types and sources of groundwater contamination are as follows.

Effluent

Effluent released in the environment may percolate underground and reach the aquifer. This is possible if the effluent is discharged into a body of water or a river which has hydraulic connection with the aquifer.

Leachates from landfills

Landfill is the cheapest and, therefore, the commonest method of solid-waste disposal. Each year trillions of tonnes of solid waste are disposed of in landfills. Although not all wastes are hazardous, there are numerous examples of toxic wastes disposed through landfills, which contaminate groundwater by leaching; this toxic liquid contaminating groundwater is called leachate (UK Groundwater Forum, 2004). See Figure 6.5.

Deep injection wells

Deep injection wells are one of the ways of disposing of wastewater. In Pennsylvania, there are such wells meant for disposal of fluids from petroleum mining, including fracking water; these wells are regulated by the Underground Injection Control Program of the US-EPA, but they are strongly opposed by environmental activists as being a source of long-term peril, especially because they function at depths not covered by regular monitoring programmes.

Mine drainage

Groundwater contamination takes place from an endless list of anthropogenic activities, among which mining is the major culprit that drains wastewater directly or indirectly into surface streams and bodies of water (Saracino *et al.*, 2002). This polluted surface water slowly percolates down to the aquifer and pollutes groundwater. Some hazardous phenomena related to mine drainage are given below. Adapted from Safe Drinking Water Forum (2008).

- **Acid mine drainage:** In sulphide rock mining, when sulphide minerals are exposed to air and water, sulphuric acid is produced. Generally, the acid is disposed into nearby streams, rivers, lakes and depressions that severely damage the aquifer water quality by percolation.
- **Heavy metal contamination and leaching:** This contamination takes place in addition to the contamination by acidification. It is caused not just by the heavy metal mining but also from a variety of ores in which heavy metals occur in trace amounts; leaching from a static source leads to enrichment of heavy metals underground, which then become a major pollutant even though the original source contains only a trace amount.
- **Processing chemicals:** Toxic chemicals (e.g. cyanides and sulphuric acid) are often used to separate ores from host rock in ore-beneficiation plants. These chemicals often spill or leak, and mix with flowing water, eventually leading to the contamination of groundwater.

Terrestrial surface water pollution from industrial solid wastes and effluents

Rivers are arguably the worst victim of water pollution because most industries are located near riverbanks. Lakes and wetlands also fall prey to solid-waste dumping and effluent discharge. When the load of pollutants exceeds the natural remediation capacity of lakes or wetlands, the ecosystems show clear signs of degradation.

Oceanic pollution from industrial solid wastes and effluents

One assumes that oceans, the largest of bodies of water occupying three-quarters of the earth' surface, are too large to be polluted. And the assumption was perhaps justified till the early decades of the twentieth century when the first major assault on the oceanic system happened during massive industrialisation triggered by World War I. Since then, as we know from intensive media coverage available in the third millennium, the variety and intensity of marine pollution have increased manifold. Today, the causes of ocean pollution are (a) oil spills from off-shore drilling and shipwreck, (b) land runoff carrying pollutants, (c) ocean mining, (d) discharge of toxic chemicals from industries and (d) littering of solid discharge and wastes (Rinkesh, no date).

6.2.14 *Water pollution identifiers*

It needs recognition that 'indicators' and 'consequences' of pollution merit separate discussions. The foremost consequence of water pollution is the severe damage to the ecosystem. Any change in the physical or chemical property of natural water affects aquatic biodiversity and the usability of water. There are ecological, social and economic identifiers.

Ecological identifier

Water pollution damages the ecosystem in many ways, affecting not only aquatic species but also wild animals and livestock. One of the commonest effects of pollution is eutrophication caused by the discharge of excessive nutrients (phosphate and nitrate) into water, which in turn causes algal bloom, choking higher aquatic life forms. Pesticides damage the reproductive systems of aquatic life. Increases in BOD and COD cause changes in the aquatic environment that affect the natural restoration ability of the ecosystem. Bio-magnification is a process in which substances like pesticides or heavy metals move up in the food chain and pass on to the higher animals (NCS Pearson, 2016). Chemicals that have a longer half-life undergo bio-accumulation (US-EPA, 2011) in the bodies of aquatic plants and animals, and reach higher toxic levels. On entering the food chain through the affected species, the effect of bio-accumulated toxins gets progressively magnified in the ecosystem; at the top of the food pyramid it eventually becomes a serious health hazard for humans. In essence, water-quality deterioration affects the quality of the ecosystem. See Figure 6.6.

Figure 6.6 Algal blooms in a eutrophic body of water.

Health effect identifier

The most hazardous as well as the most common chemical pollutants in water are arsenic and fluoride: it needs pointing out in this context that these two pollutants cause chronic health problems like fluorosis and arsenicosis. In addition to arsenic and fluorine, there are many other elements that enter into bodies of water from industrial effluents; the heavy metals such as lead, mercury, cadmium and chromium, are relatively more common. Toxicity of heavy metals varies with the element type and concentration in water (see Table 6.9 for quality guidelines on industrial effluents).

Economic performance identifier

According to the US-EPA (2016), the economic effects of water pollution are as follows:

- **Drinking water cost:** Nitrates and algal bloom in drinking water sources can drastically increase treatment costs. Nitrate-removal systems in Minnesota caused supply costs to rise from 5–10 cents per 1000 gallons to over USD 4.0 per 1000 gallons. Above all, the cleaning up of polluted bodies of water, a periodic and vital necessity, costs billions of dollars; every dollar spent on preventing pollution ultimately saves water-treatment costs.
- **Tourism losses:** Bodies of water, lakes and rivers are tourist attractions. A polluted tourist spot loses its attraction and consequently loses out on the revenue earned from tourism. In the United States, the tourism industry loses close to USD one billion each year because of nutrient pollution and harmful algal blooms in bodies of water.
- **Commercial fishing and shellfish losses:** Algal blooms and chemical toxins can kill fish and contaminate shellfish. This not only reduces the production and profit in fisheries but also inflicts consequent losses on all downstream industries in the value chain.
- **Real estate losses:** The mere presence of clean water in the vicinity, such as waterfront property, can increase value in the real estate market by at least 25%. Value of waterfront property, in short, is very high and much sought after by the rich. But it can suffer serious depreciation in value when the body of water gets polluted. A part of the East Calcutta wetland (locally known as Dhapa) was once a municipal waste dumping ground, as a result of which the real estate value of the area was extremely low, even though the area fit the description of 'waterfront property'. After relocation of the dumping site elsewhere and cleaning up of the wetland (which is now declared a Ramsar site), real estate values multiplied many times.

Water pollution affects industries in various ways. Some case studies are listed in Box 6.1.

6.2.15 Management and control of water pollution

Pollution management comes under two categories: wastewater management and solid-waste management.

Consequences of Water Pollution

BOX 6.1

1 **South African mines:** South Africa is rich in mineral resources, and hundreds of mines have been developed over the last couple of centuries. South Africa has strong environmental laws to regulate mining pollution. But recent studies have detected (Crowley, 2014) 'high quantities of uranium, arsenic, sulphuric acid and other toxic materials in streams and rivers' of South Africa around the gold mines. The presence of toxic substances in water is also reported from the coal mines of Mpumalanga Province. The Federation for a Sustainable Environment, a South African NGO, reports that huge piles of uranium containing wastes from gold mines are dumped near the residential areas of Johannesburg. The organisation has made criminal charges against the mining companies and also against the authorities for criminal noncompliance. A probe is in progress. In another incident, a coal mining company has been fined for noncompliance of environmental acts (bloomberg.com, 2016)

2 **Indian pharmaceutical companies:** Water pollution has many adverse effects on the economy. Investors may take strong exception to companies causing pollution. A recent study by a nordic investor company NORDIA stated that investors are concerned about 'water pollution caused by drug manufacturers in India' (Marriage and Grene, 2015); twelve pharmaceutical manufacturing companies in Andhra Pradesh are under the threat of closure by the order of the Pollution Control Board. These companies supply chemicals for large drug manufacturing companies throughout the world, and their closure will impact the production of large multinational pharmaceutical companies sourcing raw material from Andhra Pradesh.

3 **Health hazards in India:** Water pollution directly causes external costs to society through health risks; in their study from Karnataka, Madhusudan, Sharif and Krishnadas (2013) have shown that the impact is not only visible in the expenditure for health care but also in widespread migration of people from the polluted site causing loss of manpower to the industries.

4 **Economic loss in China:** Wen *et al.* (2013) has reported high discharge of COD by paper and pulp industries in China causing the government to take stringent action. As a consequence, many small industries faced closure. Guang (2016) estimated that the national economic loss due to water pollution in 1998 was 98.61 billion Yuan against industrial production of 2894.165 billion Yuan.

The first step in controlling water pollution is to take up a pollution management programme *from the industrial side*, failing which there are mitigation policies, regulations, guidelines and institutional frameworks from the government side to incentivise (e.g. tax relief) or compel the industries to undertake pollution management. Additionally, civil society can compel both industry and the government to work for a clean environment by forming pressure groups (e.g. Green Peace and WWF); these pressure groups derive power from the public by sensitising them to the issues at hand. In many western countries, industries are bound by law to declare their pollution prevention programmes, and periodic submission of environmental compliance reports is also mandatory. China, despite having regulatory laws in place, has not yet been able to control the enormous industrial pollution released into its rivers and bodies of water. In developing countries, the picture is rather gloomy; in India, the environmental laws are adequate, but enforcement mechanisms need strength.

Issues in pollution control

There are several issues in waste-water management and pollution control which are not only industry-specific but also country-specific. The issues are:

- Countries need to keep an account of polluting industries and maintain a register of activities related to effluent discharge and map those polluting activities under different categories, e.g. 'good', 'bad' and 'severe'.
- Identification of obsolete technology and policies that are still in practice.
- One of the most glaring issues is, of course, venal practices and noncompliance.

The above issues are alarming and require the government to take suitable measures. Some of these measures are:

- Total bans on discharging highly toxic elements (e.g. arsenic, cadmium and mercury) along with strong vigilance on the quality of discharged water.
- Impose economic pressure on the polluting industries. In the UK, industrial effluents discharged into public sewers, by agreement, are subject to a financial charge according to a formula that estimates the cost of collection and treatment; practices such as these receive public support because the expense for treatment is not transmitted to the tax payer.
- In the United States, some states have adopted the policy that 'polluters should pay'. The European Union Urban Wastewater Treatment Directive (Woodford, 2015) states that 'the *polluter-pays* has become a guiding principle among member countries', thus defining the obligation of industries to the environment and the society.
- There are multiple pressures and calls on national finances. In the past, wastewater management and water quality were not seen as a priority. Since the Rio Summit in 1982, however, it is incumbent upon national economies to factor in 'wastewater management' in budget policy.
- In many countries, the mandatory concept of 'polluters should pay' cannot be implemented because wastewater management services are not valued; or even if they are valued, they are under-priced. Implementation of the 'polluters should pay' policy requires evaluation of water-treatment services for cost recovery.
- Recognition of wastewater management and its critical role in sustainable development.
- And most importantly, stimulating political will.

Development of a public policy

Agenda 21 of the Rio Summit and the subsequent Dublin Declaration agreed that water is a finite resource and the private sector has a role in its management. Countries abiding by Agenda 21 have already framed environmental laws and guidelines for their industries. The consequent public policy framed in different countries also enumerates separate controlling mechanisms for the effluent and wastewater quality that can be discharged, there being separate guidelines for discharging into the sea, surface bodies of water and groundwater. Recalling, from Article 26 of Agenda 21, the issue of 'accountability of … private companies', it is the responsibility of governance to ensure that industries also comply with the statutes in this regard. All countries that are signatories in Agenda 21 are required to frame a public policy on water pollution and related issues. In order to lend teeth to the public policies thus framed, signatory countries have drawn up legal frameworks on ambient water quality, standards of industrial discharge, prohibited areas (e.g. Ramsar sites, Biodiversity Reserves, Heritage Sites, etc.), water-sampling and testing protocols, licensing procedures, fees and taxes; accordingly, wastewater taxes or pollution taxes are levied from industries to generate revenue for further investment in remediation processes in highly polluted areas (FAOLEX 2004).

Development of strategic action plan

Industries within the regulatory framework, functioning proactively with the government or otherwise, can develop a strategic action plan to control water pollution. The action plan includes one or more sets of goals to be accomplished within the time frame defined in the action plan. What the action plan needs to focus on is the formation of institutions and the identification of persons who will be accountable for time-bound implementation of the plan. It goes without saying that effectiveness of the action plan will depend upon a continuous information stream feeding the plan and, most importantly, adequate funds. For example, the strategic action plan to control water pollution adopted in China has identified four activities under 'voluntary corporate environmental behaviour'; namely, '(1) public access to environmental information, (2) public participation in environmental planning and decision making, (3) access to justice in environmental matters and (4) corporate environmental responsibility' (China Development Gateway, 2015).

The other important goals that the action plan needs to focus on for effectiveness are:

- Monitoring discharged water for visible improvements;
- Expansion plans are based on a conceptual framework, considering the carrying capacity of the river basin and catchment as primary hydrological units;
- Development of a standardised framework for water sampling, analyses and preparation of water-quality maps (Figure 6.7);
- Controlling human and industrial activities to protect the water environment (e.g. recharging area);
- Case-specific technological innovations supported by both government and educational institutions;
- Regulating use of water by industries through a rational pricing mechanism;
- Rationalising fees for wastewater treatment, disposal and water withdrawal (groundwater or surface water);

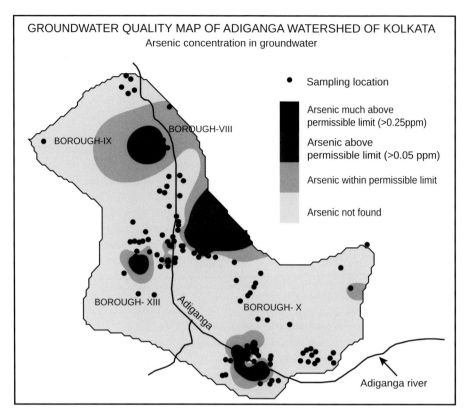

Figure 6.7 Water quality map.

- Incentivising schemes for industries for improving their pollution-control facilities;
- Institutional framework for delivering a hassle-free service to industries;
- Division of responsibilities among stakeholders;
- Strengthening public participation and social supervision through NGOs, CBOs and knowledge groups; Organising meetings and conferences to share water information with the public;
- Taking up emergency action plans as and when necessary;
- Ensuring availability of basin-wise assessment reports for programme improvement as and when necessary;
- Identify the best management practice suitable for respective industries to manage and control effluents, emission and solid-waste disposal;
- Conducting regular water audit; and
- Awareness generation and capacity building of staff and stakeholders.

6.2.16 Wastewater management

Industry generates a substantial proportion of total wastewater as sewage water and industrial effluents (Oyebode, 2015), resulting from manufacture, trade or business, or from development of natural resources, or from livestock, such as in the dairy and poultry industries. The discharged water includes contaminated storm water and leachate from solid-waste facilities. Industry is both morally and legally duty bound to release wastewater only after proper treatment. Efficient wastewater management in some countries is incentivised to encourage environment-friendly waste disposal. Investments in wastewater management generates additional benefits (Corcoran *et al.*, 2010) if the principles of (1) reducing wastewater, (2) treating before discharge, (3) reuse, and (4) recycling are adopted. This ultimately reduces the cost of pre-processing water sources and a reduction of the volume of intake water. The most cost-effective solutions are to prevent toxic substances from entering into the discharged wastewater stream and adopting a zero water discharge or a closed system.

Wastewater includes grey water, industrial wastewater and storm-water runoff. These three categories of wastewater should be managed separately. Though industrial pollution is basically point-source pollution, there are cases of diffused pollution from agriculture-based industries and illegal discharges from tankers[1] (UN-Water, 2015). Industrial pollution from point sources can be contained, controlled and treated. On the other hand, small chemical industries, metal-finishing and garment factories can cause substantial damage to the environment as they do not have any identifiable discharge system. It is estimated that '70% of industrial discharges in developing countries is dumped untreated' (Thomas, 2008). Different countries have different waste-disposal policies. There are four types of industrial effluents to be considered (Alturkmani, no date):

1 **General manufacturing effluents:** Effluents generated from the industrial process may be regular, intermittent or seasonal. In agro-based industries, generation of seasonal effluent is common. In the pharmaceutical and synthetic chemical industries, the pollution potential of effluents varies widely, depending on the specific process.
2 **Specific effluents:** Some effluents such as heavy metals from electronics or electroplating industry need special treatment.
3 **General service effluents:** These are basically grey water produced from canteen wash and nontoxic wastewater from boilers and cooling towers.
4 **Effluent overloads:** Excessive loads of effluent in the effluent-treatment plant, effluent ponds and storage tanks (called lagoons) beyond their carrying capacity often cause serious management problems. It is manifested by excessive sludge, high ammonia, high BOD and several other indicators.

6.2.17 Disposal of wastewater

As mentioned earlier, industrial pollution is generally considered point-source pollution. According to the U.S. Environmental Protection Agency (US-EPA), point-source pollutions are those which originate from a single point. For example, an industry that discharges its wastewater through a dedicated discharge outlet is readily identifiable and therefore, manageable. The volume of pollution caused by an industry depends on how the industry manages its waste or effluents prior to its disposal into the environment. Industries employ various methods to dispose of wastewater:

- **Direct disposal:** This is the commonest way of polluting the environment.
- **Disposal after treatment:** Regulatory bodies need to ensure that appropriate care is taken to maintain the standards of purification of industrial wastes.
- **Disposal through the urban sewage stream:** Industries avoid investments in water purification plants and dispose of wastewater and effluents into the urban sewage system. This adds to the environmental hazard because urban sewage systems generally do not have the capacity to purify industrial waste.
- **Disposal through storm-water drains:** Investment is also avoided, and environmental hazard is increased by the disposal of untreated wastes into storm drains. Since storm-water drains are not used as sewage, they remain dry most of the time, thus affording opportunity to unscrupulous industries to discharge their waste into storm drainage.
- **Disposal into common water-treatment plants:** A common water-treatment plant is a best practice for industrial environmental management. These plants are designed for distributing the capital cost burden of installing effluent treatment plants and are located in large industrial estates or in special economic zones (SEZ), where small, large and medium industries are clustered. In such cases, instead of having their own wastewater management installation, industries outsource effluent treatment to the common water-treatment plants.
- **Disposal into engineered wetlands:** One of the most popular ways of wastewater disposal is constructed or engineered wetlands. These wetlands perform chemical and biological processes of removing pollutants. As water flows through the wetland, it slows down, and insoluble pollutants become trapped by vegetation; soluble pollutants are also taken up by plants and are rendered inactive. Wetland plants also foster the necessary conditions conducive to growth of microorganisms. Through a series of complex processes, these microorganisms also transform and remove pollutants from the water. For example, wetland microbes can convert organic nitrogen into useable, inorganic forms (NO_3 and NH_4) that are necessary for plant growth. The by-product of this microbial process escapes into the atmosphere as methane.[2] Aquatic vegetation plays an important role in phosphorus removal and, if harvested, extends the life span of the system by postponing phosphorus saturation (Guntensbergen, Stearns and Kadlec, 1989). See Figure 6.8.

  Constructed wetlands are built outside floodplains or flood ways in order to avoid damage to natural wetlands and other aquatic resources. Wetlands are frequently constructed by excavating, backfilling, grading, diking and installing water-control structures to establish desired hydraulic flow patterns. If the site has highly permeable soil, an impervious layer is constructed at the floor of the wetland. Wetland vegetation is then planted or allowed to grow naturally.

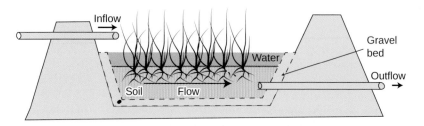

Figure 6.8 Constructed wetland.

6.2.18   Effluent treatment

Effluent treatment is both the legal and ethical responsibility of all industries (Jorgensen, 1979). In some developing countries, large volumes of industrial effluent are discharged into the sewage network, never reach a treatment plant and are eventually discharged into streams or the sea. This practice may not be illegal in some countries that lack legal guidelines, but it is certainly unethical. Eminent scholars have developed and recommended several methods of effluent treatment. This chapter highlights only the commonly adopted practices.

The main objective of effluent treatment is abatement of environmental pollution and recovery of useful materials from the wastewater. The main objectives of treatment are:

- Removal of solids;
- Removal of grease and oil;
- Removal of biodegradable organics;
- Treatment of other organics like pesticides, paints and so on;
- Treatment of acid and alkali required for neutralising wastewater; and
- Removal of toxic materials.

Effluent treatment is classified as primary, secondary and tertiary according to the grade of purification. In primary treatment, a portion of the suspended solids from the wastewater is removed; in secondary treatment, the organic matter and the residual suspended solids are removed; and in tertiary treatment, also called advanced treatment, it is ensured that the TDS of the discharged water is reduced to less than 20 mg/l and BOD is reduced to not more than 30 mg/l.

6.2.19   Treatment methods

Treatment methods are classified as *unit operation* and *unit process* (Murali Krishna, 2008). Unit operations involved in raw water treatment and wastewater treatment are shown in Table 6.15.

Unit process: Unit processes are either chemical or biological in nature. Some unit processes are described in Table 6.16.

Though there are several effluent treatment processes and methods, not all are skilfully applied by various industries. Some industry specific effluent treatment processes are shown in Table 6.17.

Effluent treatment in an industry is carried out through an effluent treatment plant (ETP) in which water-treatment processes are involved and carried out through unit operations. Before installation of an ETP the following points should be considered:

- The guidelines and standards of the country to be complied with;
- Volume of effluent;
- Contaminants and chemical quality of the effluent;
- Concentration of environmental parameters like pH, BOD, COD, and so on;
- Future plan of increase in production;
- The amount of money to be invested;
- ETP experts to be engaged; and
- Design of the ETP and its compatibility with the industry.

Table 6.15 Unit operations involved in water and wastewater treatment.

Unit Operation	Method and purpose
Equalisation	When quality and quantity of wastewater vary hourly, the water is stored in a temporary tank for equalisation.
Filtration	This is an operation of mechanical straining through a sieve (opening size varying from a molecular size to as large as 10 cm). These are used to remove suspended solids and even bacteria.
Solid removal	• **Sedimentation and settlement:** Particles sufficiently large and heavy can be allowed to settle down by detaining the wastewater in a tank for 2 to 6 hours. This method does not apply to very small particles. This is currently the most common method of reducing the suspended solids content of leachate. If the particle sizes are colloidal, it may be necessary to add a flocculent. • **Sand filtration:** Occasionally used if the solids are very fine or colloidal. Sand filtration has a high initial capital cost and requires a high degree of control. • **Dissolved air flotation:** This is sometimes used when available land does not allow the construction of settlement tanks. Leachate usually requires conditioning prior to treatment, and there are high capital costs associated with this method of treatment.
Coagulation and flocculation	Addition of aluminium sulphate up to 30 mg per litre can coagulate colloidal particles and settle them. Some dissolved substances like iron, fluoride, manganese and phosphorus are removed by this process after a pre-treatment process like oxidation and addition of suitable chemicals.
Floatation and skimming	Large floating particles which are lighter than water are removed by skimming; the method is also used to separate oil from wastewater.
Centrifugation	By swirling water in a tank, suspended particles are forced out towards the outer periphery of the tank and then removed.
Solvent extraction	This method is used to extract a particular substance from a mixture of two or more substances using a solvent in which the extractable substance is soluble and others are not.
Air stripping	In this method dissolved gases are separated from the wastewater. A stream of air is blown through the liquid for the expulsion of dissolved gases. There are two popular methods: 1 **Methane stripping:** the use of diffused air to strip out or reduce the dissolved methane content of leachate is commonly used. 2 **Ammoniacal-N removal:** depending on pH and temperature, to be effective it may be necessary to raise the pH and heat the leachate to remove ammonia. 3 **Stripping of other volatile contaminants:** dependent on the contaminants present and is unlikely to remove all contaminants completely
Evaporation	In this operation, dissolved solids are separated by drying the wastewater in the sun or blowing dry air.
Fractional distillation	In this process, two liquids with different boiling points are separated.
Adsorption	Adsorbents are used for removal of foul taste, odour, gases and toxic substances.
Ion exchange	Anions and cations are trapped by an adsorbent in this process and are separated from wastewater. The adsorbent used is a resin, natural or synthetic, comprising of giant molecules, called ion exchangers. There are separate exchangers for cations and anions.
Reverse osmosis	Osmosis is a flow of a solvent through a semi-permeable membrane from lesser to higher concentration. Reverse osmosis is just the opposite process where pressure is applied on the liquid of higher concentration to pass through the membrane and solvents of lower concentration are extracted.
Electrodialysis	Ions of dissociated salts can be filtered through molecular sieves by using an electrical field as the propelling force. The molecular sieve is impermeable to larger molecules.
Activated carbon adsorption	Many industrial wastes contain refractory organics, which are difficult or even impossible to remove by the conventional biological treatment. These are removed by adsorption on activated carbon or synthetic active-solid surface.

Adapted from Murali Krishna (2008) and Munter, (2004).

Table 6.16 Unit processes.

Unit process	Method and purpose
Neutralisation	Industrial wastes that are either acidic or alkaline are corrosive and have damaging impacts on the environment. Such wastes are neutralised before release. They are treated with alkali or acid to bring the pH within the permissible limit. In such cases, acid is added to alkaline wastewater and vice versa.
Precipitation	This is a method of separating some soluble chemicals in industrial waste by treating them with suitable chemicals. These chemicals, when added to the wastewater, cause precipitation of some dissolved salts. The precipitates are then separated by filtration. By this process, harmful chemicals are discarded, and useful materials are recovered.
Oxidation and reduction	Oxidation and reduction are the processes by which oxygen is added or removed in a chemical process to precipitate various chemicals. .
Biological oxidation	Nontoxic organic matter in sewage can be fed to bacteria of suitable type in a *reactor*, so as to convert the organic waste to carbon dioxide, water, methane, mercaptans, and so on, which escape into the atmosphere. There are two types of biological oxidation, aerobic and anaerobic. In aerobic systems, contact with oxygen is required for the survival of bacteria. The aerobic process is conducted in aerated lagoons. In the anaerobic process, the bacteria can live without oxygen. This treatment is required for wastewater having high BOD/COD ratio.
Chlorination	In this process, chlorine (Cl_2) or bleaching powder is added to water for killing certain types of bacteria.

Adapted from Murali Krishna (2008)

Table 6.17 Methods of wastewater treatment for different industries.

Industry	Methods of treatment
Acid/alkali	Neutralisation and equalisation
Coal washing	Floatation and settlement
Cyanide	Alkaline chlorination
Dairy and distillery	Aerobic biological oxidation
Sugar	Lagooning and aerobic oxidation
Slaughterhouse	Anaerobic biological oxidation
Metal finishing	Chemical precipitation, ion exchange
Paper/rubber	Biological treatment
Plating	Oxidation-reduction, precipitation, neutralisation
Steel	Neutralisation, coagulation
Textile	Neutralisation, precipitation, biological treatment
Detergent	Floatation, precipitation
Explosives/insecticides	Adsorption, chlorination at high pH.
Petrochemical	Physicochemical treatment, biological treatment, ion exchange, membrane treatment, electrocoagulation
Paper and pulp	Electrocoagulation

Figure 6.9 An effluent treatment plant. Photo courtesy Sanjay Patel.

A typical ETP has the following components:

- **Collection tank:** Waste and industrial effluent from different sectors are collected here.
- **Mixing and cooling:** Effluent cooled and thoroughly mixed to have a homogeneous chemical and physical quality.
- **Neutralisation:** PH is controlled at this stage.
- **Chemical coagulation:** $Fe_2(SO_4)_3$, $Al_2(SO_4)_3$ etc. are used for coagulation.
- **Biological oxidation tank:** Blowing air helps to allow micro-organisms to act on the waste.
- **Sedimentation and separation of sludge:** The blanket of precipitations is skimmed off to another tank, and remaining solution is removed to pressure filter.
- **Sludge chamber:** Sludge is collected in this chamber for drying. After that, it is handed over to recyclers.
- **Filtration:** The filtration layer consists of sand and rock that can filter wet sludge to extract water from it.
- **Discharge to drain:** The treated water is released in to the environment after checking the chemical quality.

An industrial ETP is shown in Figure 6.9.

6.2.20   Solid-waste management

Solid wastes from industries can be hazardous as listed in Table 6.13. The best way to manage solid waste is to reduce it. Care should be taken

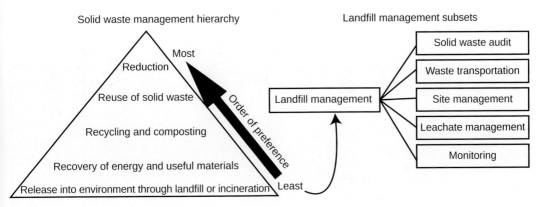

Figure 6.10 Solid-waste management hierarchy and land fill management subsets.

for hazardous solid wastes so that they do not contaminate soil and water. In Figure 6.10 the solid-waste management hierarchy is illustrated.

Landfill management measures

Schmoll, Howard, and Chilton, (2006) have suggested several management measures which fall broadly into the following categories:

1 **Site selection:** Site selection of a proposed landfill is fundamental to preventing water pollution. Broad aspects of site selection are:
   - **Legal:** The site should be cleared by competent authorities to avoid subsequent litigation; authorities are also responsible for pre-assessing the probable contamination of water resources.
   - **Natural disaster:** It must be ensured that the site is not disaster prone, especially in the context of transportation of contaminants by flood.
   - **Geographical and geological considerations:** The geological profile of the area should be verified so that shallow aquifers do not become vulnerable to pollution (by excavation, backfilling, pipeline installation, paving, building construction, etc.). Also to be taken into consideration are the issues of human settlements and their socio-economic connection with the shared water source.
2 **Monitoring and assessment:** The plant should install instruments and develop procedures to monitor contamination. Some of the important measures are:
   - Installation of a network of water quality monitoring stations for remedial actions when necessary.
   - Following installation of monitoring stations, installation of automatic or semi-automatic water-quality monitoring devices to generate a continuous temporal and spatial database.
   - A chemical laboratory within the premises equipped with precision instruments where water samples can be analysed for hazardous chemicals and heavy metals. If the industry cannot afford a dedicated laboratory, it can get samples tested in recognised laboratories.
3 **Engineering measures:** Engineering measures are highly recommended to prevent pollution. There is a wide range of construction designs that may form effective defences against water pollution. Some of the engineering measures are:
   - The contaminated liquid that leaches from the solid wastes is called a leachate. Leachates go down to the aquifer and contaminate groundwater. To prevent leaching waste, dumping ponds are constructed

Figure 6.11 Leachate evaporation pond in a landfill site located in Cancún, Mexico. Source: Victor787. https://commons.wikimedia.org/wiki/File:Leachate_ Pond.JPG. Used under CC-BY 3.0 https://creativecommons.org/licenses/by/3.0/ deed.en.

which have an impermeable lining at the base. There are several types of clay and synthetic lining materials that can prevent leaching to the groundwater. A leachate tank with polythene lining is shown in Figure 6.11.
- Appropriate alarm and monitoring systems can be installed to prevent overflow and spills, and vehicle movement should be efficiently managed to avoid accidents.
- Appropriate technology should be applied to prevent hazardous wastes and chemicals from being released into the environment. Neutralisation, solidifying and encapsulating are some of the helpful processes.
- Hazardous chemicals and wastes should be protected from rain or weather extremes like high wind, because this may lead to leaching. This can be prevented by the construction of appropriate canopy over stored materials.

4 **Management of pollution through operational procedures**
- Hazardous chemicals and materials should be handled carefully and stored in such places that they do not come into accidental contact with each other and undergo chemical reaction.
- Combustible materials should be placed in isolation.
- Prepare an operational procedure for the staff and train them how to handle hazardous materials.
- Label each material with appropriate caution signs.

6.2.21 *Management of leachate*

Solid wastes become hazardous pollutants when they are disposed of in open fields or are dumped as landfills. Leachate is the by-product of landfill waste.

Leachate treatment

Leachate treatment is essential for protecting natural water resources from pollution generated from landfills. There are a number of widely adopted processes used for treatment of leachate, either alone or in combination. The processes described in Table 6.16, 6.17 and 6.18 can be applied to leachate treatment also.

6.3 Conclusion

Knowledge of water quality parameters and quality requirements of industrial water is very important for decision making in the industrial sector. The water and beverage industries need to be vigilant over the quality of intake water as well as purified potable water. Water discharged from industries in the form of effluent is a matter of environmental concern. Government and international lobbies on the environment are very strict about maintaining the quality of industrial discharge, and industries need to follow the guidelines set by governments to avoid legal consequences.

Solid-waste disposed of by an industry is also an environmental hazard. Leachates from solid waste can contaminate both surface and groundwater resources. Leachate management is also an important aspect of industrial water management.

Notes

1. Some industries dispose of their waste water in tankers which often leak during transport and cause hazards.
2. Constructed wetlands are sources of methane, which is of environmental concern due to its potent global warming capacity. Methane emission can be managed by removing sludge and biomass regularly. Methane emission is also seasonal and increases in warm weather.

Bibliography

Agrawal G and Khare M K (2012) 'Material and EnergyWastes Minimization in a Machining System: A Review'. *Environ. Sci.* 4(2) 251–256.

Alturkmani, A. (no date) *Industrial wastewater.* Available at http://www.4enveng.com/userimages/INDUSTRIAL%20WASTEWATER.pdf (accessed: 26 May 2016).

Australian Government (2015) *Australian drinking water guidelines 6 2011 national water quality management strategy*. Available at https://www.nhmrc.gov.au/_files_nhmrc/file/publications/eh52_australian_drinking_water_guidelines_151013.pdf (accessed: 28 May 2016).

Bahadori, A. (2013) 'Water pollution control', in *Pollution Control in Oil, Gas and Chemical Plants*. Springer Science + Business Media, pp. 119–165.

bloomberg.com (2016) *WHO gathers hair to probe uranium, Johannesburg gold mines*. Available at http://www.fse.org.za/index.php/item/495-who-gathers-hair-to-probe-uranium-johannesburg-gold-mines (accessed: 11 April 2016).

Buccini, J. (2004) *The global pursuit of the sound management of chemicals*, pp. 1–67. Available at http://siteresources.worldbank.org/INTPOPS/Publications/20486416/GlobalPursuitOfSoundManagementOfChemicals2004Pages1To67.pdf (accessed: 5 October 2016).

Bureau of Indian Standards (2013) *IS 10500 (2012): Drinking water*. Available at http://cgwb.gov.in/Documents/WQ-standards.pdf (accessed: 28 May 2016).

Burr, E. (2015) Fukushima: Uranium and plutonium contamination of large areas of oceans, groundwater, soils. Available at http://www.globalresearch.ca/fukushima-uranium-and-plutonium-contamination-of-large-areas-of-oceans-groundwater-soils/5439271 (accessed: 28 May 2016).

Central pollution control board (2007) *Environmental standards, water quality criteria*. Available at http://cpcb.nic.in/Water_Quality_Criteria.php (accessed: 28 May 2016).

Chatterjee, A., Das, D. and Chakraborti, D. (1993) 'A study of groundwater contamination by arsenic in the residential area of Behala, Calcutta due to industrial pollution', *Environmental pollution (Barking, Essex : 1987)*, 80(1): 57–65.

Chemwiki (2013) *Unusual properties of water*. Available at http://chemwiki.ucdavis.edu/physical_chemistry/physical_properties_of_matter/bulk_properties/unusual_properties_of_water (accessed: 28 May 2016).

China Development Gateway (2015) *Voluntary corporate environmental behavior (China)—China Development gateway—sharing information, knowledge and tools*. Available at http://en.chinagate.cn/archives/egp/2015-07/28/content_36167887_2.htm (accessed: 11 April 2016).

China Water risk (1996) *Maximum allowable discharge concentrations for other pollutants in China*. Available at http://chinawaterrisk.org/wp-content/uploads/2011/05/Maximum-Allowable-Discharge-Concentrations-For-Other-Pollutants-in-China.pdf (accessed: 19 October 2016).

Corcoran, E., Nellemann, C., Baker, E., Bos, R., Osborn, D. and Savelli, H. (2010) *Sick water? The central role of wastewater management in sustainable development a rapid response assessment*. Available at http://www.unwater.org/downloads/sickwater_unep_unh.pdf (accessed: 26 March 2017).

Crowley, K. (2014) *Water pollution near mines prompts south African probe*. Available at http://www.bloomberg.com/news/articles/2014-05-21/water-pollution-near-mines-prompts-south-african-ombudsman-probe (accessed: 7 March 2016).

Dictionary of Sustainable management (2016) '*Unintended consequences.* 'Available at http://www.sustainabilitydictionary.com/unintended-consequences/ (accessed: 22 May 2016).

EMPA (2009) *Hazardous substances in e-waste*. Available at http://ewasteguide.info/hazardous-substances (accessed: 14 May 2016).

EPA Ireland (2001) *Environmental protection agency*. Available at https://www.epa.ie/pubs/advice/water/quality/Water_Quality.pdf (accessed: 28 May 2016).

European Union (2014) *SI122 2014*. Available at https://www.fsai.ie/uploadedFiles/Legislation/Food_Legisation_Links/Water/SI122_2014.pdf (accessed: 28 May 2016).

FAOLEX (2004) *Turkey: Water pollution control and regulation.* Available at http://faolex.fao.org/docs/pdf/tur13466E.pdf (accessed: 11 February 2016).

Fondriest (2015) *Conductivity, salinity & total dissolved solids—environmental measurement systems.* Available at http://www.fondriest.com/environmental-measurements/parameters/water-quality/conductivity-salinity-tds (accessed: 28 May 2016).

Ford, D.L., Englande, A., Eckenfelder, W.W., Engl, A.J. and Jr., W. Wesley Eckenfelder (2008) *Industrial water quality.* 4th edn. The McGraw-Hill Companies.

Frost & Sullivan (2006) *Pollution from the paper and pulp industry: Are we feeding the intellect at the expense of nature?* Available at http://www.frost.com/sublib/display-market-insight-top.do?id=83462501 (accessed: 5 October 2016).

Gary (2016) *Chemical oxygen demand.* Available at http://science.jrank.org/pages/1388/chemical-oxygen-demand.html#ixzz3xz4t2ndo (accessed: 7 March 2016).

General and special standards (2009) Available at https://webcache.googleusercontent.com/search?q=cache:MNsusBiCc4MJ:https://www.dwa.gov.za/Dir_WQM/docs/Leg_General_and_Special_Standards.pdf+&cd=1&hl=en&ct=clnk&gl=in (accessed: 19 October 2016).

Godwin, A. (2016) *Final action imminent on steam electric power generation ELGs.* Available at http://www.waterworld.com/articles/iww/print/volume-15/issue-4/departments/viewpoint/final-action-imminent-on-steam-electric-power-generation-elgs.html (accessed: 14 May 2016).

Government of Canada and Safe Environments Directorate (2014) *Guidelines for Canadian drinking water quality—summary table [Health Canada, 2012].* Available at http://www.hc-sc.gc.ca/ewh-semt/pubs/water-eau/sum_guide-res_recom/index-eng.php (accessed: 28 May 2016).

Guang, X. (2016) *An estimate of the economic consequences of environmental pollution in China.* Available at http://www.homerdixon.com/projects/state/chinaeco/pollut.htm (accessed: 11 April 2016).

Guimarães, M. E., Mascarenhas, A., Sousa, A. C., Boski, T. and Ponce Dentinho T. (2012). 'The impact of water quality changes on the socio-economic system of the Guadiana Estuary: an assessment of management options.' *Ecology and Society* 17(3): 38.

Guntensbergen, G.R., Stearns, F. and Kadlec, J.A. (1989): Wetland vegetation. In Hammer, D.A., ed. (1989): Constructed wetlands for wastewater treatment. Chelsea, MN: Lewis Publishers, 73–88.

HM Government of UK (2007) *Guidance for the treatment of landfill leachate.* Available at https://www.gov.uk/government/uploads/system/uploads/attachment_data/file/322411/Guidance_for_the_Treatment_of_Landfill_Leachate_part_1.pdf (accessed: 29 April 2016).

Jorgensen, S.E. (1979) *Industrial wastewater management.* Amsterdam: Elsevier Scientific Publishing Company.

Madhusudan, P., Sharif, M. and Krishnadas, M. (2016) 'Economic impacts of water pollution on human health and migration in Nanjangud industrial area of Karnataka state', *Environment & Ecology*, 31(2), 518–523.

Marriage, M. and Grene, S. (2015) *Pharmaceutical pollution in India is bitter pill for Nordea.* Available at http://www.ft.com/cms/s/0/b0121282-1679-11e5-b07f-00144feabdc0.html#axzz45UDIwkXD (accessed: 11 March 2016).

Ministry of Health of China (2006) National Standards of the People's Republic of China. Available at http://www.iwa-network.org/filemanager-uploads/WQ_Compendium/Database/Selected_guidelines/016.pdf (accessed 3 July 2016).

Munter, R. (2004) *Pathways of industrial effluents treatment.* Available at http://www.balticuniv.uu.se/index.php/component/docman/doc_download/

286-water-use-and-management-19-industrial-wastewater-treatment (accessed: 28 May 2016).

Murali Krishna, K.V.S.G. (2008) *Rural, municipal and industrial water resource management*. New Delhi: Reem Publishers.

The national environment (standards for discharge of effluent into water or on land) regulations, S (2008) Available at http://www.nemaug.org/regulations/effluent_discharge_regulations.pdf (accessed: 19 October 2016).

NCS Pearson (2016) *Effects on ecosystem*. Available at http://biology.tutorvista.com/environmental-pollution/effects-of-water-pollution.html (accessed: 5 October 2016).

Ncube, E.J. and Schutte, C. (2005) 'The occurrence of fluoride in South African groundwater: A water quality and health problem', *Water SA*, 31(1).

Oyebode, O. (2015) 'Effective management of wastewater for environment, health and wealth in Nigeria', *International Journal of Scientific & Engineering Research*, 6(7): 1028–1059.

Queensland Government (2016) *Water quality indicators*. Available at https://www.ehp.qld.gov.au/water/monitoring/assessment/water_quality_indicators.html (accessed: 5 October 2016).

Regional Aquatic Monitoring Programme (no date), Water Quality Indicators. Available at http://www.ramp-alberta.org/river/water+sediment+quality/chemical.aspx (accessed 21 February 2016).

Rinkesh (no date) *What is ocean pollution?* Available at http://www.conserve-energy-future.com/causes-and-effects-of-ocean-pollution.php. (accessed 12 May 2016).

Rogers, K. (2016) 'What kinds of pollution do textile factories give off?', *Small Business*. Available at http://smallbusiness.chron.com/kinds-pollution-textile-factories-give-off-77282.html (accessed: 5 October 2016).

Safe Drinking Water Forum (2008) *Mining and water pollution*. Available at http://www.safewater.org/PDFS/resourcesknowthefacts/Mining+and+Water+Pollution.pdf (accessed: 8 March 2016).

Saracino, A., Phipps, H., Sacramento, S.-K., Calif and Harter, T. (2002) *Groundwater contaminants and Contaminant sources*. Available at http://groundwater.ucdavis.edu/files/136257.pdf (accessed: 8 March 2016).

Sawyer, C.A., McCarty, P.L. and Parkin, G.F (1994) *Chemistry for environmental engineering and science*. Available at https://4lfonsina.files.wordpress.com/2012/11/chemistry_for_environmental_engineering_and_science.pdf (accessed: 7 February 2016).

Sawyer, C.N., McCarty, P.L. and Parkin, G.F. (2002) *Chemistry for environmental engineering and science*. Boston: McGraw-Hill Higher Education.

Schmoll, O., Howard, G. and Chilton, J. (eds.) (2006) *Protecting groundwater for health: Managing the quality of drinking-water sources*.1st edn. United Kingdom: Intl Water Assn.

Science daily (2006) *Clean air act reduces acid rain in eastern United States*. Available at http://www.sciencedaily.com/releases/1998/09/980928072644.htm (accessed: 5 February 2016)

Sengupta, P.K. (2014), The Chronicle of Arsenic, *Dream*, 2047, 17,1.

Shanthi, L.V. and Gajendran, N. (2013) 'The impact of water pollution on the Socio-economic status of the stakeholders of Ennore creek, bay of Bengal (India): Part I', *Indian Journal of Science and Technology*, 2(3): 66–79.

Thomas, A. (2008) *Message 4: Sanitation helps the environment*. Available at http://www.unwater.org/wwd08/docs/sanitation-environment.pdf (accessed: 11 April 2016).

UK Groundwater Forum (2004), *The risk from industrial chemicals* (2004) Available at http://www.groundwateruk.org/downloads/industrial_and_urban_pollution_of_groundwater.pdf (accessed: 8 March 2016).

UN Water (2015) *Wastewater management A UN-water analytical brief analytical brief.* Available at http://www.unwater.org/fileadmin/user_upload/unwater_new/docs/UN-Water_Analytical_Brief_Wastewater_Management.pdf (accessed: 11 April 2016).

US-EPA (2000) *Technology transfer network air toxics web site.* Available at http://www3.epa.gov/ttn/atw/hlthef/cadmium.html (accessed: 28 May 2016).

US-EPA (2007) *Drinking water standards and health advisories table.* Available at https://www3.epa.gov/region9/water/drinking/files/dwsha_0607.pdf (accessed: 28 May 2016).

US-EPA (2011) *Bioaccumulation/biomagnification effects.* Available at https://www.epa.gov/sites/production/files/documents/bioaccumulationbiomagnificationeffects.pdf (accessed: 5 October 2016).

US-EPA (2016) *Table of regulated drinking water contaminants.* Available at http://www.epa.gov/your-drinking-water/table-regulated-drinking-water-contaminants (accessed: 28 May 2016).

USGS (2016a) *Contaminants found in groundwater, USGS water science school.* Available at http://water.usgs.gov/edu/groundwater-contaminants.html (accessed: 28 May 2016).

USGS (2016b) *PH: Water properties, from the USGS water-science school.* Available at http://water.usgs.gov/edu/ph.html (accessed: 28 May 2016).

USGS (2016c) *Hardness in water, USGS water science school.* Available at http://water.usgs.gov/edu/hardness.html (accessed: 28 May 2016).

USGS (2016d) *Physical water properties.* Available at http://water.usgs.gov/edu/waterproperties.html (accessed: 28 May 2016).

Wastewater discharge standards in Latin America (2016) in *Wikipedia.* Available at https://en.wikipedia.org/wiki/Wastewater_discharge_standards_in_Latin_America (accessed: 19 October 2016).

Waterloo Hydrogeologic (2015) *Contaminant transport from landfill sites.* Available at http://www.waterloohydrogeologic.com/resources/visual-modflow-flex-resources/modeling-groundwater-contamination-from-landfill-sites (accessed: 22 April 2016).

Welsch, F.W. and Lieber, M. (1955) 'Groundwater pollution from industrial wastes [with Discussion]', *Sewage and Industrial Wastes*, 27(9): 1065–1072. doi: 10.2307/25032871.

Wen, C., Yinhua, M., Mingyong, L. and Xiujian, P. (2013) *Economic and environmental effects of water pollution abatement policy.* Available at https://www.gtap.agecon.purdue.edu/resources/download/6457.pdf (accessed: 15 June 2016).

Woodford, C. (2015) *Water pollution: An introduction to causes, effects, solutions.* Available at http://www.explainthatstuff.com/waterpollution.html (accessed: 28 May 2016).

WHO (2004) *Uranium in drinking-water background document for development of WHO guidelines for drinking-water quality.* Available at http://www.who.int/water_sanitation_health/dwq/chemicals/en/uranium.pdf (accessed: 28 May 2016).

WHO (2009) *Guidelines for drinking-water quality.* 3rd edn., incorporating the first and second addenda. Available at http://www.who.int/water_sanitation_health/dwq/fulltext.pdf (accessed: 28 May 2016).

WHO (2011) *Cadmium in drinking-water background document for development of WHO guidelines for drinking-water quality.* Available at http://www.who.int/water_sanitation_health/dwq/chemicals/cadmium.pdf (accessed: 28 May 2016).

7

Water Abstraction, Purification and Distribution

7.1 Overview

It is time for big companies to ensure a fair distribution of water in most countries. This is crucial to sustainable business practices in general, and for water-related industries in particular. There are five stages in water distribution process in an industrial environment; a single industry may require one or more of the five stages. The stages are:

1 **Water sourcing:** Water sourcing deals with the obtaining of water from (a) natural sources, i.e., surface water and groundwater, or (b) agencies that sell water, e.g. municipalities. According to FICCI (2011), surface water is the major source of water for Indian industries (41%) followed by groundwater (35%) and municipal water (24%).
2 **Conveyance:** Conveyance is about designing connection systems that link water sources to users. A water conveyance system may be an open canal or drain or pipelines of various diameters depending on the required flow of water.
3 **Water treatment:** Treatment involves making raw water qualitatively suitable for (a) human consumption, and (b) industrial use. The treatment processes are specific to the type of use.
4 **Water storage:** In large industries with townships, water is stored in underground sumps and in overhead tanks; sump size and location, height and size of overhead tanks are designed in accordance with the quantified water requirement at various user points. Water storage systems are also called 'service reservoirs'.
5 **Water delivery:** Effluent water from industries is required by law to be treated before it is discharged. Delivery of discharged water covers two broad aspects: (a) recycling one part of the water for use, and (b) diverting the other part for replenishing natural sources. Figure 7.1 shows the cycle of water withdrawal and use.

Industrial Water Resource Management: Challenges and Opportunities for Corporate Water Stewardship, First Edition. Pradip K. Sengupta.
© 2018 John Wiley & Sons Ltd. Published 2018 by John Wiley & Sons Ltd.

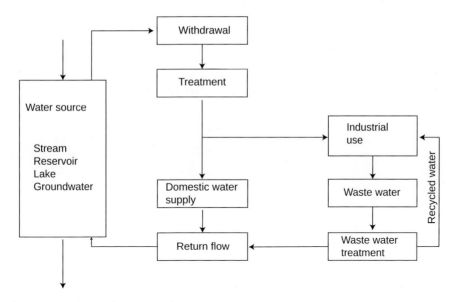

Figure 7.1 Industrial water cycle.

7.2   *Water sourcing by industries*

There is no such source of water which can be specifically earmarked for industries. Industries can use any source of water if sufficient water is available and the water regulations allow the industries to do so. The major sources from which an industry generally draws water are described in Table 7.1.

Considerations for locating and planning water intake are discussed below:

1 **Availability:** Before planning for water abstraction, its availability at the source should be ascertained. If the source is a river, the assessment is done through long-term data collection on river-flow to quantify the response of the river to seasonal changes. For assessing groundwater availability, pre- and post-monsoon water-level data are collected and processed for quantification of utilisable resource. If the source is a lake, then lake hydrology and bathymetric studies are conducted. Detailed methods of source-water estimation are discussed in Chapter 8.
2 **Licence:** Generally, a licence is required from an appropriate authority to withdraw groundwater or surface water. Groundwater licences are awarded by groundwater authorities while surface water licences are awarded by environment agencies or river basin authorities of the country. The licensing authority examines the purpose and evaluates the possible impact on the environment. If it is found that the withdrawal may adversely impact the ecosystem, the authority may impose certain conditions or prescribe measures to prevent environmental degradation. The authority may impose site-specific operating rules for managing abstractions (Environment Agency, Government of UK, 2013). For example, in a water-stressed condition, the authority may ask for detailed water recycling plans before granting

Table 7.1 Sources of water and methods of abstraction.

Sources		Methods of abstraction
Surface water	Stream, lakes, ponds, canals, reservoirs, impounded reservoir	Water is lifted through industrial pumps and delivered to the water purification plant.
Groundwater	Open well, deep well/tube well, artesian well, spring, infiltration gallery	Centrifugal or submersible pumps are used to lift water from open wells or tube wells. The pumping method and position of the pump depend on the character of the aquifer and the depth of water table/piezometric head.
Marine sources	Seawater	Seawater is withdrawn mainly for cooling or hydraulic fracturing. In water-stressed countries, drinking water is also sourced through desalination of seawater.
Rainwater	Rainwater catchment and storage system	Rainwater is an alternative source of freshwater for industrial use. Rainwater is collected in suitable rainwater harvesting structures and used.

a licence or may ask the industry to limit per day abstraction of natural water to a certain quantity.

3 **Quality:** The quality of source water is important because the technology and capacity of water-treatment plants depend on the quality of the input water. Periodic monitoring of source-water quality will help managers to adopt appropriate water-treatment facilities.

4 **Protection from disturbances and hazards:** The intake point should be free from disturbance and hazards like public resistance, climate extremes, erosion, and so on. The water intake structure should be made secure against external impacts.

5 **Accessibility:** The source of water should be easily accessible so that water withdrawal systems like pumps, storage, and so on. can be properly maintained and serviced.

6 **Distance:** In the absence of proximal water sources, the industry will require long pipelines and heavy-duty pumps to maintain the hydraulic head. They will also require consent regarding their rights to take the pipeline through privately owned land. This is a costly and a long-drawn-out affair.

7 **Durability:** The intake structure should be designed to withstand local immanent forces arising from natural hazards (as ascertained from long-term meteorological data on the area), local fauna and weathering effects.

7.3 Surface water abstraction

For surface water abstraction, appropriate engineering studies should be conducted at the proposed intake point so that the source is free from hazards or insufficient flow. The intake structure housing encapsulates (1) the pump, (2) the storage structure, where required, and (3) its connection to the conveyance system; protective measures should be designed accordingly. The design of intake structures depends on hydraulic head, pump capacity, type of source (river, lake, reservoir, etc.) and distance from the intake point. Seasonal fluctuation of water level at the source is also taken into consideration when designing intake structure. There are three types of intake structures for surface water abstraction: (1) intake structure in a reservoir (2) intake structure in a lake and (3) intake structure for a river. These structures may be fixed or mounted on a wheeled/floating platform for mobility.

One major issue for surface water abstraction is the licensing process. The licensing authority may sound out corporate bodies on the baseline hydrological character of the river basin before granting licence; the most commonly asked for information includes:

- Detailed report on hydrological investigation;
- Duration and purpose of withdrawal;
- Detailed break-up of the proposed abstraction quantity;
- Flow duration curve (see Box 7.1);
- Baseflow data of the river;
- Hydrographic data; and
- Impact analyses of impact of river water withdrawal.

7.3.1   Reservoir intake

Water level in a reservoir fluctuates seasonally while water quality varies with depth and seasonality. Reservoir water contains floating materials like plankton, dead organisms and other impurities. Generally, a depth of one metre or more from the water surface is considered the best intake level.

A reservoir intake structure is a concrete or masonry construction located at the upstream toe of the dam (Figure 7.3); the main part of the structure is a well-type construction where several intake gate valves at different depths

---

Flow Duration Curve

The flow-duration curve (FDC) is a cumulative frequency curve that shows the percentage of time during which specified discharges were equalled or exceeded in a given period (Searcy, 2011).

In Figure 7.2, a flow-duration curve of a hypothetical river is shown. In that curve, the x-axis represents percentage of time, and the y-axis represents the flow (Q). If the required flow is 100 m³/sec, then it can be seen from the curve that the required flow is available for only 40% of the calendar year, that is for 145 days.

Source: Searcy, J.K. (2011) *Flow-duration curves manual of hydrology: Part 2. Low-flow techniques Geological Survey Water-Supply Paper 1542-A.*

BOX 7.1

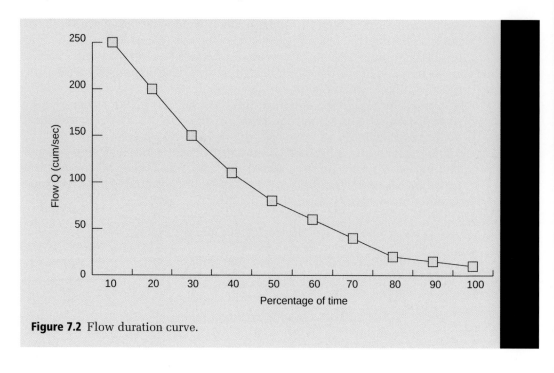

Figure 7.2 Flow duration curve.

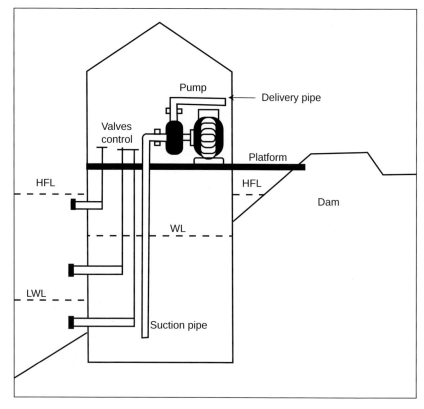

Figure 7.3 Reservoir intake structure.

Figure 7.4 Cincinnati water intake point. By Unknown, [Public domain], via Wikimedia Commons.

are built into the submerged part of the well. Gate valves are protected by screens to prevent entrance of such materials that can damage the pump. The gate valves are kept open or closed according to the water level of the reservoir. The pump is kept on a platform, and an intake pipe is attached to the pump that draws water from the well. The delivery conduit is attached to the delivery end of the pump.

7.3.2   River and lake intakes

Three types of structures are generally constructed to abstract water from a river or lake. The first one is similar to the reservoir intake (Figure 7.4), constructed in deep rivers. In another system, the pump house is constructed on the riverbank, and the intake pipe is immersed below the lowest recorded water level of the river (Figure 7.5); appropriate screening measures are taken so that debris does not enter into the pump. The third intake system applies where the water level varies seasonally; in such cases truck/trailer/boat mounted portable intake mechanism is used.

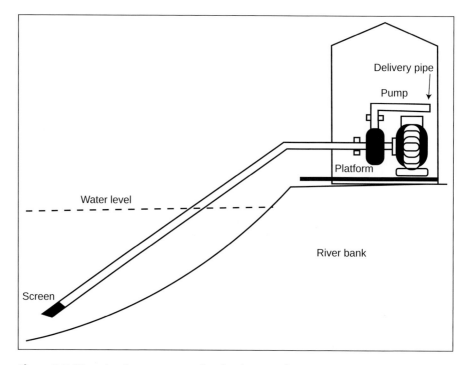

Figure 7.5 River intake structure, riverbank pump house.

7.3.3 Impacts of surface water abstraction

Direct abstraction from streams requires construction of dams, lifting infrastructure and diversion of flow, all of which are detrimental to the ecology of the river basin, unless remedial measures are installed alongside water abstraction installations. Adverse effects of surface water abstraction are:

- Direct abstractions exacerbate low levels of water in rivers (particularly during the summer). This also reduces the ecological status of tributaries and hydrologically connected wetlands (Scottish Environment Protection Agency, 2007).
- Abstraction causes human-induced variation of water levels in lakes and reservoirs that has adverse effects on the growth of aquatic plants and fish spawns.
- Dams cause barriers to fish migration and often result in shrinkage of their spawning area.
- Dams change the natural siltation pattern on the upstream and downstream sides, which in turn radically changes the local habitat, notably by reducing nutrition supply to the downstream segment.
- Reduced flow effects natural dilution of pollution by the streams and may result in a depletion of DO level in water.

7.4 Methods of groundwater abstraction

Groundwater is abstracted from three types of underground sources: (1) base flow of a river bed, (2) shallow aquifer and (3) deep aquifer. To those uninitiated in hydrogeology, the terms 'shallow' and 'deep' apparently suggest depth of the aquifer. It therefore, needs clarifying that the water level of a shallow or deep aquifer depends on the depth of the total head from the ground level. For instance, the water table of a shallow unconfined aquifer of great thickness may have a water table much below the centrifugal pumping limit. On the other hand, a deep confined aquifer may have its static water level (piezometric level) very close to the surface or even above the ground level in artesian condition.

7.4.1 Abstraction of baseflow

There are conditions in which a stream bed is dry but there is water flowing beneath the stream bed (it is also a source of drinking water in water-scarce areas in the dry season); stream flow is, thus, generated from two types of flows: (1) overland flow, which is by far the most common type of stream flow, and (2) 'baseflow' i.e. the portion of stream-flow that results from seepage of water from the ground into a channel slowly over time. Baseflow originates at locales where water is flowing beneath the stream bed but the water table is still above the stream bed (Figure 7.6). In dry seasons, baseflow contributes to downstream overland flow in the channel, especially in flood plains bordering catchment. A good amount of baseflow is often available for abstraction.

Structures required to abstract water from the river bed are called collector wells (Figure 7.7). Collector wells are masonry structures constructed on the riverbank or on the river bed. These wells are connected with one or

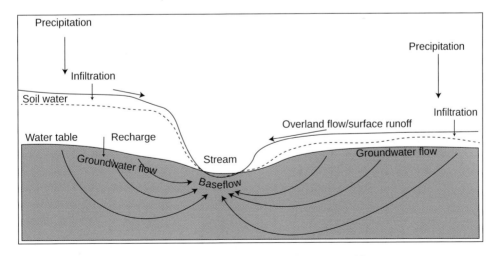

Figure 7.6 Overland flow and base flow along longitudinal river profile.

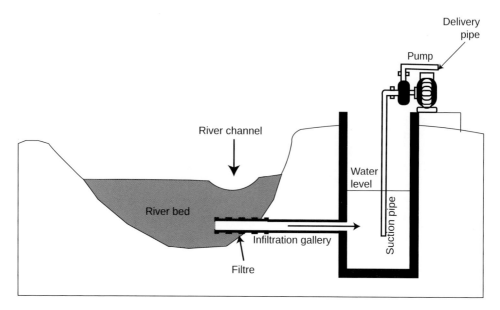

Figure 7.7 Collector well and infiltration gallery.

more horizontal conduits, called infiltration galleries, to draw water from the river bed. These conduits are attached with the tube well strainers of appropriate size that pre-filter the water from suspended particles. Water moves from the river bed through conduits to the collector well from which water is pumped out. Licences for such withdrawal are generally awarded by groundwater departments, ideally, in consultation with the river basin authority.

7.4.2 Abstraction of groundwater from aquifer

Groundwater is one of the most preferred industrial water sources. If a high-yielding aquifer is available within the factory premises or close to the factory, cost of water withdrawal and transport will be much less than surface water abstraction; additionally, purification cost for groundwater is often lower than that for surface water. Naturally therefore, manufacturing industries that consume large quantities of fresh water for selling low-cost products prefer groundwater. Groundwater abstraction needs a well or a tube well, pumping arrangement and water purification system (if required). Capacity of lifting devices (pumps) depends on the depth of aquifer and the capacity of the aquifer to deliver water. Parameters on which yield depends are (1) specific yield or aquifer storativity, (2) transmissivity of the aquifer and (3) depth of static water level or water table.

For groundwater abstraction two types of wells are constructed:

1 **Dug wells or open wells:** Dug wells are common in shallow aquifer tapping. Water yield of dug wells is generally low and not suitable for industrial use. But where water demand is low, dug wells, being low-cost water sources,

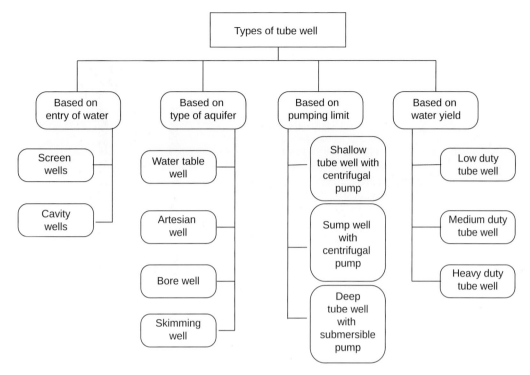

Figure 7.8 Types of tube well. Adapted from Mohammed (2016).

become commonplace. For instance, tea gardens along the Himalaya foothill belt source domestic water from dug wells.

2 **Tube wells:** There are several types of tubewell structures: Shallow tube wells, deep tube wells, low-duty, medium-duty and heavy-duty tube wells (Figure 7.8). Shallow or deep tubewells are often used by industries, depending upon water requirements and aquifer condition. There are several types of terminologies related to tube wells. A list of such terminologies with their descriptions is given in Table 7.2.

7.4.3 Construction of a tube well

The steps involved in the construction of a tube well are:

Site selection

Site selection is the first and most vital task for locating tube wells. The aspects of site selection are: (1) hydrogeological feasibility, (2) social acceptance and (3) convenient location. Most industries will prefer to install the tube well within its premises if it is hydrogeologically feasible. To get a good idea of the proper site selection, a pilot hydrological study is required. The study will recommend (a) probable depth of drilling, (b) methods of drilling and (c) best location in terms of probable yield and transportation of drilling equipments and pumping devices.

Table 7.2 Description of different types of tube wells.

Type of Tube Well	Description
Screen well	Entry of water is made through screens, filters or slotted pipes.
Cavity wells	Water enters through the bottom opening of the casing pipe.
Shallow tube well	A tube well is called shallow if the water level is within the centrifugal/vacuum pumping limit (<10.3 metre below ground level).
Deep tube well	If in a tube well the pumping water level is below the centrifugal/vacuum pumping limit, it is called a deep tube well. In a deep tube well, the water is lifted either through a submersible pump or a jet or compressor pump.
Low-duty tube well	If the yield of a tube well is less than or equal to $30\,m^3$/hour, it is called a low-duty tube well.
Medium-duty tube well	If the yield of a tube well is between 30 and $100\,m^3$/hour, it is called a medium-duty tube well.
Heavy-duty tube well	A heavy-duty tube well yields more than $100\,m^3$/hour
Water table well	Tube wells constructed in unconfined aquifers are called water table wells.
Artesian well	The wells constructed in a confined aquifer are called artesian wells. Artesian wells often develop auto-flow.
Bore well	Bore wells are types of wells which are generally drilled in hard rock where the bore hole itself acts as the production well.
Skimming well	Skimming wells are constructed in aquifers where the water in the lower level is brackish or saline (Mohammed 2016) and in the upper part is relatively fresh. Only the upper part is tapped in this type of wells.

Groundwater licence

There was a time not too many years ago when industries could drill at will; consequences followed. Now, in most countries, it is illegal to drill for groundwater without obtaining a licence. The licence application must contain detailed information on a variety of parameters like location (including land ownership documents or tenure), depth of proposed drilling, expected yield, purpose of drilling a well (drinking, process, etc.), and so on. Even after providing this information, the licensing authority may ask for a detailed hydrogeological report from the applicant.

Pilot study

There is more to groundwater exploration than water divining. A preliminary feasibility study at the selected site conducted through vertical electro-resistivity sounding (VES) or through pilot boring has its advantages. In an alluvial country with shallow aquifer, a pilot study may be skipped. But in rocky terrain where drilling is costly, a slim-hole boring or VES goes a long way towards ensuring the reliability of the aquifer, provided of course that the feasibility study is carried out by an experienced hydrogeologist. The hydrogeological report will not only quantify sustainable groundwater abstraction (i.e. required pump capacity, per day running hours and total maximum water withdrawal in a day) but will also facilitate regulatory compliance as well; for example, the spacing of wells (distance between two wells or the distance of the proposed well from any existing well) to avoid interference.

Community mobilisation

Social consent to operate a tube well is necessary, even if there is a licence to operate. Local people should be taken onboard and assured that their established lifestyle will not be adversely affected by the proposed well.

Commencement of drilling

Drilling methods range progressively from hand-drilling to heavy-duty rig-drilling; accordingly drilling costs increase progressively from the inexpensive hand-drilling to the expensive heavy-duty rig-drilling. The drilling method is chosen by the aquifer type and requirement. See Box 7.2 for popular drilling methods. A typical drilling rig is shown in Figure 7.9.

Logging and preparing a borehole log chart

As drilling proceeds, sand, silt, clay or rock fragments come up to the surface. These materials are sampled and arranged sequentially for preparation of a borehole log. The borehole log chart (also called a litholog) reflects depth, thickness and texture of different layers encountered in drilling.

Popular Tube Well Drilling Methods

Rotary drilling is a method of drilling that employs a sharp, rotational drill bit to make its way through the earth's crust. This system has four parts:

1 Hoisting equipment used to hold the drill rod and drill bit in the required position.
2 Rotating equipment that rotates the drill rod and pushes it through the earth and drills a hole.
3 A circulatory system that circulates drilling fluid (generally a mixture of water and mud) through the drill rod and the hole.
4 A drill bit fitted at the bottom of the drill rod

Percussion drilling (also called cable tool drilling) is suitable in soft and fissured rock formations. In this technique, a drill bit attached to a rope or cable is repeatedly raised and lowered, hammering the rock, to make a hole in it.

Auger drilling applies to drilling a well in shallow aquifers; it is a drilling device that usually includes a rotating helical screw-type blade that rotates and moves through the earth forming a hole in it.

Hand driving is popular in alluvial areas, especially where cheap labour is available. The technology is identical to rotary drilling with the difference that the drill rod is rotated manually.

BOX 7.2

Figure 7.9 Photograph of a drilling rig in operation in Chittagong in Bangladesh. Photo source: Earth Care Products & Services, Kolkata, India.

In soft sediments or highly weathered rocks, the drilling fluid washes away the drilled material, creating gaps in the litholog; in the event of significant material loss, electrical logging and photo logging may have to be conducted to complete the litholog. The appropriate aquifer zone (i.e. the tapping zone) is selected on the basis of borehole log chart; for heavy-duty tube wells, aquifer selection is best done by experienced hydrogeologists.

Assembly design

After analysis of the bore-hole log, a tube well assembly is designed by a hydrogeologist, and the tube well is constructed according to the prescribed design. A tube well assembly has the following parts from top to bottom: (1) the housing pipe, also called 'casing'; (2) blank pipe (above strainer); (3) strainer or slotted pipe or filter pipe; (4) blank pipe (below strainer); and (5) bottom plug. The diameter of the tube well depends on the yield of the well; diameter of a low-duty tube well is 50 to 100 mm, while a heavy-duty tube well may be up to 300 mm in diameter. The pipes of the tube well assembly are made of steel, galvanized iron or PVC. The strainer may be made of PVC or brass or stainless steel. The housing pipe diameter is a little larger than that of the blank pipe. The electrical log, lithological log and assembly design of an imaginary tube well are shown in Figure 7.10.

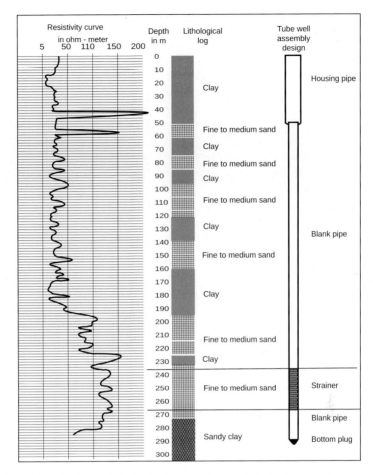

Figure 7.10 Electrical log, lithological log and assembly design of a tube well.

Strainer

Strainers are perforated pipes or slotted pipes covered by wire mesh that allows water to come into the well. The strainer should be made of durable material that can withstand rough handling (Punmia *et al.*, 2009) and is resistant to the corrosive electrolytic action of water. There are several types of strainer:

1 **Cook strainer:** A brass tube on which slots are made with cutting machine. Slot size varies from 0.15 to 0.4 mm.
2 **Brownlie strainer:** A slotted steel pipe with steel mesh surrounding it.
3 **Ashford strainer:** Perforated steel tube with a wire mesh surrounding it.
4 **Phoenix strainer:** Mild steel tube with slots grooved on its wall.
5 **Layne and Brownlie strainer:** Wedge-shaped steel wire is wound round slotted steel or iron pipe.

Formerly, metal strainers were used, but now corrosion-proof high-density PVC strainers are becoming increasingly common.

Lowering of tube well

Following design, the tube well assembly is lowered gradually into the borehole. The diameter of the bore hole is a little larger than the diameter of the pipes, so that there is a clear annular space around the blank pipes. The annular space, or cavity, is then packed with fillers like coarse sand or gravel. The grain size of the filler material is slightly larger than the pore size of the strainer or slotted pipe to prevent the inflow of filler material into the tube well. Any deviation from pipe verticality affects flow; consequently, while lowering the tube well part by part, verticality tests are to be carried out for each attached segment.

Development of tube well

This is the most important stage because the life of the tube well depends on it. In this process, a compressor/submersible pump is fitted with the tube well, and groundwater is pumped out for several hours until the discharged water is free from silt and suspended particles. Compressor development has two functions: it breaks down the mud envelop on the wall of the bore hole and cleans the tube well from mud silt and fine materials accumulated in the hole during the drilling. See Figure 7.11.

Figure 7.11 A production tube well is being developed with a submersible pump. Photo source: Earth Care Products & Services, Kolkata, India.

Pumps:

The pump is installed in the tube well after development. There are different types of pumps for different types of tube wells. Pump selection criteria are:

- Depth and diameter of the tube well;
- Depth of water table and possible drawdown;
- Quality of water;
- Desired discharge;
- Power requirement (fuel, electric, solar, wind); and
- Capital and maintenance cost of the pump

Centrifugal pumps are used if the water table is within the centrifugal pumping limit. Other types of pumps are used where the water level is beyond the centrifugal pumping limit, as in very deep aquifers; a list of such pumps with their working principle is given in Box 7.3. In Figure 7.12 different types of pump assembly are shown.

Different Types of Pumps Used in Deep Tube Wells

1. Vertical turbine pump: These are heavy-duty pumps. There are four components: a vertical spindle electrical motor, water flow pipe, a vertical turbine pump and a vertical shaft. The shaft and the delivery pipe are attached to the pump at one end, and the other end reaches the lowest drawdown level of the tube well. The pump is attached at the lower end of the shaft; multi-stage pumps can be attached to develop very high total head. A vertical turbine pump is capable delivering up to 2000 lps. These pumps are used for large irrigation projects, municipal water supply and industrial water supply.

2. Submersible motor pump: In this type, both pump and the motor are submerged in well water. Its function is similar to that of a centrifugal pump but consists of a series of impellers and diffusers. As both the pump and motor are submerged in water, there is no suction head, and water is delivered under pressure. Submersible pumps are very popular in irrigation fields and have industrial and domestic use as well. They can be installed in very low-diameter (50 mm) wells. Electrical power is supplied to the pump by means of submersible cables.

3. Jet pump: Jet pumps are small-duty single-stage centrifugal pumps in which the foot valve of the suction pipe reaches the lowest water level of the well. A stream of water is injected by the pump into the suction pipe through a constricted nozzle placed just above the foot valve, creating a Venturi effect to suck up and push out water along the delivery pipe.

4. Air lift pumps: In this system, compressed air is injected into the well to force water into the delivery pipe. The mouth of the well is sealed air-tight to maintain the high pressure in the well. This system is not generally used in production wells but is used in well development and testing.

BOX 7.3

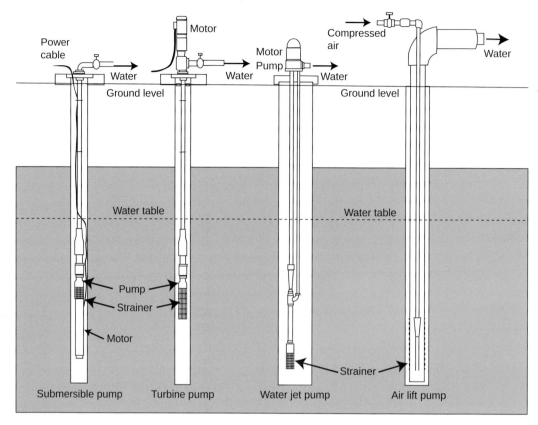

Figure 7.12 Different types of pumps used for lifting water from deep tube wells.

7.4.4 Impacts of groundwater abstraction

Groundwater abstraction has geological, socioeconomic, hydrological and ecological impacts. Unconfined aquifers have hydraulic connection with surface water streams and bodies of water. If the withdrawal rate is more than the recharge rate at the point of abstraction, a cone of depression is generated around the well (discussed in Chapter 3). If the area of the cone extends up to a surface body of water, the body of water will lose water to the aquifer.

Geological impact

Geological impacts are identified by intensification of water-rock interaction (Klimas and Gregorauskas, 2002) manifest in an intensified erosion of karst landscape, development of sinkholes and land subsidence. Due to the drying up of aquifer, large cracks often form on the land surface.

Socioeconomic impact

Depletion of water level is commonplace in areas where water abstraction is not mitigated by artificial recharge. Depletion of the water table and the piezometric head requires installation of higher-capacity pumps, which are not only more expensive to buy but also more expensive to run. This additional expenditure to maintain the same pumping output effectively means higher cost per litre of water, which is bad for business. And flawed business policy invariably has cascading effects on ecosystem.

Hydrological impact

Hydrological impact can be categorised as:

- **Reduction of surface runoff:** If the groundwater level around the water-pumping station drops below the level of the local stream beds, the stream loses water, causing drying out of small tributaries, wetlands and reduced river base flows during periods of low rainfall. A recent study in Bangladesh has reported that net recharge to the aquifers has increased by 5 to 15 mm/year between 1985 and 2007 (Shamsudduha *et al.*, 2011) due to an extensive withdrawal of groundwater, which consequently reduced the surface runoff in small rivers.
- **Impact on surface storage:** Wetlands are a vital source of livelihood for the surrounding population, especially marginal people. High groundwater withdrawal causes drying up of wetlands. In the Suti area of Murshidabad, India, a large perennial body of water covering about fifty square kilometres has been reduced by 90% due to extensive pumping of groundwater from shallow tube wells for irrigating the summer paddy crop.
- **Seawater intrusion:** Coastal aquifers generally have interfaces with saline aquifers or the sea. There is always a possibility of saline water intrusion into the aquifer if the water table of the freshwater aquifer goes below the level of saline water. Saline water finds its way to replace the fresh water in such cases. In the southern Adriatic coast of Italy, saltwater has invaded the coastal aquifer as the freshwater tables are now below sea level (Antonellini *et al.*, 2008).

Quality impact

Drawdown of the water table can alter the geochemical environment in the drawdown zone, which can lead to a release of toxic elements trapped in aquifer material, creating a serious health hazard. For example, arsenic is present in the Holocene aquifer zone of the Hanoi area; so Hanoi Water Works now draws groundwater from the deeper Pleistocene aquifer (Berg *et al.*, 2008). High withdrawal has caused drawdown in the Pleistocene aquifer, which affects the overlying Holocene aquifer, reducing those conditions favourable for mobilisation of arsenic.

Ecological impact

Reduced surface storage and surface flow due to groundwater withdrawal causes loss of aquatic habitats. Some tree species extend their roots to the aquifer to draw water; a depressed water table may result in the wilting of such tree species. For example, a large shallow aquifer in Western Australia is under abstraction for municipal water supply (Groom, Froend and

Mattiske, 2000). An impact analysis of the ongoing abstraction reveals that the plant genus *Banksia*, in woodlands around the abstraction well, is dying extensively, and the ecology of the area is under severe threat.

7.5 Water abstraction from the sea

Abstraction involves pumping seawater, using it for industrial processes and then returning it to the sea. Most abstracted water is used for cooling. There is a limit placed on the temperature at which water can be returned. In the UK, the majority of coastal water abstractions, by volume, are for power stations (98% or 20.2 Mm^3 per day). The remainder 2% or 0.4 Mm^3 per day is to be used by the fish-processing and manufacturing industries. A large power station may abstract from coastal or estuarine sources at the rate of 200 m^3/second or more during peak load for cooling. In most cases, non-evaporative cooling is used where the steam turbine exhausts are cooled via the main condensers before the water is discharged back to the sea. The evaporative cooling process is often used by power stations located inland which use cooling towers to recycle the water. Other uses include fish farms and fish-processing factories, the manufacture of food, wood and chemical products, agriculture, marine animal sanctuaries and navigation abstractions to maintain water levels in impounded docks. According to a study in 2013, Qatar and Kuwait are 100% reliant on desalination for their domestic and industrial freshwater needs, while Oman and Saudi Arabia source freshwater to the tune of 85% and 70% respectively from seawater.

7.5.1 Environmental impact of seawater withdrawal

The impact depends on the use and purification process of seawater. A recent study on the impact of seawater abstraction for the Kapar Power Plant in Malaysia reveals that due to elevated temperature and reduced pH, the marine environment has been harmed by eutrophication and an increase in abundance of small-bodied copepods at the expense of previously dominant large-bodied species (Chew *et al.*, 2015). In Table 7.3 the major uses of seawater, purification process and environmental impact are listed.

7.6 Conveyance system

After withdrawal, water is conveyed to the point of purification and storage. The objective is to provide the desired through-put from the source to the user area. Water is first conveyed from the source to the treatment plant.

Table 7.3 Major uses of seawater, purification process and environmental impact.

Uses	Purification process	Environmental impact
Extraction of crude petroleum and natural gas	No purification	High adverse impact when wastewater is disposed into sea or nearby body of water.
Power plant	No purification	Increase in temperature adversely affects the environment where it is discharged.
Agriculture, drinking and domestic use	Desalination	Highly saline slush created from the desalination and high temperature of the brine can destroy marine flora and micro-organisms if discharged into the sea.
Inland pisciculture	Diluted	Low impact

7.6.1 Conveying water from the source to the treatment plant

The conveyance system requires hydraulic head between the source and the destination. If the source is at a higher altitude than the destination facility, hydraulic head is created by gravity; if not, the required head is created by deploying a 'point pressure system' at the source. Thus, depending on the available hydraulic head, there are two types of conveyance systems: (1) free-flow system and (2) pressure system.

In the free-flow system, the constructed conduit follows the topographic gradient, so that water flows through the conduit under gravitational pull. Common free-flow systems are canals, flumes, grades, aqueducts and grade tunnels. See Figure 7.13.

Pressure conduits are essentially closed conduits; water is forced by pressures higher than atmospheric pressure to flow against gravity. The bed or 'invert' of the conduit system is designed to follow the path of least resistance to optimise construction cost; such conduits are also called 'pressure pipes'. A wide variety of materials is used for construction of pressure pipes such as PVC, wrought iron, cast iron, steel, copper, brass, reinforced cement concrete Hume steel, vitrified clay, and reinforced plastic pipes.

7.7 Water purification

Raw water must be treated and purified before it is supplied for domestic, industrial or any other use. Treatment processes depend on the quality of raw water and type of use. Efficiency of a water-treatment plant, apart from engineering considerations, depends on two geographical considerations:

- **Selection of site:** Site selection is the foremost consideration in installation of water-treatment plants; it has direct consequences in terms of construction

Figure 7.13 The Central Arizona Project: Aqueduct transfers untreated water. Source: wikimedia.org, public domain.

cost and delivery efficiency. Relevant factors are topography, geology and land use, so that construction of the treatment plant and delivery system does not require any major change in the landscape.
- **Hydraulic profile:** Hydraulic profile is the other major consideration. Treatment plants with delivery, that is powered by gravity are most cost effective both in terms of construction and maintenance cost. Otherwise, cost-intensive and environmentally damaging pumping station(s) will be required. Head loss occurs due to friction in (a) pipes/conduits/canals (b) bends and (c) sluice gates and (d) valves between entrance and exit points (imperceptible frictional head loss also takes place due to volume expansion caused by hot locales in the delivery route). Before construction, a high-resolution land survey will do much to achieve cost-efficient gravity flow.

Raw water passes through several stages of purification to remove impurities like suspended particles, floating material, oil, grease and bacteria. Raw water-treatment processes are shown in Figure 7.14.

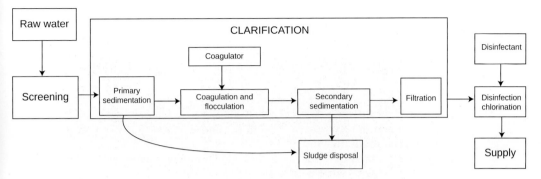

Figure 7.14 Processes of raw water treatment.

7.7.1 *Primary screening*

The eliminates floating material in water like leaves, sticks, plastic, and so on. The Primary Screen is a mesh of required size through which water is passed leaving behind all floating materials.

7.7.2 *Clarification*

This removes suspended particles that cannot be removed by primary screening. In this process, water passes through different stages:

- **Primary sedimentation:** In this process, water is stored in large tanks, called 'settling tanks', for a while so that most of the suspended particles (up to clay-sized grains) settles at the bottom of the tank.
- **Coagulation:** Particles finer than clay, that is, colloidal particles, do not settle even in still-water conditions due to Brownian movement, that is, particles in thermal balance with water and therefore, unresponsive to the force of gravity. Chemical coagulants added to water induce particles in Brownian motion to coagulate and form particles of larger size that can settle by gravity. The most commonly used coagulator is alum or aluminium sulphate.
- **Flocculation:** Water treated by a coagulator is taken to the 'flocculation basin' where it is agitated by paddles attached to a rotating axle. Agitation creates multiple nucleation centres causing further coagulation and formation of 'flocs' augmenting clustering of colloids into larger sizes; flocs also trap bacteria.
- **Secondary sedimentation:** In the secondary sedimentation process, water is again detained in tanks, that is, still-water sedimentation basins, to allow settling of clustered colloids. Secondary sedimentation basins may be circular, square or rectangular. Typical sizes are:
  - Circular: 5 to 100 metre in diameter and 2 to 5 metres in depth;
  - Rectangular: Dimension depends on sludge removal mechanism; or
  - Square: 10 m to 70 metres in width and 2 to 5 metres in depth.
  Precipitates at the bottom of secondary sedimentation tanks are removed as sludge and clean water is transferred to the 'filtration chamber'.

- **Filtration chamber**: Water from the secondary sedimentation tank is finally passed through slow sand filter bed where all suspended material in water is completely removed. Unlike coagulation and flocculation, this is a physical separation process in which water flows through layers of sand by gravity (Figure 7.15). Component parts of gravity-aided *rapid sand filter* are:
  - Chamber: filter tank or filter box;
  - Filter media (sand);
  - Under drain system with manifold; and
  - Wash water troughs
- **Backwashing of rapid sand filter:** When raw water passes through the filter, particles accumulate in the filter media and after a few days of operation, the filter gets clogged. To keep the filter in working condition it must be cleaned periodically by a process called 'back wash' in which the treated water from the storage tank is pumped through the backwash inlet and passed through the filter in the reverse direction. This process cleans accumulated impurities lodged in the inter-granular space of the filter; back wash water from reverse filtration is removed through the backwash drain.

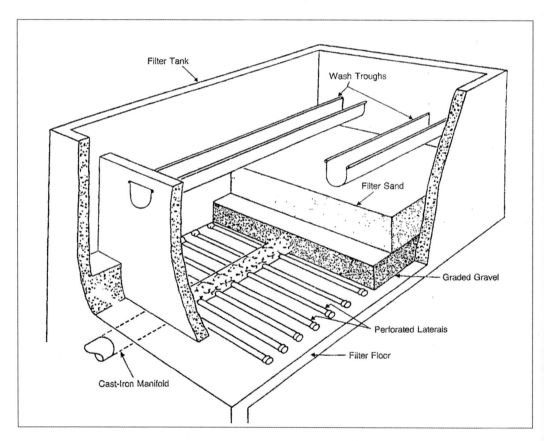

Figure 7.15 Gravity aided rapid sand filter. By US Environmental Protection Agency, Public domain, via Wikimedia Commons. https://commons.wikimedia.org/wiki/File:Rapid_sand_filter_EPA.jpg.

7.7.3 Disinfection

Disinfection is a selective destruction of pathogens. It is the ultimate purification process of drinking water and must be done for all water supply schemes. The processes of disinfection are:

- **Lime treatment:** Bacterial content in water is killed by high alkalinity. Filtered water is passed through lime to raise the pH of water to a level unfavourable for bacteria. Highly alkaline water is not suitable for public water supply; subsequent treatment is required for lime removal.
- **Ozone treatment:** Ozone in water breaks down into nascent oxygen. Ozone is thus a powerful oxidising agent that kills bacteria.
- **Chlorination:** Chlorination also kills bacteria by breaking down into enzymes. Ozone treatment and chlorination also reduce alkalinity.
- **Ultraviolet disinfection:** Bombardment by ultraviolet (UV) rays damages the DNA structure of micro-organisms. UV lamps of germicidal spectrum frequency are used for further disinfection. This process kills about 99% of germs in water.
- **Boiling:** Boiling kills all bacteria and water-borne viruses. But boiling is cost-intensive and generally used for pharmaceuticals and health care products.

7.7.4 Desalination

Desalination can be defined as any process that removes salts from water, (Krishna, 2004), especially seawater. Desalination has applications in municipal, industrial and commercial water supply. It is used as the main source of water in some Middle Eastern countries. Desalination plants are widely installed in developed countries for industrial use; 20% of desalination plants of the world are in the United States. As new technologies are emerging, desalination is becoming more cost effective and gaining popularity to meet growing water needs.

Desalination processes are energy intensive; the relevant technologies are categorised as 'thermal technology' and 'membrane technology'. Three commonly used thermal technologies are as follows:

Multi-stage flash distillation

In multi-stage flash distillation, saline water is heated by steam and then passed through heat exchangers under high temperature and pressure. After that, it is diverted to a low pressure chamber, which results in an immediate boiling of the saline water (now called brine) without additional heat. The steam thus formed is passed through demisters and condenses into fresh water. This process is used in large desalination plants associated with thermal power plants. See Figure 7.16.

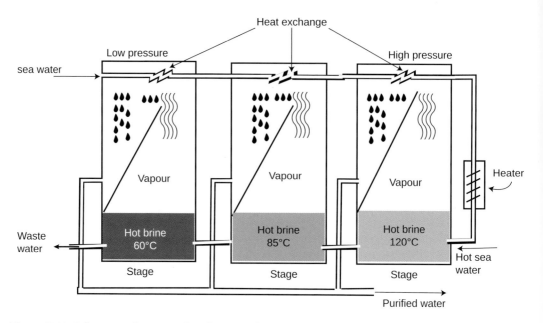

Figure 7.16 Schematic diagram of multi stage flash distillation process.

Vapour compression distillation

In this process, salt water is evaporated at atmospheric pressure. The vapour is then compressed in such a way that the temperature and pressure of the compressed steam becomes about 105 °C and 0.2 kg/cm² respectively. The compressed hot steam is then passed through brine and the brine is boiled to produce more water vapour. While heating the brine, hot water vapour loses heat and condenses into fresh water. The system is schematically shown in Figure 7.17.

Multi-effect distillation

In this process, steam is used to heat seawater to produce steam. The source of the initial steam is the spent steam at a slightly elevated pressure exiting from a steam-operated power station. The secondary vapour produced is used to generate tertiary steam at a lower pressure. The process is repeated several times or stages, and at each stage some steam is condensed to produce water. A schematic of this process is shown in Figure 7.18.

7.7.5   Membrane technologies

Three types of membrane technologies are widely used: electrodialysis (ED), electrodialysis reversal (EDR) and reverse osmosis (RO).

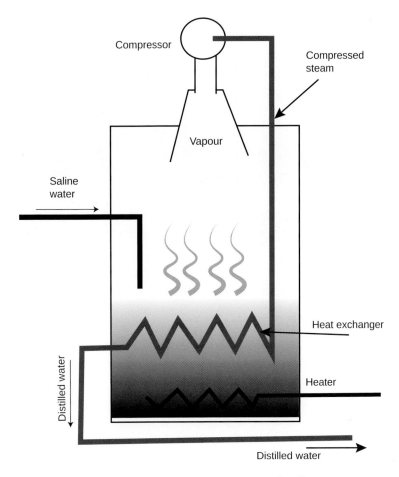

Figure 7.17 Schematic diagram of vapour compression distillation process.

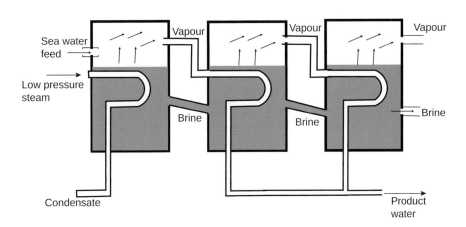

Figure 7.18 Multi effect distillation process. Adapted from BARC (2010).

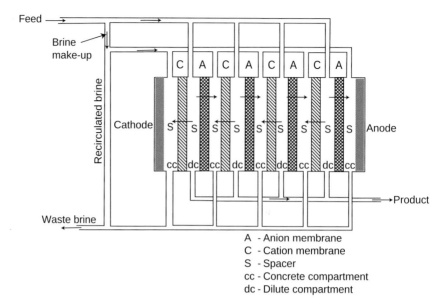

A - Anion membrane
C - Cation membrane
S - Spacer
cc - Concrete compartment
dc - Dilute compartment

Figure 7.19 Schematic diagram showing electrodialysis process. Adapted from Takenaka, Chen and Miele (1975).

Electrodialysis (ED)

In ED, saline water or wastewater, after removal of suspended particle and organic matters, is transported through stacks containing alternate anion and cation selective membranes. Cation-selective membranes are negatively charged materials, which reject negatively charged ions and allow positively charged ions to flow through. On the other hand, anion selective membranes allow negatively charged ions to flow through. By placing positively charged and negatively charged membranes alternately in a row, both types of ions can be removed from wastewater. Thus in ED, water in one column becomes free from ions, which is delivered to the supply line; while in the other column, water becomes more concentrated with ions and is rejected into the wastewater stream. See Figure 7.19.

Electrodialysis reversal (EDR)

The working principle of EDR is similar to that of ED. Ions released from water often form scales on the membranes. To remove the scales, the polarity of the electrodes is reversed so that the electrodes can repel the scales. EDR is suitable for treating brackish water with TDS not above 8000 mg/l. See Figure 7.20.

Reverse osmosis

Reverse Osmosis (RO) is a process in which water is deionised or demineralised by passing water through a semi-permeable membrane called the RO membrane. Reverse osmosis, as the name implies, is the opposite of osmosis.

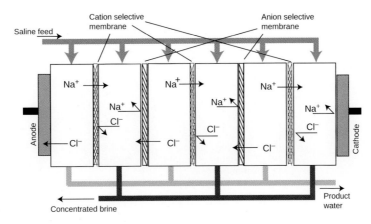

Figure 7.20 Electrodialysis reversal process.

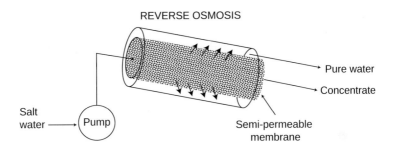

Figure 7.21 Reverse osmosis process.

In osmosis, liquid with lower salt concentration migrates towards liquid with higher concentration of salts when separated by a semi-permeable membrane, thus diluting salt concentration of the target liquid. In reverse osmosis, water with higher salt concentration is forced through the RO membrane under high pressure; the membrane allows water to pass through but not the dissolved solutes and other impurities. Up to 99% of dissolved salts, particles, colloids, bacteria and pyrogens are removed by reverse osmosis. The reverse osmosis processes is illustrated in Figure 7.21.

Ion exchange

Ion exchange removes unwanted ions from water using resins. Alkaline resins are used for removing anions and acidic resins for cation removal. Ion exchange is used for different water-treatment applications:

- **Softening (removal of hardness):** Natural water containing calcium and magnesium ions (with other cations like strontium, barium in minor amounts) is called hard water. Hard water forms scales when boiled and as such, is unsuitable for industrial applications that require heated water; softening is essential in most industries, notably where boilers are used.

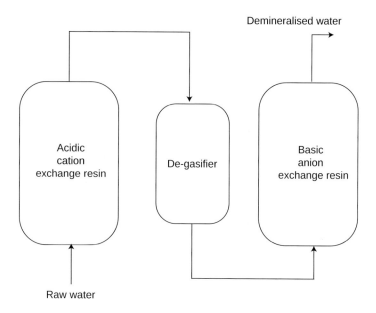

Figure 7.22 Ion exchange processes in a demineralisation plant.

- **De-alkalisation (removal of bicarbonate):** Weakly acidic cation resins are used in this process to remove both hardness and alkalinity of water. This process is used in breweries and low-pressure boilers. It also reduces dissolved salts in water.
- **Decationisation (removal of all cations):** In this process, strongly acidic cation exchange resins are used to remove all cations from water.
- **Demineralisation (removal of all ions):** For any industry using steam in the production process, the removal of all ions is essential. Water is treated with cation, and anion exchange resins to prevent scaling/corrosion in the steam manufacturing unit. Breakdown of bicarbonates during demineralisation produces carbon dioxide which is removed by degasifiers. Industrial application of this process is in high-pressure boilers in thermal and nuclear power stations, electronic industries, textile and paper industries, laboratories and pharmaceutical industries. See Figure 7.22.

7.8 Water supply and distribution

Water distribution systems carry drinking water or industrial water from treatment plants to consumers. A water distribution system is a complex combination of pipes, pumps, valves, storage tanks, reservoirs, meters, fittings, and other hydraulic appurtenances supplying suitably treated water, that is, potable water to employees and non-potable water to machines. Small industries often source drinking water from the municipal supply or other water works.

7.8.1 *Pipes*

Pipes are categorised by (a) dimension, there being a vast range of diameters, and (b) purpose as in transmission mains, distribution mains, service lines, and premise plumbing (*Drinking water distribution systems*, 2006). Pipes are made of various materials, such as ductile iron, pre-stressed concrete, polyvinyl chloride (PVC), reinforced plastic and steel.

Transmission mains

These pipelines carry large volumes of water over long distances. The diameter of such pipes are 61 cm or higher. Transmission mains deliver water to storage tanks.

Distribution mains

These pipelines convey water from storage tanks to end-users and generally follow city streets or connecting roads in industrial townships. Distribution mains are generally 10.0–30.5 cm in diameter.

Service lines

Service lines carry water from distribution mains to buildings or industrial units. For residential buildings, the diameter of the service line is generally 2.5 cm; for industrial units, the diameter varies according to the requirement of the flow of water.

Premise plumbing

In premise plumbing, pipelines carry water from the service line to the kitchen tap, shower, flushing system, water heater and various other utilities.

Layout of water distribution system

Hydraulically the flow of water in a distribution system is pipe flow, as the flow is always under pressure. In a distribution system, pipe flow rate is generally between 0.9 to 1.3 m s^{-1}. To maintain the flow, appropriate gradient or pressure head should be maintained in the system. There are many good discussions on the water distribution hydraulics, which is beyond the scope of this book. See Figure 7.23. In this figure two types of layout of water distribution networks are shown.

7.8.2 *Storage system*

An efficient water system catering to a large demand for water needs a storage system. A storage system can be located underground, on the ground or overhead.

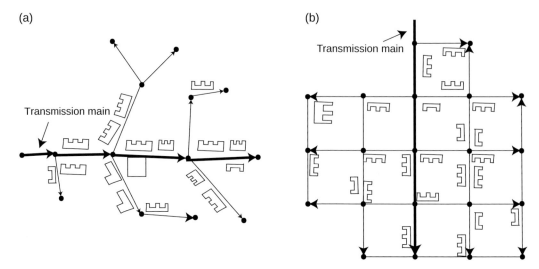

Figure 7.23 Two basic configurations for water distribution systems: (a) Branched distribution system, (b) Looped network distribution system.

- **Underground:** Underground storage reservoirs (also called sumps) are generally used to store water where there is a shortage of space. Initial storage of purified water in areas of large demand use sumps from where water is pumped to overhead tanks. Sumps are also used for storing harvested rainwater.
- **Surface tanks:** Surface tanks are constructed where high hydraulic head is not required. These tanks serve the purpose of both initial storage near the water intake point and of storing purified water. Surface tanks are also used for storing harvested rainwater.
- **Overhead tanks:** Overhead tanks are constructed at a height to gain hydraulic head required for distributing water.

A storage system is essential in the following contexts:

- **Store water to cover fluctuation in demand:** Where water demand is not uniform in the course of a day, the supply system is designed to (a) meet the peak demand hours with pumped water, and (b) to store water in overhead tanks to meet the demand during the peak hours.
- **Smooth out variations in supply:** Pumps do not run 24/7. A good amount of pumped water is stored in overhead tanks to meet demand during non-pumping hours.
- **Provide water security in case of supply interruptions or disaster:** A storage tank can be used to store water for emergency and supply-disorder.
- **Storing water for fire fighting:** It is a statutory requirement that every industry store water for fire fighting. The capacity of a fire-fighting storage tank depends on the size of the industry.
- **To provide adequate hydraulic head:** Water is generally distributed under gravitational force to avoid energy costs. Overhead storage tanks provide the hydraulic head.

Water is stored for distribution in 'service reservoirs'. The design criteria for service reservoirs are as follows:

Location of water tanks

A water tank is located in such a way that it can be used in the best possible manner. Size, location and elevation of a storage tank depend on the water demand, the location of the pumping station and the magnitude of the supply network. In an industrial situation, there are requirements for separate storage tanks for industrial water supply, potable water supply and rainwater storage. A service reservoir, as a rule of thumb, should be capable of storing 25 to 40% of peak daily water demand (Trifunovic, 2002). It should also be situated at a higher altitude than the distribution area, but close to it. In hilly areas, such elevation is available but in a flat topography, an overhead tank is required. The other location criteria are stability of soil, aesthetics and security.

- **Site feasibility:** Site feasibility considerations are as follows:
    - The site should have adequate space for construction and subsequent maintenance installations.
    - It should be located as close to the service location as possible.
    - Existing ground surface elevation and site drainage (University of Idaho, 2008).
    - Feasibility in terms of geotechnical assessment.
    - Power source availability.
- **Tank Shape:** Tanks can be designed in many shapes according to capacity, stability and aesthetics factor. In general, square, rectangular, conical and cylindrical tanks are used.
- **Tank Material:** Water storage in reservoirs is a traditional household practice, and various materials from clay and terracotta to reinforced concrete are used as storage tank materials. In industries, four types of materials are popularly used: PVC, brick and mortar, reinforced concrete and galvanised iron.
- **Maintenance:** the following maintenance criteria should be observed:
    - Tanks and reservoirs should be cleaned periodically.
    - Check for any leak, crack or other damage in the tank as well as in the inlet and outlet pipes.
    - Conditions of approach roads, ladders, and so on for approaching the tank should be free from obstructions.
    - Hard standing platform, parapet walls, fencing, manhole covers, indicating plates, and so on should be properly maintained.

7.9 Water delivery and distribution software

Water distribution and management software is widely used in industries to model water flow in a water distribution system. One of the most popular programmes is EPANET 2.0 developed by US-EPA. The salient features of EPANET as sourced from US-EPA (2016) are described below. In Figure 7.24, a screenshot of the software is given.

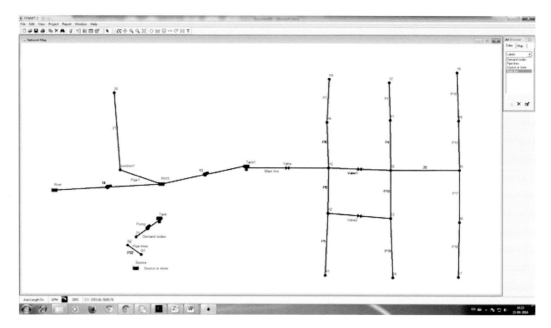

Figure 7.24 A typical water distribution model in EPA.

7.9.1 Overview

US-EPA states, 'EPANET is a software programme that models water distribution piping systems. It is public domain software that may be freely copied and distributed. It is a Windows 95/98/NT/XP program. EPANET performs extended period simulation of the water movement and quality behaviour within pressurized pipe networks. Pipe networks consist of pipes, nodes (junctions), pumps, valves and storage tanks or reservoirs. EPANET tracks:

- The flow of water in each pipe;
- The pressure at each node;
- The height of the water in each tank;
- The type of chemical concentration throughout the network during a simulation period;
- Water age;
- Source; and
- Tracing.' (US-EPA, 2016)

7.9.2 Capabilities

According to US-EPA, 'EPANET's Windows user interface provides a visual network editor that simplifies the process of building piping network models and editing their properties and data. EPANET provides an integrated computer environment for editing input data. Various data-reporting and visualisation tools are used to assist in interpreting the results of a network analysis. These include:

- Colour-coded network maps;
- Data tables;
- Energy usage;
- Reaction;
- Calibration;
- Time series graphs;
- Profile plots; and
- Contour plots.

EPANET provides a fully equipped, extended-period hydraulic analysis package that can:

- Simulate systems of any size;
- Compute friction head loss using the Hazen-Williams, the Darcy-Weisbach, or the Chezy-Manning formula;
- Include minor head losses for bends, fittings, and so on;
- Model constant or variable-speed pumps;
- Compute pumping energy and cost;
- Model various types of valves, including shutoff, check, pressure regulating and flow control;
- Account for any shape storage tanks (i.e. surface area can vary with height)
- Consider multiple demand categories at nodes, each with its own pattern of time variation;
- Model pressure-dependent flow issuing from sprinkler heads; and
- Base system operation on simple tank level, timer controls or complex rule-based controls.

In addition, EPANET's water quality analyser can:

- Model the movement of a non-reactive tracer material through the network over time;
- Model the movement and fate of a reactive material as it grows (e.g. a disinfection by-product) or decays (e.g. chlorine residual) over time;
- Model the age of water throughout a network;
- Track the percentage of flow from a given node reaching all other nodes over time;
- Model reactions both in the bulk flow and at the pipe wall;
- Allow growth or decay reactions to proceed up to a limiting concentration;
- Employ global reaction rate coefficients that can be modified on a pipe-by-pipe basis;
- Allow for time-varying concentration or mass inputs at any location in the network; and
- Model storage tanks as being complete mix, plug flow, or two-compartment reactors.'

7.9.3 Applications

The applications of EPANET can help water utilities to maintain and improve the quality of water delivered to consumers and can be a helpful tool to estimate water quality at any stage of a system. 'It can be used to:

- Design sampling programs;
- Study disinfectant loss and by-product formations;

- Conduct consumer exposure assessments;
- Evaluate alternative strategies for improving water quality, such as altering source use within multi-source systems;
- Modify pumping and tank-filling/emptying schedules to reduce water age;
- Use booster disinfection stations at key locations to maintain target residuals;
- Plan and improve a system's hydraulic performance;
- Assist with pipe, pump, and valve placement and sizing;
- Energy minimisation;
- Fire flow analysis;
- Vulnerability studies; and
- Operator training.' (US-EPA, 2016)

7.10   Conclusion

Industrial water management greatly relies on efficient management of water procurement, purification and distribution of water. There are several useful books and papers on this topic. This chapter gives an overview of the system so that water managers may find necessary guidelines regarding installing any water infrastructure, either for industrial purpose or for delivering water stewardship.

Bibliography

Antonellini, M., Mollema, P., Giambastiani, B., Bishop, K., Caruso, L., Minchio, A., Pellegrini, L., Sabia, M., Ulazzi, E. and Gabbianelli, G. (2008) 'Salt water intrusion in the coastal aquifer of the southern Po plain, Italy', *Hydrogeology Journal*, 16(8): 1541–1556.doi: 10.1007/s10040-008-0319-9.

BARC (2010) *Desalination & water purification technologies*. Available at http://www.barc.gov.in/publications/eb/desalination.pdf (accessed: 8 July 2016).

Berg, M., Trang, P.T.K., Stengel, C., Buschmann, J., Viet, P.H., Van Dan, N., Giger, W. and Stüben, D. (2008) 'Hydrological and sedimentary controls leading to arsenic contamination of groundwater in the Hanoi area, Vietnam: The impact of iron-arsenic ratios, peat, riverbank deposits, and excessive groundwater abstraction', *Chemical Geology*, 249(1-2): 91–112. doi: 10.1016/j.chemgeo.2007.12.007.

Chew, L.L., Chong, V.C., Wong, R.C.S., Lehette, P., Ng, C.C. and Loh, K.H. (2015) 'Three decades of seawater abstraction by Kapar power plant (Malaysia): What impacts on tropical zooplankton community?', *Marine Pollution Bulletin*, 101(1): 69–84. doi: 10.1016/j.marpolbul.2015.11.022.

Drinking water distribution systems (2006) Washington DC: The National Academies Press.

Environment Agency, Government of UK (2013) *Managing water abstraction*. Available at https://www.gov.uk/government/uploads/system/uploads/attachment_data/file/297309/LIT_4892_20f775.pdf (accessed: 30 March 2016).

FICCI (2011) *Water use in Indian industry survey*. Available at http://ficci.in/SEDocument/20188/water-use-indian-industry-survey_results.pdf (accessed: 30 May 2016).

Groom, B.P.K., Froend, R.H. and Mattiske, E.M. (2000) 'Impact of groundwater abstraction on a Banksia woodland, Swan Coastal Plain, western Australia', *Ecological Management and Restoration*, 1(2): 117–124. doi: 10.1046/j.1442-8903.2000.00033.x.

Klimas, A. and Gregorauskas, M. (2002) 'Groundwater abstraction and contamination in Lithuania as geoindicators of environmental change', *Environmental Geology*, 42(7): 767–772. doi: 10.1007/s00254-002-0554-7.

Krishna, H. (2004) *Introduction to desalination technologies*. Available at http://www.twdb.texas.gov/publications/reports/numbered_reports/doc/r363/c1.pdf (accessed: 23 June 2016).

Mohammed, E. (2016) *Types of water wells*. Msc thesis. Available at https://www.academia.edu/5403958/TYPES_OF_WELLS_TYPES_OF_WATER_WELLS_ARBA_MINCH_UNIVERSITY_INSTITUTE_SCHOOL_OF_POST_GRADUATE_STUDENT_DEPARTMENT_OF_WATER_RESOURCE_AND_IRRIGATION_ENGINEERING_COURSE_OF_GROUND_WATER_HYDROLOGY_TYPES_OF_WELLS (accessed: 1 April 2016).

Punmia, B.C., Lal, P.B.B., Jain, A.K. and Jain, A.K. (2009) *Irrigation and water power engineering*.16th edn. New Delhi: Lakxmi Publication Private Ltd.

Scottish Environment Protection Agency (2007) *An introduction to the significant water management issues in the Scotland river basin district*. Available at https://www.sepa.org.uk/media/38319/an-introduction-to-the-significant-water-management-issues-in-the-scotland-river-basin-district.pdf (accessed: 15 April 2016).

Searcy, J.K. (2011) *Flow-duration curves manual of hydrology: Part 2. Low-flow techniques geological survey water-supply paper 1542-A*. Available at http://pubs.usgs.gov/wsp/1542a/report.pdf (accessed: 30 March 2016).

Shamsudduha, M., Taylor, R.G., Ahmed, K.M. and Zahid, A. (2011) 'The impact of intensive groundwater abstraction on recharge to a shallow regional aquifer system: Evidence from Bangladesh', *Hydrogeology journal*, 19(4): 901–916. doi: 10.1007/s10040-011-0723-4.

State of Michigan (2010) *Water well drilling methods*. Available at http://www.michigan.gov/documents/deq/deq-wb-dwehs-gwwfwim-section5_183030_7.pdf (accessed: 31 March 2016).

Takenaka, H.H., Chen, C. and Miele, R.P. (1975) *Demineralization of wastewater by electrodialysis*. Cincinnati, OH: US-EPA.

Trifunovic, N. (2002) *TP40 21 water distribution*. Available at http://www.samsamwater.com/library/TP40_21_Water_distribution.pdf (accessed: 24 June 2016).

US-EPA (2016) *EPANET*. Available at https://www.epa.gov/water-research/epanet (accessed: 23 June 2016).

8 Water Resource Assessment

8.1 Introduction

The subject matter of this chapter calls for a definition of the word 'highfalutin' at the risk of offending readers, but it is a risk worth taking, given the seriousness of the subject matter in hand. The online Merriam-Webster's Learner's Dictionary defines '"highfalutin" as seeming or trying to seem great or important'. Water resource assessment is highfalutin terminology because desert tribes all over the world have been practicing it for millennia, such as the Kalahari bush people. It is the privileged that need to be educated on source-water assessment, and if they do not learn *suo moto* they learn it the hard way; for instance, the Californians, who are suffering 25% water cut since 2015. Freshwater resources have come to such a pass that its availability cannot be taken for granted anymore; it is now essential to have an idea where the water is coming from and how much of it can be abstracted without making life difficult for our posterity. This awareness is especially important for groundwater conservation.

This chapter discusses qualitative and quantitative estimation of source water. The term 'source water' encompasses rivers, streams, natural springs, constructed canals, groundwater aquifers or any other freshwater source tapped by people. Water consumption planned on the basis of source-water assessment (SWA) is a moral duty for the Third Millennium parental generation to ensure that future generations are spared the misery of water starvation. While discussing SWA, the World Bank (2007) document on 'Strategic Environmental Assessment and Integrated Water Resources Management and Development' laments that SWA does not figure prominently in 'strategic environmental assessment' (SEA): '...relatively few SEAs have been applied in the water sector, especially in developing countries.' And thereby hangs a tale of cop-out because multinational corporations

Industrial Water Resource Management: Challenges and Opportunities for Corporate Water Stewardship, First Edition. Pradip K. Sengupta.
© 2018 John Wiley & Sons Ltd. Published 2018 by John Wiley & Sons Ltd.

(MNC) are major water consumers in 'developing countries', which is a euphemism for countries where a major part of the population is poverty stricken and lives on a few buckets of water from ponds and/or wells/tube wells (i.e. untreated water). Unplugged bathtubs worldwide drain away enough treated water in a day to provide bush people with a perennial oasis. Be it understood that SWA applies more to water-guzzling MNCs and to a lesser extent to the bathtub brigade than to the man on the street.

In the United States, water departments of the various states periodically conduct SWAs and publish reports. This assessment is mainly based on the qualitative aspect of the source water with the objective to protect the source from pollution and contamination. In almost all other countries, there are government departments that periodically assess their water sources and publish reports and maps. The results of such assessment are not uniform globally, and there are governance and management issues in some countries that often hamper adequate data generation and report publishing. Moreover, data collected by government departments are often done at the macro level and give only impressions about regional water status. Water data of small watersheds and local aquifers, streams and so on are not mentioned in their studies.

Industrial water requirements are generally sourced from local basins, either from groundwater aquifers or from surface bodies of water. Whether or not such sources are capable of delivering sustainable yield is a matter of in-depth study to be conducted on a case-to-case basis. According to Sutcliffe and Lazenby (1990), industrial water demand is larger in scale than domestic supply and needs full assessment of the yield of sources. In a free market economy, the government or its appointed agencies cannot possibly monitor all tapped sources of water. It is therefore, incumbent upon industries to develop their own network of monitoring stations to safeguard business interests without falling afoul of the society at large.

Water resource assessment (WRA) is required to assess the quality and available quantity of exploitable water in a basin or catchment. It serves a variety of purposes of the industry, starting from building up resilience under climate uncertainties to corporate water disclosure. Therefore, the drivers of industrial water resource assessment are as follows (see Figure 8.1):

- **Legal requirements:** The regulatory body may ask the industry to conduct periodical water resource assessment for environmental compliance.
- **Water policies and environmental integration:** Industries can conduct WRA as a functional part of their water and environmental strategy.
- **Participatory management:** WRA can be a part of activities under IWRM.
- **Climate change:** WRA can monitor the impacts of climate change.
- **Water profiling and disclosure:** WRA is required for water profiling and disclosure.
- **Impact assessment:** WRA is a tool for monitoring environmental impact assessment (EIA).

Measuring the different parameters to quantify water resources, qualitative analysis of water and estimating current water uses by different sectors

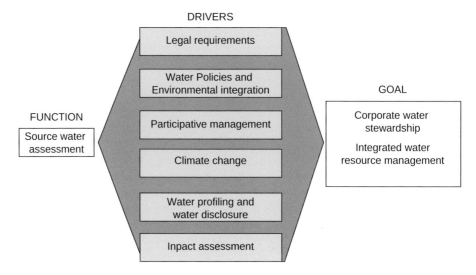

Figure 8.1 Drivers and goals of WRA.

in a geographical area is known water resources assessment (WRA). WRA requires two stages: (1) before commissioning of the project, which is also a requirement for EIA studies where baseline information is gathered for assessing the future impact on environment; and (2) periodical monitoring and assessment of the source water to understand the impact of water withdrawal on the environment. In both cases, the assessment should be done in terms of quality and quantity. Water resource assessment has a broader perspective and is useful for a range of Cross sectoral management needs. See Table 8.1.

In a corporate environment, various water assessment tools may be developed which facilitate decision support system (DSS) for water management and corporate water stewardship. See Table 8.2.

Source-water assessment is also a part of the source-water protection plan (SWPP) in which different stakeholders are involved. Source-water protection is a set of activities for maintaining ecosystem services in the area; participation of industries is crucial to effective SWA and SWPP.

8.2   Water resource assessment tools

Global Water Partnership has developed a water assessment tool (WAT) which is intended to evaluate 'water resources in relation to a reference frame' (Global Water Partnership, 2013). The frame may be a social, administrative or natural frame like a river basin, a watershed, an aquifer or an ecosystem. In case of corporate water assessment, the reference frame should be a natural hydrological unit whose physical, chemical and biological features are taken into account in the WRA tool.

Table 8.1 Cross-sectoral management needs of WRA.

Understanding	Application	Tasks
Current status of water resources at different scales, including inter-and intra-annual variability	Initial assessment for EIA, Compliance of statutory norms, baseline study	Assessment of quality, quantity in the catchment and use of water by different sectors.
Current *water use* (including variability), and the resulting societal and environmental trade-offs	Decision support for water management and CSR, baseline scenario development	Socio-economic survey, water use survey
Scale-related externalities, especially when patterns of *water use* are considered over a range of temporal and spatial scales	Modelling future use; defining projected scenario.	Catchment-related water resource assessment, long-term data generation on water; storm-water management
Social and institutional factors affecting access to water and their reliability	Conflict resolution, equitable access	Policy engagement
Opportunities for saving or making more productive, efficient and/or equitable	Decision support for economic measure and cost reduction	Water audit and water footprint assessment
Efficacy and transparency of existing water-related policies and decision-making processes	Water governance	Stakeholders' participation
Conflicts between existing information sets, and the overall accuracy of government (and other) statistics	Decision support system	Modelling and calibration

Adapted from Keller (no date).

Table 8.2 Sub-domains and components of WRA.

Sub-domains of water assessment	Components of the assessment	Target/goal of assessment
Demand and availability assessment	Assessment of quality, quantity in the catchment and use of water by different sectors. This tool visualises the balance between available resources and demand.	• Water management in the industry. • Compliance with regulatory authority. • Compliance with requirements of financial institutions.
Water assessment as a part of EIA	EIA is a dedicated tool for assessing the social and environmental impact of any project. It is a regulatory requirement of which WRA is an important part.	Compliance with requirements of regulatory authorities.
Assessment of social appropriation of water resources	This tool assesses the impact of water withdrawal on society and on the livelihood of people residing within the catchment or physical boundary of impact.	Compliance with requirements of the management and the society.
Assessment of risk or vulnerability	Assessment of industrial resilience to potential disasters, natural or social. Major component of this assessment is water.	Planning for future water needs, reduce possibility of conflict with society.
SWA for development and source protection	Identifying possible contamination source of and developing methods for protection.	Designing water processing infrastructure.

Adapted from Global Water Partnership (2013) toolbox.

8.3   General scenario

Industrial water demand is in general 'larger than domestic supply and may require larger water sources' (Sutcliffe and Lazenby, 1990). In India and Bangladesh, small industries that demand less than 500 kl/day depend on groundwater. According to a report of NGWA (2016), 18% of the total groundwater withdrawal is consumed by industries in Russia, China 26%, Japan 48%, Thailand 26% and Italy 10%. Thermal power plants and oil refineries are major water consumers; in order to conserve groundwater, these industries seek locations on or close to major river-banks. If large rivers are absent in the vicinity, small check dams, Gabion weirs and so forth are constructed across small perennial streams for steady water supply.

8.4   WRA basics

8.4.1   Conceptual and policy framework

There is no fixed template for WRA, the exercise being location specific. In a region where water is abundant, assessment may be done when deemed necessary. But, in water-stressed regions, WRA at regular intervals (gener-ally annually) is a priority task for evaluating water-supply capacity of the source.

Reliable water-resource estimation involves:

- Preparing a data collection format for water balancing;
- Estimating annual water balance from quantified (1) total intake, (2) distribution to various outlets in the industrial establishment and (3) discharge;
- Data-supported delineation of command area and formulation of command area development plans;
- Cross-checking primary data against randomly collected secondary data; and
- Presenting water-balance report to stakeholders.

WRA has three parts: (1) primary field data collection (2) secondary data collection and (3) data processing. External experts may be engaged ini-tially, but it is advisable for industries to train their own staff in data collec-tion and data processing. WRA is built upon a set of spatial and temporal data on water availability. While collecting data protocols established by regulatory bodies, the following should be maintained. WRA protocol involves:

- **Data:** Data constitute all measurable parameters (e.g. groundwater level fluctuation, monthly rainfall, distribution points, user population) used in tabulation, generating maps and feeding computer models. Satellite imagery, a commonly used tool in WRA, is called digital data.
- **Information:** Information constitutes analysed and interpreted data sets that provide answers to specific questions. For example, the average annual

rainfall is 'information' in that it cannot be measured; it is computed from monthly data.

- **Knowledge:** Information that provides insight into multi-parameter processes is knowledge.

WRA derives from the information-flow system that builds up knowledge. Figure 8.2 shows the framework of information flow system.

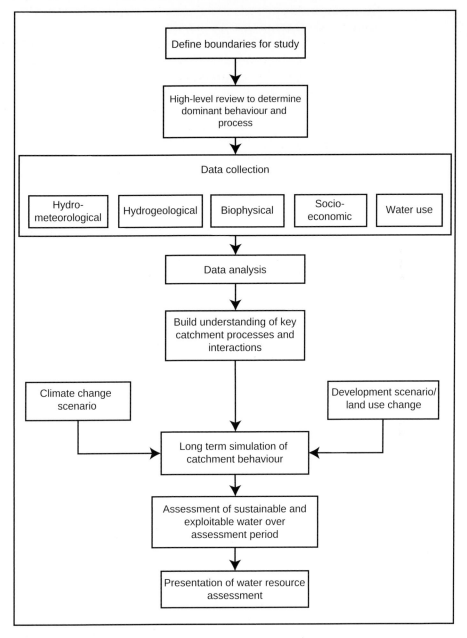

Figure 8.2 Outline schematic of the water resource assessment development process. Printed with permission from World Meteorological Organization (2012).

8.4.2   Defining a research agenda

Defining a research agenda (Huitema *et al.* 2009) requires knowledge of (1) analytical and statistical tools, (2) GIS applications, and (3) modelling software; the knowledge base thus created is the tangible outcome of research which can then be presented as a comprehensive report for policy formulation.

8.4.3   Defining the physical boundary

Water resource assessment undertaken by an industry entails mapping the zone affected by industrial activity. The exercise thus begins with a delineation of the hydrological boundary within which industrial activities are located. Scale selection for boundary delineation is crucial because major rivers typically cross through state/national/international boundaries. The boundary, pertinent to a given industry, is delineated in local watershed scale, clearly incorporating withdrawal and discharge points in the map. The difference between small-scale and large-scale maps is explained in Figure 8.3. The map of Nakusp (1:20,000) on the left (8.3a) is a *large-scale* map in relation to the *small-scale* map (8.3b) of southern British Columbia (1:5,000,000) on the right (Kuschk, 2010).

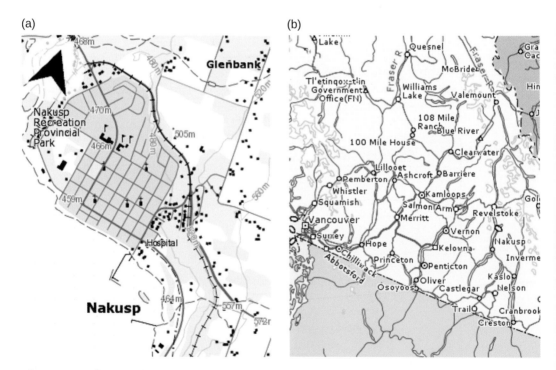

Figure 8.3 (a) large-scale map and (b) small-scale map. Sourced from (Kuschk, 2010). http://basementgeographer.com/large-scale-maps-vs-small-scale-maps/. Used under CC BY 2.0 https://creativecommons.org/licenses/by/2.0/.

8.5 WRA data generation

WRA data are collected from different sources; broadly, there are two types of data: temporal and spatial. Spatial data include land parcels, land-use pattern, villages, water wells, water divides, stream network configurations and their flood plain geometry, rocky outcrops, and so forth, which have fixed locations and can be plotted on a map. Temporal data include location-specific variable data like rainfall, stream flow, groundwater level, end user population, and, catastrophic changes brought about by natural disasters (earthquakes, tsunamis, storm surges, floods etc). Fixed spatial data are to be generated only once and can often be sourced from the public domain. Temporal data are to be generated periodically in hourly/daily/monthly/annual basis. Data can be collected from the field or from experimental setups. Such types of data are called primary data. Sometimes data available from government, research institutes, published research papers and so forth are so-called secondary data. For example, monthly rainfall data are regularly published by the meteorological department of a country and are secondary data.

8.5.1 Secondary data collection

In most countries, water assessment is periodically carried out by government agencies or state or regional regulatory authorities. These assessment boundaries are generally large catchments or administrative boundaries such as districts, municipalities or some local administrative entities. In some countries like the United States, there are several water authorities that regularly collect data for assessment; in similar federal structures, such as India, there are both national and state-level authorities who regularly collect hydrological data from an array of data-monitoring stations. But these are essentially macro-level (i.e. small-scale) data covering entire districts, states or even countries, in which data density is low, say, one data point in several km². In contrast, industries concentrate on the limited area comprising their industrial site and its environs to meet with the statutory requirements of environmental impact assessments (EIAs) and environmental management plans (EMPs).

Sanctioning authorities evaluate environmental sensitivity of industries on the basis of EIA and EMP reports; consequently, water-resource assessments for industries typically require data with say, one data point per km², that is, data density far higher than what is available in macro-scale surveys carried out by government agencies. Macro-level data generated by government agencies are useful in site selection, but after site selection, industries need to generate site-specific primary data for submission of EIA and EMP reports. It helps the cause of obtaining/renewal of business licence if corporate managers are conversant with norms and protocols of secondary data generation. In problematic areas, governments have already developed infrastructures to generate very high quality data on a regular basis (e.g. weather stations, river gauging stations, observation wells, etc.). For example, in arsenic-contaminated areas of Bangladesh, a large amount of

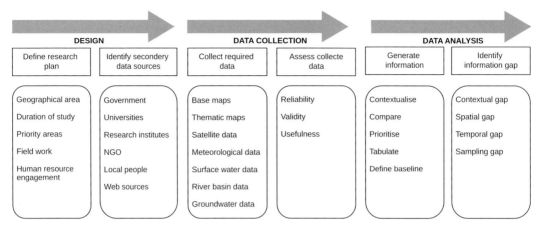

Figure 8.4 Data flow pattern in water resource assessment. Adapted from Acaps (2011).

groundwater-related data has been collected by many national and international agencies, including the British Geological Survey. The data generated by researchers will be of great help to industries; it needs emphasising that existing data do not absolve industries from their responsibity to generate site-specific data. Useful data generated by government agencies is like a free lunch, and industries should make the best of it. Reviewing existing data helps industries to optimise their data-collection plan.

8.5.2   Primary data generation

Primary data generation depends on the data gaps, reliability and relevance of secondary data. Industries can develop their own workforce for primary data collection or can outsource it to reliable organisations. Different types of data that are required for water-resource estimation and water-balance study are given in Figure 8.4.

8.5.3   Biophysical data

Terrain, surface water and groundwater recharge zones constitute an interactive trio. Visualisation of this interactive process is vital to corporate water stewardship. Biophysical data include data related to topography, soil, geology, vegetation and land use (Sheng, 1990). These data are required to calculate the volume of components in the water balance of a watershed and also in preparation of different thematic maps for visualisation. Types of biophysical data include the following:

General data

Watershed names, locations, boundaries, size, elevation, streams, tributaries, sub-watersheds and so on are general data and are required for geo-coding spatial data units (Piestrak, 2010). A good topographic map, that is, a toposheet, is very helpful in this regard. Procuring toposheets often, involves red tape; however, Google Earth has become an excellent source for obtaining general data.

Topographical data

Toposheets are basic to terrain visualisation because they show both physiography (i.e. contours and drainage) and artefacts (i.e. roads, railways, habitations etc.). Topographical maps are available in different scales. Selection of the appropriate scale is crucial to terrain visualisation; for small areas, larger-scale maps are required and vice-versa. Topographic data are essential for (1) planning field investigations, and (2) providing a base for thematic maps, in this case, water-resource assessment maps. A list of information available from a toposheet is given below.

- Elevation and relief shown by contour lines or spot heights marked on the map;
- Drainage patterns and bodies of water;
- Infrastructure, such as roads, bridges, rail lines, power transmission lines, etc.,
- Administrative boundaries like village, district, state and international boundaries;
- Broad land use, like settlement, forest, barren land, rocky outcrop; and
- Geographical coordinates.

Figure 8.5 shows how a cadastral map is used as base map and a thematic map is then generated from it.

Apart from toposheets, there are good resources from spatial digital data on the internet, which can be downloaded (sometimes free) to generate watershed maps, drainage maps, slope maps and land-use maps using appropriate GIS software. In Figure 8.6 (b), an example of a watershed map is shown, which is generated from elevation data. Digital elevation models (DEM) of watersheds are very helpful in visualising the land geometry and drainage configuration. While collecting spatial data, the ground resolution and scale should be kept in mind. In small watersheds with water-stressed conditions, large-scale data may be required. Sources of DEM are:

- USGS GTOPO30 is a global digital elevation model (DEM) with horizontal grid spacing of 30 arc seconds (approximately 1 km) (USGS, 2015a);
- Shuttle Radar Topography Mission (SRTM) 1 Arc-Second Global data (USGS, 2015b); and
- ETOPO5 and ETOPO2 are global elevation databases (National Geophysical Data Center, 2016).

Air photos and satellite imagery precede DEM and have been widely used; topographical maps, prepared from ground surveys, precede air photos, and satellite imageries have been widely used; these 'ancient' tools are still useful in validating DEM data.

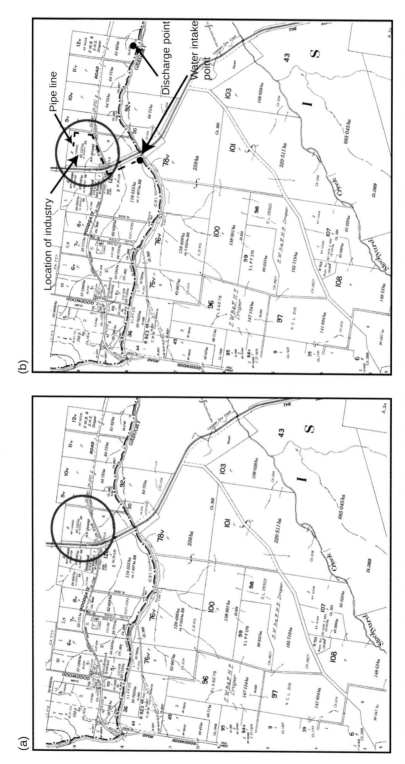

Figure 8.5 A cadastral map is used as base map, and a hypothetical thematic map is generated on the said base map. Adapted from ICSM (2016). Under CC BY license.

Land use and land cover

Land-use and land-cover data are forms of spatial data and can be mapped. Data can be collected from field work and plotted on a cadastral map. A land-use map can also be prepared through satellite image processing. A land-use data sheet expresses all land uses within the watershed by area covered and by percentages of area covered.

Watershed boundary and drainage

Watershed boundaries can be delineated manually by connecting points of highest elevation surrounding a river system in a contour map. Manual drawing can now be replaced by processing DEM data using appropriate software. The advantage of DEM data over surveyed contour maps is that software has the flexibility to generate watershed boundaries at the required scale. Figures 8.6a and 8.6b show two watershed maps, one drawn manually and the other from DEM (using SRTM data and Global Mapper 13.2 software).

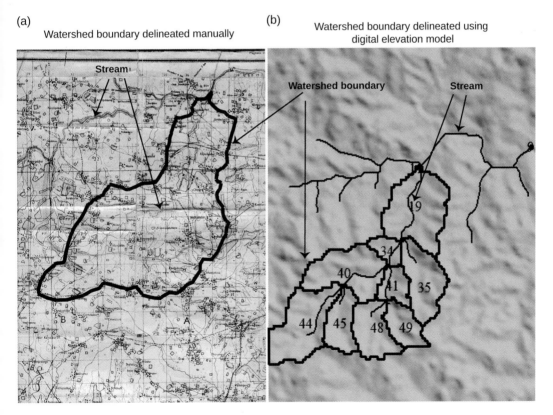

Figure 8.6 Watershed map showing (a) watershed boundary delineated manually on a toposheet, (b) watershed boundary delineated using SRTM data, in which sub-watersheds are shown by number.

Geology

Surface geological data collected from exposed rocks (supplemented by laboratory data on rock samples collected from the field) are plotted on topographic sheets of appropriate scale to generate geological maps showing distribution of rock types and their configuration; for hydrological applications, less accurate, but fairly reliable surface geological maps are generated from satellite imagery validated by limited ground checking. Subsurface rock types and configurations are reconstructed from exploratory drilling and geophysical studies; in hydrology, application of sub-surface geology is to reveal the aquifer geometry.

8.5.4   Hydrometeorological data

A list of required data methods of data generation is given in Table 8.3.

The advent of automatic weather stations (AWSs) has made the task of meteorological data-collection easier, especially from remote areas. The AWS is an automated system that collects weather data in close intervals

Table 8.3 List of hydrometerological data required for a WRA.

Parameter	Method of data acquisition	Frequency
Rainfall	A rain gauge is used for measuring the amount of rain that reaches the ground surface during a storm. The rain gauge is the most common method for measuring rainfall. Digital rain gauges with data acquisition systems are also available.	Daily/hourly
Evaporation	Evaporation data is collected from a evaporimeter pan. The pan is kept in open air; the depth of water evaporated from the pan during a unit of time (say, one day) is recorded.	Daily
Temperature	Measured with maximum-minimum conventional thermometers or digital thermometers.	Daily
Evapo-transpiration	Evapo-transpiration is the sum total of surface-water evaporation, capillary fringe evaporation from water table and transpiration. Evapo-transpiration is a function of temperature, sunshine hour, relative humidity and average wind speed.	Daily
Relative humidity (RH)	Relative humidity is the ratio of water vapour in air and the maximum amount of moisture that air can hold at a given temperature and pressure. Relative humidity is measured with a hygrometer.	Daily
Wind speed	The instrument for wind speed measurement is called an anemometer; both analog and digital metres are available.	Daily/hourly
Runoff	Estimation of runoff is given in Section 8.7.1	Daily/hourly
Stream discharge	Stream flow is measured with the help of a river-gauging system. The method is discussed in Section 8.9.	Daily
Pond (lake/reservoir) storage	Lake or reservoir storage is measured at any given time by multiplying the surface area of the pond/reservoir by the average depth of the pond. This is further elaborated in Section 8.11	Monthly
Soil moisture	Soil moisture of a plot of land is measured with a soil moisture metre; both analog and digital metres are available.	Daily/weekly

Figure 8.7 An automatic weather station at Discovery Park, Seattle, USA.

and keeps records in digital format that can be retrieved in a spreadsheet. Figure 8.7 shows an automatic weather station located at Discovery Park, Seattle.

8.5.5   Data table

A simple data recording table of hydrometeorological data is given in Table 8.4 that records meteorological parameters essential for WRA. This table is valid for manually collected data. AWSs deliver data in tabular format through data-acquisition systems.

8.5.6   Hydrogeological data

- **Groundwater level:** Groundwater level is measured from an open well, a tube well or an observation well. Groundwater provides information on (1) the depth of water table/peizometric surface below ground level and (2) the elevation of water table/piezometric surface above mean sea level. A typical data table is shown in Table 8.5.

A detailed method of groundwater level measurement is shown in Box 8.1

Table 8.4 Standard hydrometeorological data table.

Location			Altitude		M		

Month			Year				

Latitude			Longitude				

Date	Max. temp (°C)	Min. temp (°C	Humidity %	Precipitation (mm)	Evaporation (mm)	Sunshine hour	Average wind speed (km/h)
1							
2							
3							
4							
...							
...							
...							

Table 8.5 Water-level data recording table.

Sl No	Location	Latitude	Longitude	Date of collection	RL of GL AMSL (m)	DWL BGL (m)	RL of WL AMSL (m)
(a)	(b)	(c)	(d)	(e)	(f)	(g)	(f-g)

RL of GL = reduced level of ground level of the measuring well. AMSL = above mean sea level, DWL BGL = depth of water level below ground level, RL of WL = reduced level of water level.

- **Groundwater flow pattern:** Groundwater level data are plotted as contour lines or isopleths generating 'depth to water level' maps. These maps can be drawn manually or can be generated through computer programmes to determine groundwater flow patterns at the time of measurement.
- **Aquifer parameters:** Aquifer parameters like specific yield, storage coefficient and transmissivity do not change with time. These data are collected through aquifer tests and are used in estimation of groundwater recharge and well designing.
- **Groundwater quality:** This entails the sampling and analysis of water from tube wells or dug wells. There are prescribed water-sample collection protocols. Methods of water-sample collection have been discussed in section 8.11.

Method of Groundwater Level Measurement

Groundwater level is measured from dug wells, tube wells and observation wells. For all measurements, a fixed reference point must be established at the well head. This point is usually the top of the casing or the access port in water-supply wells. The reference point is accurately surveyed to determine elevation above mean sea level, the accuracy level being 0.01 ft; the same measuring point must be used in successive measurements. Measurement of depth to free ground water surface is done by the following methods (method accuracy is given in the FPS system also because the material is sourced from US Government agencies):

- **Chalked steel tape method** (Mills, 2005): The most accurate measurement is obtained with chalked steel tape (accuracy ± 0.01 ft or 0.005 m). This method utilizes a graduated tape with a weight attached to its end; lead weights were used formerly, but present concerns about water quality require that the weight be brass or stainless steel. The lower 3–4 ft (or 1.5 m) of the tape is coated with carpenter's chalk and lowered into the well until the weight hits the water surface, which can be heard. The tape is held at a reference point ('hold' position) and the tape position recorded. The depth to the water level is determined by subtracting the length of wet tape (indicated by wet chalk) from the total length of tape lowered into the well. To lessen the possibility of computation errors, measurement should be repeated to ensure accuracy. Steel tape measurements are usually done where horizontal gradients are very low. For tube wells with submersible pumps, particularly small-diameter (<6 inches or 100 mm) domestic wells with pitless adapters, the tapes are used without weight to avoid entanglement with pump wiring. See Figure 8.8.
- **Electronic water level indicators:** These consist of a spool of dual conductor wire with a probe attached to the end and an indicator (light and/or buzzer). When the probe comes in contact with the water, the circuit is closed, and the indicator signals contact. Penlight or 9-volt batteries are normally used as power sources. Measurements should be made and recorded to the nearest 0.01 feet (US-EPA, 2013).

BOX 8.1

8.5.7 Socioeconomic data

- **Population and demography:** Populations of people living in a river basin or the assessment area can be directly surveyed along with socioeconomic survey. Population data and growth rates are also available from census reports for a particular census year. Population size in the required year is estimated from the growth rate.
- **Industry:** While evaluating the environmental impact of industries, their socioeconomic contributions have to be kept in mind, which is precisely what governments do. And yet governments are vilified for their efforts: A classic example is freshwater protection plans of state and union governments in India; these are elaborate plans designed to function in collaboration with industries. Implementation of these plans works effectively, in that the rampant pollution prevalent up to the 1980s has been largely contained, but the restoration process, understandably, is slow.

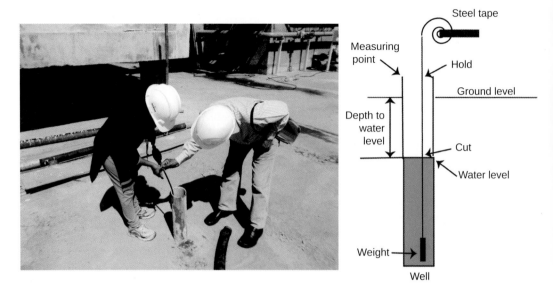

Figure 8.8 Measuring groundwater level in an industry by chalked steel tape method.

- **Agriculture:** Agricultural data are collected from the agriculture department and also from field studies in different seasons, where required. The collected data provide information on cropping area, types of crop, irrigation water requirements and the magnitude of livelihood from agriculture.
- **Recreation:** Recreational facilities of a watershed are also a marker of the socioeconomic status of the project area.
- **Infrastructure:** All infrastructure-related information, such as transport facilities, water infrastructures, educational facility, hospitals, government offices and so on should be collected.

8.5.8   Water use and discharge

The parameters and methods of collection of water use data are shown in Table 8.6.

8.6   Water balance

Water balance estimation is a tool for estimating the status of water availability from different sources over a specific time period. Actual water availability can be visualised from water-balancing data, which is crucial to decision making in water management. The monthly water balance equation for a watershed is:

$$P = Q + ET + EV + \Delta S \tag{8.1}$$

Where P = precipitation, Q = stream discharge/runoff, ET = evapo-transpiration, EV = evaporation from reservoir, and the parameter ΔS is total change in

Table 8.6 Parameters and methods of collection of water use data.

Parameter	Method of data acquisition	Frequency of measurement	Unit
Irrigation draft	Data-collection method is complicated. In general, it is collected from the owners of irrigated land or from the agency providing irrigation.	Monthly or annual	m³
Industrial draft	Field data involve (1) the number of intake points where water is sourced from surface bodies of water, (2) number of tube wells for groundwater withdrawal, (3) rate of withdrawal and (4) daily working hour. The formula for quantifying daily groundwater withdrawal is: $GW_d = Q_{gw} \times H \times n$ Where, 'GW_d' is withdrawal *from one well*, 'Q_{gw}' hourly pumping capacity of that well, 'H' is daily working hour and 'n' is the number of working days of the month; if water is drawn from multiple tube wells, data from each tube well are summed up. Likewise, for surface water withdrawal the formula is: $SW_d = Q_{sw} \times H \times n$ Where 'SW_d' is water withdrawal *from one intake point* and 'Q_{sw}' is hourly pumping capacity at each intake point.	Monthly or annual	m³
Water abstraction for domestic use.	The formula for total domestic consumption is simple: (average per capita water use) × (population within the catchment). The data collection, however, needs field survey.	Monthly or Annual	m³
Discharged water	Water discharged from industries, return flow from agricultural fields and municipal wastewater should be measured. Agricultural return flow is estimated from field measurements. Data on industrial wastewater is furnished by the industry. Domestic wastewater can be quantified through household surveys and measuring municipal drainage outfall.	Monthly or annual	m³

storage, that is the sum of changes in reservoir storage (ΔR), change in aquifer storage (ΔGw) and change in soil moisture (ΔSm). A positive value of aquifer storage change is considered to be groundwater recharge.

$$\Delta S = \Delta Sm + \Delta Gw + \Delta R \qquad (8.2)$$

Methods of measuring different parameters are given Table 8.7.

8.7 Estimation of surface runoff

Runoff is defined as the portion of the precipitation that makes its way towards rivers or oceans as surface or sub-surface flow. After infiltration and other losses, the excess water on land surface flows out along ephemeral rivulets into main drainage channels. Infiltrated rainwater beneath sloping ground

Table 8.7 Methods of estimating parameters under water-balance estimation (volumetric).

Parameters	Data requirement	Method of calculation
P	Area of the watershed and precipitation depth	Precipitation volume = (Precipitation depth) × watershed area. Data is available from local meteorological offices. Industries may install their own rain gauge for more frequent data.
ET	• PET (potential evapo-transpiration) • Crop coefficient • Land use • Cropping data	Actual evapo-transpiration (ET) can be calculated from the potential evapo-transpiration (PET) using the following equation: ET (mm) = PET (mm) × Crop Coefficient (k_c) for each crop. Note: It is difficult for industries to estimate the PET and K_c of an area. It is, therefore, a secondary data input available from local meteorological or agriculture office.
EV	• Monthly rate of pan evaporation	EV = (Reservoir surface area) × (Monthly evaporation). Evaporation rate is measured by pan evaporimeter.
Q	• Wetted area of river cross section • Average flow rate	River discharge is measured at stream gauging stations. Q = (wetted cross section) × (flow rate measured at the gauging station). Measurements can be taken daily or weekly and monthly rate of flow can be estimated.
ΔSm	• Monthly average change in soil moisture depth • Area of each soil class	This parameter is the difference of average soil moisture by volume of the current month and average soil moisture of the previous month.
ΔGw	• Monthly groundwater level fluctuation (Δh) • Assessment area (A) • Specific yield (S_y)	Monthly groundwater level fluctuation is the difference between the water level of the current month and that of the previous month. For unconfined aquifer groundwater recharge[1] is, $\Delta Gw = A \times \Delta h \times S_y$
ΔR	• Surface area of each pond • Average change in depth of each pond	ΔR = Pond volume in the last month minus pond volume of the current month. Pond volume should be measure on a fixed date each month.

[1] Generally, groundwater recharge is estimated annually. In that case, the difference between pre- and post-monsoon water levels is considered for estimating annual water level fluctuation.

moves parallel to the land surface as sub-surface flow and reappears on the surface at certain other points; such flows are called 'interflow' or 'through-flow' which joins surface flows where the slope levels out. The infiltrated water that does not emerge out as throughflow, moves downward and merges with groundwater, inducing sub-surface flow or groundwater flow.

Runoff can be estimated by various methods. These can be classified under the following headings:

• Runoff estimations based on empirical formulae and tables
• Runoff estimations based on land use and land treatment

There are several methods for estimation of runoff in a catchment area, of which two very popular ones are discussed here: (1) Khosla's method, and (2) the runoff curve number (CN) method.

8.7.1 Khosla's Formula

Khosla analysed the rainfall, runoff and temperature data for various catchments in India and the USA to arrive at an empirical relationship between runoff and rainfall. The time period is taken as a month.

$$Rm = Pm - Lm$$

Where,

Rm = Monthly runoff in cm.
Pm = Monthly rainfall in cm.

Lm is monthly evapo-transpiration (ET) loss in cm; $Lm = 0.48 \times Tm$ (for $Tm > 4.5\,°C$).
Where

Tm = Mean monthly temperature (°C) in catchment.
Annual Runoff = $\sum Rm$ (i.e., sum total of Rm in a calendar year).

The Khosla formula is indirectly based on the water-balance concept, and the mean monthly temperature is used to reflect losses due to ET. Runoff estimated from the mean monthly temperature and precipitation in Uthala watershed of India is shown in Table 8.8.

8.7.2 Estimation of rainfall runoff by SCS curve number (CN) method

The method, also known as the Hydrologic Soil Cover Complex Method, was developed by the Soil Conservation Service (SCS) of the US Department of Agriculture (USDA) for use in rural areas. It is a versatile and widely used procedure for runoff estimation. The requirements for this method are land use, rainfall amount and curve number. The curve number is based on (1) hydrologic soil group, (2) land use treatment and (3) soil moisture condition.

SCS soil scientists classify soil into four hydrologic groups A, B, C and D. See Table 8.9.

Land use and treatment classes

The commonly used land use and treatment classes are briefly described below. These classes are used in determining hydrologic soil-cover complexes, which are used in one of the methods for estimating runoff from rainfall.

- **Cultivated lands:** These include all crop fields such as maize, sugarcane, paddy, wheat and so on.
- **Fallow lands:** These are lands that have been cropped and kept fallow after harvest. The period over which land is kept fallow is not less than one year and not more than five years.

Table 8.8 Runoff of Uthala watershed according to Khosla's Formula.

Parameter	Jan	Feb	Mar	Apr	May	Jun	Jul	Aug	Sep	Oct	Nov	Dec	Total
Pm (cm)	1.9	3.3	2.5	2.3	6.4	21.7	32.5	32.5	1.4	9	1.5	0.4	115.4
Tm (C)	19	22	27	32	34	31	28	28	28	27	22	20	
Lm (cm)	0.91	1.58	1.2	1.10	3.07	10.42	15.6	15.6	0.67	4.32	0.72	0.19	55.39
Rm (Pm-Lm)	0.98	1.71	1.3	1.19	3.32	11.28	16.9	16.9	0.72	4.68	0.78	0.21	60.01
Runoff volume (MCM)	1.36	2.37	1.79	1.65	4.59	15.56	23.30	23.30	1.00	6.45	1.08	0.29	82.74

Table 8.9 Hydrological soil groups.

Soil group	Description
A	Soils that have low runoff potential and high infiltration rates even when thoroughly wetted. These are thick sand/gravel beds with high water transmission rates (>0.30 inch/hour).
B	Moderately fine textured soil with moderate infiltration when thoroughly wetted; water transmission rate being 0.15 to 0.30 inch/hr.
C	Soils with low infiltration rates when thoroughly wetted; consisting chiefly of soils with a layer that impedes downward water movement and/or soils with moderately-fine to fine textures. These soils have a low rate of water transmission, ranging from 0.05 to 0.15 inch/hr.
D	Soils having high runoff potential with very low infiltration rates when thoroughly wetted; these are clay soil with high swelling potential, soils soaked in permanently high water table, soils with clay pan or clay layer at or near the surface, and shallow soils over impervious bed rocks; water transmission ranges from 0 to 0.05 inch/hr.

Source: USDA (1989). The units are in FPS system because it is sourced from an US agency.

- **Uncultivated lands** include:
  - Permanent pastures and other grazing lands.
  - Cultivable waste, which are lands available for cultivation but not used or, cultivated and abandoned after a few years for one reason or another. Land once cultivated but lying fallow in excess of five years in succession shall also be included in this category.
- **Forest area** (state-owned or private) includes all land classed as forest by legal enactment or maintained as potential forest land.
- **Tree crops** include woody perennial plants that reach a mature height of at least eight feet and have well defined stems and a definite crown shape.
- **Lands put to non-agricultural use**, such as buildings, roads, railroads and so on.
- **Barren and uncultivable lands**, such as mountains, deserts and so on.

Antecedent Moisture Condition (AMC)

This is an indicator of watershed wetness and soil moisture content prior to a storm. AMC can have a significant effect on runoff volume. Recognizing its significance, Soil Conservation Services (SCS) have developed a guide for adjusting the curve number (CN) method to AMC based on the total rainfall in the five-day period preceding a storm. Three levels of AMC are used in CN method: AMC-I for dry conditions, AMC-II for normal and AMC-III for wet conditions. Table 8.10 gives seasonal rainfall limits for these three antecedent moisture conditions.

Runoff curve number (CN)

This is an empirical parametric value of a land parcel and a coefficient that reduces the total precipitation to runoff potential deducting losses through evaporation, absorption, transpiration and surface storage; the higher the

Table 8.10 Rainfall Limits for Antecedent Moisture Condition.

AMC	Five-Day total antecedent rainfall (cm)	
	Dormant Season	Growing Season
I	<1.25	<3.5
II	1.25 to 2.75	3.50 to 5.25
III	>2.75	>5.25

Source: USDA (1989).

CN value the higher the runoff potential. SCS has developed a table of CN values based on land use, land treatment and hydrologic soil group. See Tables 8.11a, 8.11b and 8.11c. These tables are sourced from USDA (1989).

8.7.3 Runoff calculation

The formulas for runoff calculation are given below in Equations 8.3 to 8.7. In this method, the CN of the land is first determined from Table 8.12a or 8.12b. Then, the value of S is determined from Equation 8.4 and one-day precipitation, and then Q or runoff is calculated from Equation 8.5 or 8.6 or 8.7.

Rainfall-runoff relation is given below:

$$Q = \left[(P-I_a)^2 \right] / \left[(P-I_a) + S \right] \tag{8.3}$$

Where, Q is the actual runoff in mm, S the potential maximum retention in mm, and I_a, the initial abstraction during the period between the beginning of rainfall and runoff in equivalent depth over the catchment in mm $Q=0$ if $P \le I_a$

In areas covered by black soils having Antecedent Moisture Conditions (AMC) II and III, I_a in the equation is equal to 0.1S, whereas in all other regions, including those with black soils of AMC I, I_a is equal to 0.3S. In order to show this relationship graphically, 'S' values are transformed into 'Curve Numbers (CN)' using the following equation (for metric system):

$$CN = 25400 / (254 + S) \tag{8.4}$$

Using the above equation, the following equations have been developed:

$$Q = \left[(P-0.3S)^2 \right] / \left[(P+0.7S) \right] \tag{8.5}$$

$$Q = \left[(P-0.1S)^2 \right] / \left[(P+0.9S) \right] \tag{8.6}$$

According to USDA (NRCS, 1986),

$$Q = \left[(P-0.2S)^2 \right] / \left[(P+0.8S) \right] \tag{8.7}$$

Table 8.11a Runoff curve number for urban areas.

Cover type and hydrological condition	Average percent impervious area	Curve number for hydrologic soil group			
		A	B	C	D
Fully developed urban areas (vegetation established)					
Open space (lawns, parks, golf courses, cemeteries, and so on:					
Poor condition (grass cover <50%)		68	79	86	89
Fair condition (grass cover 50 to 75%)		49	69	79	84
Good condition (grass cover > 75%)		39	61	74	80
Impervious areas:					
Paved parking lots, roofs, driveways, and so on (excluding right of way)		98	98	98	98
Streets and Roads					
Paved, curbs and storm sewers (excluding right of way)		98	98	98	98
Paved; open ditched (including right of way)		83	89	92	93
Gravel (including right of way)		76	85	89	91
Dirt (including right of way)		72	82	87	89
Western Desert Urban Areas					
Natural desert landscaping (pervious areas only)		63	77	85	88
Artificial desert landscaping (impervious weed barrier, desert shrub with 1–2 inch sand or gravel mulch and basin borders)		96	96	96	96
Urban Districts					
Commercial and business	85	89	92	94	95
Industrial	72	81	88	91	93
Residential Districts by Average Lot Size					
1/8 acre or less (town houses)	65	77	85	90	92
1/4 acre	38	61	75	83	87
1/3 acre	30	57	72	81	86
1/2 acre	25	54	70	80	85
1 acre	20	51	68	79	84
2 acre	12	46	65	77	82
Developing urban areas:					
Newly graded areas (pervious areas only, no vegetation)		77	86	91	94

Source: USDA (1989).

Equation 8.5 is applicable to all soil regions of India except black soil areas in the Hydrological Soil Groups column. Equation 8.6 applies to black soil regions. This equation should be used with the assumption that cracks, which are typical of these soils when dry, have been filled. Therefore, Equation 8.3 should be used where AMC falls into groups II and III. In cases where the AMC falls in group I, Equation 8.4 should be used. For practical purposes, consider watersheds to be AMC II, which is essentially an average moisture condition.

It is recommended that for a given watershed where there are many land classes and land uses a weighted average CN for the watershed should be estimated. Area-weighted composite curve numbers for various > conditions of land use and hydrologic soil conditions are computed as follows:

Table 8.11b Runoff curve number for cultivated agricultural land.

Cover description			Curve number for hydrologic soil group			
Cover Type	**Treatment**	**Hydrologic Condition**	**A**	**B**	**C**	**D**
Fallow	Bare soil		77	86	91	94
	Crop residue cover (CR)	Poor	76	85	90	93
		Good	74	83	88	90
Row crops	Straight row (SR)	Poor	72	81	88	91
		Good	67	78	85	89
	SR + CR	Poor	71	80	87	90
		Good	64	75	82	85
	Contoured ©	Poor	70	79	84	88
		Good	65	75	82	86
	C + CR	Poor	69	78	83	87
		Good	64	74	81	85
	Contoured and treated (C&T)	Poor	66	74	80	82
		Good	62	71	78	81
	C&T + CR	Poor	65	73	79	81
		Good	61	70	77	80
Small grain	SR	Poor	65	76	84	88
		Good	63	75	83	87
	SR + CR	Poor	64	75	83	86
		Good	60	72	80	84
	C	Poor	63	74	82	85
		Good	61	73	81	84
	C + CR	Poor	62	73	81	84
		Good	60	72	80	83
	C&T	Poor	61	72	79	82
		Good	59	70	78	81
	C&T + CR	Poor	60	71	78	81
		Good	58	69	77	80
Close seeded or broadcast legumes or rotation meadow	SR	Poor	66	77	85	89
		Good	58	72	81	85
	C	Poor	64	75	83	85
		Good	55	69	78	83
	C&T	Poor	63	73	80	83
		Good	51	67	76	80

Source: USDA (1989).

$$CN = (CN_1 \times A_1) + (CN_2 \times A_2) + (CN_3 \times A_3) + \cdots\cdots (CN_n \times A_n) / A \qquad (8.8)$$

Where A1, A2, A3, ..., An represent areas of polygon having CN values CN1, CN2, CN3,.....,CNn respectively and A is the total area.

An example of estimating weighted average CN of a watershed is given Table 8.12.

Table 8.11c Runoff curve number for Arid and semiarid rangelands.

Cover description			Curve numbers for hydrologic soil group			
Cover type	Hydrologic condition	A	B	C	D
Herbaceous—mixture of grass, weeds, and low-growing brush, with brush the minor element	Poor	–	80	87	93
	Fair	–	71	81	89
	Good	–	62	74	85
Oak-aspen—mountain brush mixture of oak brush, aspen, mountain mahogany, bitter brush, maple, and other brush	Poor	–	66	74	79
	Fair	–	48	57	63
	Good	–	30	41	48
Pinyon-juniper—pinyon, juniper, or both; grass understory	Poor	–	75	85	89
	Fair	–	58	73	80
	Good	–	41	61	71
Sagebrush with grass understory	Poor	–	67	80	85
	Fair	–	51	63	70
	Good	–	35	47	55
Desert shrub—major plants include saltbush, greasewood, creosotebush, blackbrush, bursage, palo verde, mesquite, and cactus.	Poor	63	77	85	88
	Fair	55	72	81	86
	Good	49	68	79	84

Poor: < 50% ground cover or heavily grazed with no mulch; Fair: 50–75% ground cover and not heavily grazed; Good: > 75% ground cover and light or only occasionally grazed. Source: USDA (1989).

Table 8.12 Estimating weighted average CN of a watershed.

Sl No.	Land Use	Area (ha)	Soil group	CN
1	Open space > 75% grass cover	50	A	39
2	Bare soil	150	A	77
3	Fallow (good hydrological condition)	300	B	87
4	Paved parking lot/roof	30	A	98
5	Total area	530		
6	Weighted average CN			80

Example

In a hypothetical watershed, a single-day rainfall is 105 mm. The land use and soil group under AMC III are given in Table 8.12. Accordingly, the peak runoff for the watershed is:

Value of S from equation 8.2, i.e., $CN = 25400/(254 + S)$
$S = (25400/CN) - 254$ or $S = (25400/80) - 254 = 63.5$ mm

From equation 8.7 we get $Q = [(P - 0.2S)^2]/[(P + 0.8S)]$

Or $Q = [(105 - 0.2 \times 63.5)^2]/[(105 + 0.8 \times 63.5)] = 54.68$ mm

8.8 Estimation of stream discharge

Stream discharge can be measured using (1) volumetric gauging, (2) float gauging, and (3) current metering. The methods are as follows.

8.8.1 Volumetric gauging

Stream discharge measures (in $m^3 s^{-1}$) the 'spot discharge', that is, flow rate at a particular place and time. In small streams, the flow rate can be measured directly with volumetric gauging. In this method, stream flow is diverted through a small notch (a V notch), and water is collected in a graduated bucket; flow rate is estimated by the time taken to fill a graduated bucket. For example, if the bucket is of 100 litre capacity and the time taken to fill it is 10 seconds then the rate of flow is $100/10 = 10 m^3 s^{-1}$.

8.8.2 Float gauging

In this method, a cross section of the stream is determined (or taken from secondary data), and the average velocity of streamflow is measured. If 'v' is the velocity and 'A' the area of cross section, then discharge 'Q' is given by $Q = A \times v$. For example, if 100 sq metres is the cross section and velocity 0.5m/sec, then stream flow is $100 \times 0.5 = 50$ $m^3 s^{-1}$. Stream velocity is generally measured with a 'flow meter'. The float method is applied in the absence of flow meters; a conspicuously visible floating object is dropped in the river, and the time needed to travel between two marked points is measured; average velocity is calculated from several measurements taken at different points between banks; averaging is done because flow is not uniform; velocity increases from banks to midstream.

Stream cross section is measured as follows (see Figure 8.9):

1 An orthogonal line is selected, along which the cross section will be measured.
2 The width is measured with measuring tape in small streams; for wide rivers, a bridge comes in handy.
3 The selected cross section line is divided into several sub-sections.
4 Depth of the stream is measured at several points along the selected cross section line with graduated poles, weighted ropes or electronic sounding.
5 Area of each subsection is estimated by multiplying depth with width.
6 Flow measurement is taken at each subsection.
7 Stream flow is calculated by averaging sub-section flow rates.

8.8.3 Current metering

The meter has rotating blades (impeller) driven by stream current. The rotor axle is coupled with a suitably calibrated galvanometer that directly gives the flow velocity.

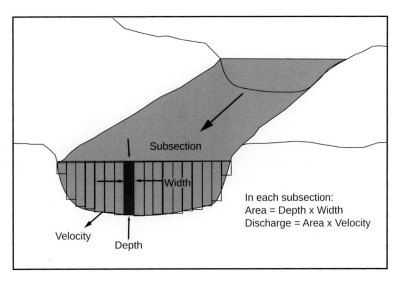

Figure 8.9 Stream discharge measurement. Adapted from USGS (2016).

8.9 Estimation of renewable groundwater resource

Quantification of groundwater resources is often critical, and no single comprehensive technique has yet been identified which is capable of accurate groundwater assessment (CGWB, 2008). However, a close estimation of groundwater resources is possible if sufficient data are available. Groundwater resources can be divided in two parts, (1) static reserve and (2) dynamic resource. Static reserve is the amount of groundwater that permanently rests in the aquifer; it is the amount of water stored in the aquifer prior to commencement of monsoon recharge. Here we shall discuss how the dynamic resource is estimated.

Dynamic resource (also called renewable groundwater resource) is the amount of groundwater recharged annually. Recharging results in rise of groundwater levels in wells, which can be measured. There are several experimental methods of estimating renewable groundwater resource. In this chapter, the two most widely used methods are discussed.[1]

8.9.1 Water level fluctuation method

The groundwater level is at its lowest during the pre-monsoon period. At the onset of monsoon, the rain groundwater level starts to rise, and towards the end of monsoon season it reaches its highest level. The difference of groundwater level between the pre- and post-monsoon periods is the annual water-level fluctuation. Similarly, the difference in water levels between the beginning and the end of a month is called the monthly water level fluctuation. Groundwater levels are monitored monthly and/or during pre- and post-monsoon period. There are four sources for data collection:

1 Observation wells installed by the government (permission required);
2 Observation wells installed by industries;
3 Privately owned tube wells; and
4 Wells installed by the industry for community water supply.

Groundwater level data is collected from several monitoring wells in a watershed and then averaged. If the average groundwater-level fluctuation over a period of time (say, one month) is 'Δh' in a watershed of area 'A' and specific yield 'S_y' then groundwater recharge is expressed by the equation:

$$\Delta Gw = A \times \Delta h \times S_y \tag{8.9}$$

In practice groundwater levels are collected before and after the rainy season, and the change in water level is considered as annual water-level fluctuation. This value is used in calculating annual groundwater recharge.

Specific yield

Specific yield depends on the permeability of aquifer material and is estimated by pumping tests. Specific yield data can be obtained from the groundwater department of the country; alternatively, if the aquifer material is known, it can also be obtained from a standardised table (Table 8.13).

Groundwater-level data should be collected from several tube wells (>3) in the watershed; tube well location and water-level fluctuation data are plotted on a base map to generate different thematic maps. These thematic maps show seasonal depth of water level, annual water level fluctuation

Table 8.13 Specific yields of various materials [Rounded to nearest whole percentage], (Johnson, 2011)

Material	Specific Yield in %		
	Maximum	Minimum	Average
Clay	5	0	2
Silt	19	3	8
Sandy clay	12	3	7
Fine sand	28	10	21
Medium sand	32	15	26
Coarse sand	35	20	27
Gravelly sand	35	20	25
Fine gravel	35	21	25
Medium gravel	26	13	23
Coarse gravel	26	12	22

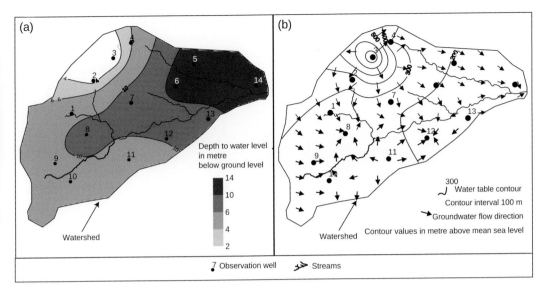

Figure 8.10 (a) Depths to water level zone map and (b) water table contour map showing groundwater flow direction of a hypothetical watershed.

and groundwater flow direction. In Figure 8.10, a depth to water level map and a water table contour map are shown.

8.9.2 Rainfall infiltration method

The rainfall infiltration method is adopted when monitoring wells in a watershed are absent or insufficient in number. The rate and amount of groundwater recharge is a function of annual rainfall and aquifer material. For example, if the aquifer material is semi-consolidated sandstone and annual rainfall 1000 mm, then the minimum rainfall recharge is 100 mm (10%). Rainfall infiltration percentages depend on several factors like annual precipitation, surface material, aquifer depth and so on. Commonly used rainfall infiltration factors are given in Table 8.14.

8.9.3 Soil water balance method

This is one of the most widely used methods of estimating annual groundwater recharge. This method was developed by Thornthwaite (1948) and was later revised. It is based on inflow and outflow of water in a system using the formula:

$$GW = P - ET - \Delta Sm - Q \tag{8.10}$$

Where GW is groundwater recharge, P is annual precipitation, ET is actual evapo-transpiration, ΔSm is change in soil water storage and Q is runoff. All parameters are quantified by volume (m^3) or by depth (mm).

Table 8.14 Rainfall infiltration factors of different types of hydrogeological situation.

S. No	Hydrogeological situation	Rainfall infiltration factor (percentage of normal rainfall)
1	Sandy alluvium	20–25%
2	Clay alluvium	10–20%
3	Semi-Consolidated Sandstones (friable and highly porous)	10–15%
4	Hard rock area	5–10%

In this method, parameters on the right-hand side are estimated according to the methods discussed in Sections 8.6 and 8.7.

8.10 Estimation of pond/reservoir storage volume

Pond volume measurement is essential for (1) water balance calculation, and (2) framing pond management plans and most importantly to (3) determine the precise amount of aquatic herbicide(s) needed for a pond. A watershed may have several types of ponding structures such as dam reservoirs, large and small impounding structures for water harvesting, excavated lakes for landscaping or recreational use, natural horseshoe lakes and wetlands. The basic calculation is: Pond volume (V) = pond area (A) × pond depth (h). Calculation of pond area 'A' uses simple formulae if the pond has a regular geometry (e.g. length × width for rectangular pond, $2 \pi r^2$ for circular pond etc), which cannot be done for irregularly shaped ponds.

8.10.1 Area calculation irregularly shaped ponds

Three methods are described here which are arranged according to increased accuracy.

1 **Average length and width method:** Take numerous measurements to determine the average length and average width as shown in Figure 8.11. Make sure that both the longest and shortest distances are measured while calculating the average length, and the widest and narrowest distances for determining the average width. The accuracy of this method depends on the number of measurements taken, the more the better. Pond area = Average length × Average width.

2 **Hand-held global positioning systems (GPS):** Recent technology like using a hand-held GPS system can be used to measure the area of a pond. Use the GPS to mark several waypoints on the shoreline of the pond. Enter the data in GIS software and accurately measure the area.

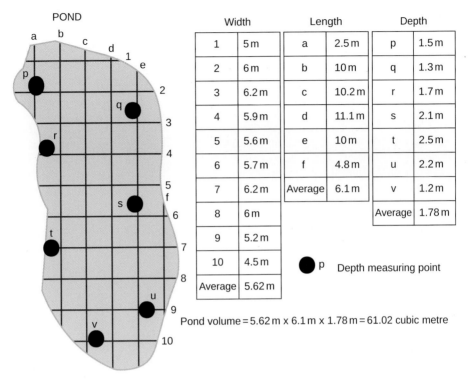

Figure 8.11 Method of pond depth and surface area calculation.

3 **Google Earth and similar applications:** Web-based mapping applications like Google Earth or Bing Maps, which use satellite imagery to display a map of your pond or lake. Use the web tool to draw an outline of the pond, transfer the data into GIS software and measure the area accurately.

8.10.2 Pond depth and volume estimation

Pond depth is measured at various points using measuring poles and water-level measuring tapes and then averaged; a small boat may be necessary if the pond is large. Pond volume = Pond area × Average depth.

8.11 Estimation of source-water quality

One of the most important concerns of epidemiology is water quality. Agencies that supply drinking water are aware of this concern and do their best to supply treated water. But water treatment, regardless of whether it is for human consumption or for industry, is not an arbitrary process; designing a water-treatment facility is specific to the source-water quality. There are two components to source-water quality estimation: (1) water sampling and (2) analysis.

USGS Guideline on Sample Collection

Field personnel must be alert to conditions that could compromise the reliability of sample.

- Collect a representative sample. Use appropriate methods and quality-assurance measures to ensure that the field sites selected and the samples collected accurately represent the environment intended for study and can fulfil data-quality objectives.
- Think contamination! To ensure the integrity of the sample, be aware of the possible sources of contamination. Contamination introduced during each phase of sample collection (and processing) is additive and usually is substantially greater than contamination introduced elsewhere in the sample-handling and analysis process. Therefore, collect a sufficient number of quality-control samples, appropriately distributed in time and space, to ensure that data-quality objectives and requirements are met.

Source: USGS (2006)

BOX 8.2

8.11.1   Water sampling

Knowledge of sampling objectives, procedure and storage is crucial to the reliability of the sample, and, needless to say, the sample location has to be noted meticulously so that the same sampling site can be revisited when required. Collected samples are stored in hermetically sealed containers specially manufactured for the process. In general, sampling bottles are made of sterile plastic or hard glass bottles with double capping and sealing systems. The size of sampling bottles varies from 100 ml to 1000 ml, depending on the number of parameters to be analysed. A guideline on sample collection is given in Box 8.2. Figure 8.12 shows the methods of water sample collection from a tube well.

Surface water sampling

Surface water can be either flowing or still bodies of water. It has been mentioned before that sample location must be accurately determined and plotted on a map. Hand-held GPS is best because location data can be downloaded into the database, both for flowing water (e.g. river, perennial/ephemeral stream, canal, flumes) and still water (ponds, lakes, reservoirs). A generic guideline on selection of flowing/still water sites follows:

- Consider the study objectives, types of data and quality of data needed, equipment needs, and sampling methods.
- Obtain all available historical information.
- Consider physical characteristics of the area, such as size and shape, land use, tributary and runoff characteristics, geology, point and nonpoint

Figure 8.12 Methods of water sample collection from a tube well.

sources of contamination, hydraulic conditions, climate, water depth, and fluvial-sediment transport characteristics.
- Consider chemical and biological characteristics of the area (aquatic and terrestrial).
- Note the types of equipment that will be needed.

Groundwater sampling

Collecting samples of groundwater that accurately represent aquifer conditions requires sampling at appropriate wells and using equipment and methods that maintain the integrity of the sample with respect to the physical, chemical and biological characteristics of interest.

Procedures for collecting raw or filtered groundwater samples into bottles, sample preservation and other sample-processing and handling activities are important; all precautions to ensure integrity of the sample should be maintained. Water samples are collected from either monitoring wells or water supply wells.

Monitoring wells are observation wells that are installed principally for the collection of water samples to assess the physical, chemical and biological characteristics of formation (aquifer) water. Samples from monitoring wells are collected either with portable, low-capacity pumps or with slim buckets that can be lowered through the well.

Water-supply wells are wells that are installed primarily for a supply of public, domestic, irrigation, commercial or industrial water and usually are equipped with a dedicated pump or water-lifting device. Water is pumped out from these wells and collected in sample bottles.

8.11.2   Water analysis

After the task of water sample collection is over, the samples are sent to a laboratory for analysis. The laboratory may be an integral part of the industry or the analysis may be outsourced to a reputable laboratory. The essential qualities of a laboratory are:

- It is accredited by the government or appropriate authority.
- It has proper instruments that meet appropriate standards and specifications and are capable of analysing the required parameters.
- It has adequately trained personnel to conduct the required analyses.

Testing procedures and parameters may be grouped into physical, chemical, bacteriological and microscopic categories.

- **Physical tests** indicate properties detectable by the senses.
- **Chemical tests** determine the amounts of mineral and organic substances that affect water quality.
- **Bacteriological tests** show the presence of bacteria, notably those characteristic of faecal pollution.

Details of chemical, physical and biological parameters that are crucial for determining quality of water have been discussed in Chapter 6 of this book.

8.12   Aquifer test

Aquifer tests are conducted to estimate aquifer parameters such as storativity and transmissivity of an aquifer. It has already been discussed that there are different types of aquifers, such as confined aquifer, unconfined aquifer, leaky aquifer and so on; hence, there are different formulae and methods of estimating aquifer parameters. Typical arrangement of a pumping well and observation wells is given in Figure 8.13.

An aquifer test arrangement needs a pumping well and one or more observation wells from which water-level data are collected at specified intervals. Aquifer tests typically have three main components: (1) design, (2) field observations and (3) data analysis. There is a basic rationale that is common to all aquifer tests, and each test has a specific set of conditions as described below USGS (no date):

1 The geology of the site is known (driller's logs, geophysical logs, mapping).
2 The construction of the wells is known (width, depth, materials used and development).
3 Test design is consistent with measuring important aquifer properties.
4 All measuring devices (time, discharge and water level) are calibrated and verified.
5 Discharge is accurately monitored.
6 Pre-pumping variations in water levels are defined and explained, as are atmospheric and climatic effects on water levels during the time of the test.
7 Data are plotted during the test.

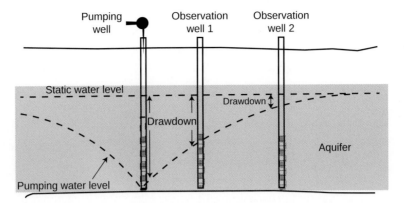

Figure 8.13 Typical arrangements of wells in a pumping test. r=distance of observation well from pumping well and b=depth of pumping well.

8 The testing rationale is flexible to accommodate changing conditions or anomalous responses.
9 Data from the recovery portion of the test are compared to the pumping portion.
10 Data are analyzed completely to fulfil objectives.

8.12.1   Field procedures

Well thought out field procedures and accurate monitoring equipment are the keys to a successful aquifer test. The following three sections provide an overview of the methods and equipment for establishing a pre-test baseline condition and running the test itself. The contents of a data collection form are (a) date, (b) temperature, (c) discharge rate, (d) weather, (e) well location, (f) well number, (g) owner of the well, (h) type of test (drawdown or recovery), (i) description of measuring point, (j) elevation of measuring point, (k) type of measuring equipment, (l) radial distance from centre of pumped well to the centre of the observation well, (m) static depth to water, (n) person recording the data and (o) pagination. In addition, the forms should have columns for recording (1) elapsed time since pumping started, shown as t; (2) elapsed time since pumping stopped, shown as t'; (3) depth to water level in metres; (4) drawdown or recovery of the water level in metres; (5) the time since pumping started divided by the time since pumping stopped, that is t/t'; (6) discharge rate in gallons per minute; (7) a column for comments to note any problems encountered, weather changes (i.e. barometric changes, precipitation), natural disasters, or other pertinent data (Osborne, 1993). A sample form as sourced from Osborne (1993) is given in Table 8.15.

8.12.2   Test procedures

Immediately before pumping is to begin, static water levels in all test wells should be recorded.

Table 8.15 Aquifer test field data sheet.

Pumped well no.	Date
Observation well no.	Location
Owner	Weather
Observer	Ambient temperature °C
Height of measuring point	Discharge rate of pumping well (lpm)
Static water level (m bgl)	Total number of observation wells
Distance from pumping well (m)	Water measurement technique
Recorded by	

Field data

Clock time	Elapsed time since pump started/stopped (min)	Depth to water level (m)	Drawdown[1] or recovery (m)	Discharge or recharge (lpm)	t/t'	Comments

[1] Drawdown is the difference between the static water level and the water level during pumping. mbgl = metre below ground level.
Source: Adapted from Osborne 1993.

Measuring water levels during test

If drawdown is expected in observation well(s) soon after testing begins and continuous water level recorders are not installed, an observer should be stationed at each observation well to record water levels during the testing. If there are numerous observation wells, a pressure transducer/data-logging system should be considered to reduce manpower needs. Recommended time interval for an aquifer test is given in Table 8.16.

Monitoring discharge rate

During the initial hour of the aquifer test, well discharge in the pumping well should be monitored and recorded as frequently as practical.

Water level recovery

Recovery measurements should be made in the same manner as the drawdown measurements. After pumping is terminated, recovery measurements

Table 8.16 Maximum recommended time intervals for aquifer test water level measurements.

Time since pumping started/stopped	Time interval
0 to 3 minutes	every 30 seconds
3 to 15 minutes	every minute
15 to 60 minutes	every 5 minutes
60 to 120 minutes	every 10 minutes
120 min. to 10 hours	every 30 minutes
10 hours to 48 hours	every 4 hours
48 hours to shut down	every 24 hours

should be taken at the same frequency as the drawdown measurements listed above in Table 8.16.

Length of test

The amount of time the aquifer should be pumped depends on the objectives of the test, the type of aquifer, location of suspected boundaries, the degree of accuracy needed to establish the storage coefficient and transmissivity, and the rate of pumping. The test should continue until the data are adequate to define the shape of the type curve sufficiently so that the parameters required are defined. It is recommended that pumping be continued for as long as possible and at least for twenty-four hours. Recovery measurements should be made for a similar period or until the projected pre-pumping water level trend has been attained.

Water disposal

Water being pumped must be disposed of legally within applicable local, state and federal rules and regulations. This is especially true if the groundwater is contaminated or is of poor quality compared to that at the point of disposal. During the pumping test, the individuals carrying out the test should perform water quality monitoring as required by the test plan and any necessary disposal permits. This monitoring should include periodic checks to ensure that water disposal procedures are following the test design and are not recharging the aquifer in a manner that would adversely affect test results. The field notes for the test should document when and how monitoring was performed.

Record-keeping

All data should be recorded on the forms prepared prior to testing (see Appendix 2). An accurate recording of the time, water level, and discharge measurements and comments during the test will prove valuable and necessary during subsequent data analyses.

8.12.3 *Pumping test data reduction and presentation*

All forms required for recording the test data should be prepared prior to starting the test and should be attached to a clip board for ease of use in the field. A portable computer located on-site with appropriate spreadsheets and graphics packages enables easier manipulation of test data. There are circumstances where computer storage by itself may not suffice; in such cases, properly catalogued files containing hard copies should be maintained.

8.12.4 *Analysis of test results*

Data analysis involves the processing of raw field data to calculate estimated values of hydraulic properties. If the design and field-observation phases of the aquifer test are conducted successfully, data analyses should be routine and successful. The method(s) of analysis utilised will depend, of course, on particular aquifer conditions in the area (known or assumed) and the parameters to be estimated. Semi-log paper plotting of 'drawdown' versus 'time' for each observation well is carried out while the test is going on. Curves plotted semi-log paper and then matched with type-curves specific to the aquifer type (e.g. confined, unconfined, leaky, infinite, homogeneous, etc.). The matching point value is used for estimating hydraulic properties of the aquifer. At present, there are many good analytical software in the market that plot the data on graphs, calculate and return the aquifer parameters without having manually match the curves. Figure 8.14 shows a time vs. drawdown curve.

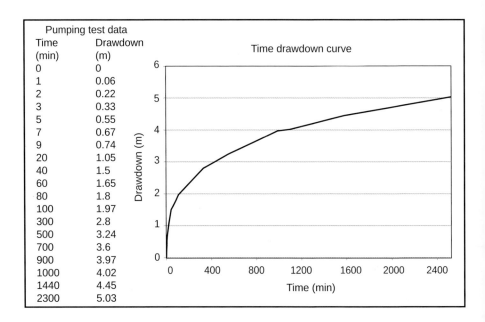

Pumping test data	
Time (min)	Drawdown (m)
0	0
1	0.06
2	0.22
3	0.33
5	0.55
7	0.67
9	0.74
20	1.05
40	1.5
60	1.65
80	1.8
100	1.97
300	2.8
500	3.24
700	3.6
900	3.97
1000	4.02
1440	4.45
2300	5.03

Figure 8.14 Aquifer test plot.

8.12.5   *Calculations and aquifer test results*

All calculations in data analyses must accompany the final report. The calculations should be referenced with the appropriate tables and graphs used for a particular calculation. Any assumption made in any part of the analysis should be noted in the calculation section. This is especially important if the data were corrected to account for barometric pressure changes, off-site pumping changes or other activities which affect the test. On completion of data analyses, aquifer test results provide the basis for estimating groundwater resource and designing tube wells.

8.13   Build understanding of key catchment processes and interaction

Preparation of thematic maps on geology, hydrogeology, land use, water level fluctuation, drainage pattern and surface bodies of water helps in visualisation of the spatial distribution of water resources and their interrelationships. Quantitative assessment of renewable water resource (e.g. renewable groundwater, exploitable surface water, etc.) helps managers to estimate the amount of water available from the catchment for industrial purposes. In short, the assessment methodologies will answer the following questions:

1 How much water is available from streams and other surface bodies of water?
2 What is the quality of surface water?
3 How far out does the aquifer continue, and what is the optimum yield of the aquifer?
4 How much total water is available?
5 What are the sustainability issues in the watershed?

8.14   Long-term simulation of catchment behaviour

One of the major applications of water resource assessment and water balance study is developing scenarios for coming years where the visible changes in land use, demography and water demand might take place. Scenario development helps in long-term simulation of catchment behaviour which is helpful in advance planning for different land use and climate change situations.

8.15   Assessment of sustainable and exploitable water over assessment period

An assessment of exploitable water is carried out through the water budget approach, which several countries have adopted in their water policies. The basic principles of estimating exploitable water resource are:

- Surface water flow or stream discharge should meet the environmental flow requirement before withdrawal.
- Groundwater withdrawal must be regulated by the sustainable yield policy of the country. Sustainable yield is defined by the European Union as 'available groundwater resource' which means that the long-term annual average recharge is less than the long-term annual rate of flow required to achieve the ecological quality objectives for associated surface waters. Indian National Water Policy, 2002 states that exploitation of groundwater resources should be so regulated as not to exceed the recharging possibilities, and also to ensure social equity.
- Water withdrawal from surface water reservoirs and ponds should not exceed the annual replenishing capacity of the ponds. It should be kept in mind that water withdrawal should not disturb the ecological health of the surface bodies of water.

8.16 Presentation of water resource assessment

Water resource assessment is vital for several reasons because it has applications in greater perspectives like input in national water databases, land and water management in various scales, political decision making and for corporate requirements as listed below:

- **Linkage with legal requirement:** Water assessment reports, as has already been mentioned, are statutory requirements for obtaining water allocation from the regulatory authorities. They are also required for environmental regulatory compliance.
- **Linkage with corporate policy:** A water resource assessment report is an important input in influencing corporate policy. Corporate bodies and their financial institutions can make informed decisions while investing in water infrastructure or water-conservation measures. It will also serve the purpose of corporate sustainability document preparation.
- **Linkage with IWRM:** IWRM is completely dependent on water resource assessment reports.
- **Linkage with water stewardship:** A major component of water stewardship is to know the local water resource to educate stakeholders. A water assessment report is a vital input in that direction.
- **Linkage with water disclosure:** A water assessment report is vital for estimating water stress of a watershed and a vital input in water disclosure document.

These requirements can be met if a technical report and a non-technical summary have been produced and presented in an appropriate forum for discussion.

8.17 Conclusion

Water resource estimation is a crucial task under industrial water resource management. Industries need to be vigilant about the water resources of the basin where the industry is located. It is more important in a water-stressed area where the sustainability of business is under risk. A systematic assessment of water resources and a periodical monitoring of water flow in streams and groundwater levels will help the industry in planning their water management activities and water stewardship initiatives.

Note

1 There are several other methods of groundwater recharge estimation. A good discussion is available at http://ces.iisc.ernet.in/energy/water/proceed/section7/paper5/section7paper5.htm.

Bibliography

Acaps (2011) *Summary secondary data review and needs assessment.* Available at http://www.reachresourcecentre.info/system/files/toolkit/acaps_secondary_data_review.pdf (accessed: 12 April 2016).

CGWB (2008) *Report of the groundwater resource estimation committee groundwater resource estimation methodology, Ministry of Water Resources, Government of India.* Available at http://cgwb.gov.in/documents/gec97.pdf (accessed: 29 June 2016).

Global Water Partnership (2013) *Water resources assessment (C1.02).* Available at http://www.gwp.org/en/ToolBox/TOOLS/Management-Instruments/Water-Resources-Assessment/Water-resources-assessment (accessed: 5 August 2016).

Huitema, D., Mostert, E., Egas, W., Moellenkamp, S., Pahl-Wostl, C. and Yalcin, R. (2009). 'Adaptive water governance: assessing the institutional prescriptions of adaptive (co-)management from a governance perspective and defining a research agenda'. *Ecology and Society* 14(1): 26.

ICSM (2016) *Fundamentals of mapping.* Available at http://www.icsm.gov.au/mapping/maps_cadastral.html (accessed: 7 October 2016).

Johnson, A.I. (2011) *Compilation of specific yield for various materials prepared in cooperation with the California Department of Water Resources.* Available at https://pubs.usgs.gov/of/1963/0059/report.pdf (accessed: 19 October 2016).

Keller, S. (no date) *Water resources assessment.* Available at http://www.sswm.info/content/water-resources-assessment (accessed: 13 April 2016).

Kuschk (2010) *The basement geographer.* Available at http://basementgeographer.com/large-scale-maps-vs-small-scale-maps/(accessed: 7 October 2016)

Mills, P.C. (2005) *Chapter 4: Groundwater-data collection measuring and mapping groundwater levels in wells.* Available at http://il.water.usgs.gov/pubs/ofr01-50_chapter4_4.pdf (accessed: 25 June 2016).

National Geophysical Data Center (2016) ETOPO1 *global relief.* Available at https://www.ngdc.noaa.gov/mgg/global/global.html (accessed: 24 April 2016).

NGWA (2016) 'Facts about global SM groundwater usage'. Available at http://www.ngwa.org/Fundamentals/Documents/global-groundwater-use-fact-sheet.pdf (accessed: 21 March 2017).

NRCS (1986) *Urban hydrology for small watersheds.* Available at http://www.nrcs.usda.gov/Internet/FSE_DOCUMENTS/stelprdb1044171.pdf (accessed: 9 July 2016).

Osborne, P.S. (1993) *Suggested operating procedures for aquifer pumping tests.* U.S.EPA. Office of Research and Development, EPA Groundwater Issue, p. 23.

Piestrak, J. (2010) *Mann library GIS Tutorial Geocoding address data using ArcGIS.* Available at http://www.westfield.ma.edu/uploads/cbraun/GIS2_geocode_addresses.pdf (accessed: 24 April 2016).

Sheng, T.C. (1990) '4. Biophysical data collection and analysis', in *Watershed management field manual.* Rome: FAO.

Sutcliffe, J.V. and Lazenby, J.B.C. (1990) '*Hydrological data for surface water resources assessment*', Beijing: IAHS Publ. Available at http://hydrologie.org/redbooks/a197/iahs_197_0251.pdf (accessed: 12 April 2016).

USDA (1989) *Runoff curve numbers1.* Available at http://www.wcc.nrcs.usda.gov/ftpref/wntsc/H&H/training/runoff-curve-numbers1.pdf (accessed: 25 June 2016).

US-EPA (2013) *Groundwater level and well depth measurement.* Available at https://www.epa.gov/sites/production/files/2015-06/documents/Groundwater-Level-Measurement.pdf (accessed: 25 June 2016).

USGS (2006) *National field manual for the collection of water-quality data.* Available at https://water.usgs.gov/owq/FieldManual/chapter4/pdf/Chap4_v2.pdf (accessed: 7 July 2016).

USGS (2015a) *Global 30 Arc-Second elevation (GTOPO30).* Available at https://lta.cr.usgs.gov/GTOPO30 (accessed: 24 April 2016).

USGS (2015b) *Shuttle radar topography mission (SRTM) 1 Arc-Second global.* Available at https://lta.cr.usgs.gov/SRTM1Arc (accessed: 24 April 2016).

USGS (2016) *How streamflow is measured. Part 2: The discharge measurement: USGS water science school.* Available at http://water.usgs.gov/edu/streamflow2.html (accessed: 30 June 2016).

USGS (no date) *Groundwater-data collection.* Available at http://il.water.usgs.gov/pubs/ofr01-50_chapter4_6.pdf (accessed: 9 July 2016).

World Bank (2007) *Strategic environmental assessment and integrated water resources management and development.* Available at http://siteresources.worldbank.org/INTRANETENVIRONMENT/Resources/ESW_SEA_for_IWRM.doc (accessed: 12 April 2016).

World Meteorological Organization (2012) *Technical material for water resources assessment.* Available at http://www.wmo.int/pages/prog/hwrp/publications/Technical_report_series/1095_en_4_Web.pdf (accessed: 5 October 2016).

9 Corporate Water Accounting and Disclosure

9.1 The context

The greatest challenges to economic growth in the Third Millennium are uncertainty in water availability and external factors in the value chain. Financial institutions and international business organisations are searching for tools to (1) quantify risks and to develop mitigating responses, and (2) to develop strategic plans for water accounting and disclosure. Investors are keen to know future strategies and ongoing water performances of industries in water-stressed areas. Investors also expect to be informed by companies about potential water risks in business and to be appraised about mitigation plans. Initiatives taken by industries to quantify water resources, risks and opportunities are called 'water accounting'. Water accounting and disclosure are complementary to each other. Water accounting is based on location-specific parameters. The corporate water disclosure document (which must incorporate all parameters of water accounting) is prepared to apprise the investor on the state of water in the industry.

A water footprint study generates the background for developing water sustainability documents. A water audit is the process of estimating water loss in an industry due to leakage and unauthorised use. A water footprint, by contrast, estimates water consumption in the value chain. Before discussing water disclosure, some basic terms need to be defined: water risk, water scarcity, water stress and water intensity.

9.1.1 Water Risk

The CEO Water Mandate has defined water risk as the challenges that a business entity experiences. The World Resource Institute (WRI) has further divided water risk into (1) physical risk with regard to quantity, (2) physical

Industrial Water Resource Management: Challenges and Opportunities for Corporate Water Stewardship, First Edition. Pradip K. Sengupta.
© 2018 John Wiley & Sons Ltd. Published 2018 by John Wiley & Sons Ltd.

risk with regard to quality, (3) regulatory risks and (4) reputational risk (Reig, Shiao and Gasser, 2013).

Physical water risk with regard to quantity

When water withdrawn from a basin exceeds the natural recharge rate, it constitutes a physical water risk to industries with regard to quantity of water; this risk may arise from unfavourable climatic condition, hydro-geological condition, basin geometry and land-use pattern. It may be a temporary or chronic phenomenon endemic to that region. For example, there are some areas where rainfall is good, but due to high runoff storms, water largely drains off leaving only a small amount of water for storage in bodies of water or aquifers; in such cases, water stress increases in summer but during monsoon season the condition is eased; whereas there are areas where water scarcity is perennial; there is no monsoonal relief.

Physical risk with regard to quality

Physical risk with regard to quality refers to regions where water is a constant health hazard. More than 30% of water in the Third World countries is polluted. In some rivers of China, the water quality is so poor that it cannot be used even for cooling towers. In Asia, Africa and Latin American countries, poor people drink unhealthy water collected from river beds (Figure 9.1). These are some of the examples of quality-related physical risks. Physical risk with regard to water quality has two measurable parameters: (1) intake water quality as a measure of upstream activities, and (2) grey-water footprint of the industry, which is a measure of the industry's impact on the environment.

Regulatory risks

Noncompliance with regulations, if detected by regulatory authorities, can lead to the closure of industries; the foremost requirement for industries is therefore to ensure compliance. Regulatory risks are also experienced by industries when the regulatory authority modifies laws, licence procedures and allocation policies, without sufficient advance notice to industries; it becomes difficult and cost-intensive for industries to implement compliance at short notice. Industries also suffer from bureaucratic delays which they need to surmount with the help of lobbyists.

There are instances where regulatory requirements are imposed only upon industries, but no restrictions are placed on the agricultural sector. For example, in West Bengal, India, there is no restriction of groundwater withdrawal for irrigation in the 'safe category blocks' whereas in the same safe category block industries are required to go through multilevel regulatory procedures; multilevel bureaucratic procedures constitute a risk for industries because of complexities involved.

Figure 9.1 Mwamongu Village water source, Tanzania. By Bob Metcalf [Public domain], via Wikimedia Commons.

Reputational risk

Loss of reputation is arguably the most damaging business risk (Deloitte, 2014). This is also related to corporate ethics and efficient delivery of corporate social responsibilities. The only way to avoid reputational risk is through successful corporate water stewardship (CWS); CWS is a multi-component system in which societal participation, that is, 'social licence' is the most important factor. In the absence of social licence, there can be backlash from local people, which invariable attracts media attention and causes loss of reputation; both established and start-up industries must ensure social licence or face reputational risks. For example, gold mines in South Africa faced tremendous reputational risk when the media focussed attention on the pollution potential of their mining operations. Like unwanted media attention, uprising among the general population also causes threats of severe reputational risks; numerous industries have faced public ire because it was perceived that the industry was drawing excess water, thus denying common people of their share; the only way to negotiate such reputational risk is to take on board a representative section of the society and explain the industry's water policy.

9.1.2 Water stress

'Water stress occurs when demand for water exceeds the available amount during a certain period or when poor quality restricts its use. Water stress causes deterioration of freshwater resources in terms of quantity (aquifer over-exploitation, dry rivers, etc.) and quality (eutrophication, organic matter pollution, saline intrusion, etc.).' (European Environment Agency, no date)

9.1.3 Water intensity

Water intensity is a measure of either (1) the volume of water used per manufactured unit; or (2) the value of the product, say, per 1000 litres of water consumed. Water intensity is thus expressed either as volume/unit production (e.g. 4000 litres per megawatt of electricity) or by economic value created/unit volume of water (e.g. it takes 2200 litres of water per kg of paddy, which costs say Rupees 55.00 in India; so, water intensity is Rupees 23/1000 litre). With regard to 'water intensity', the industry should (1) minimise the volume of water consumed per unit production and (2) increase the value of products manufactured per 1000 litres of water. Water intensity is thus a valuable tool to evaluate the success level achieved in sustainable water resource management.

9.2 Methods of assessing water risk

It is difficult for an industry to develop expertise on all aspects of water management; it has to rely on external expertise for water assessment, accounting and disclosure. Water risk assessment can be done by (1) water risk assessment tools and (2) data generation and internal assessment.

9.2.1 Water risk assessment tools

Several global organisations including organisations under UNO have developed tools for assessing water risk for the corporate sector. Some web-based tools also have the capacity to generate maps to show the status of water risk of a region; these web-based tools not only help industries to assess water risk but also enable industries to contribute to the global database on water. Some web based tools are described in Table 9.1

The applicability of the above-mentioned tools is jointly validated by the United Nations Environment Programme (UNEP) and the CEO Water Mandate (Morrison, Schulte, and Schenck, 2010). According to their validation study, water accounting tools should contain the following abilities:

- Assess both social (e.g. accessibility, affordability) and natural (e.g. hydrology, soil) dimensions that help industries in delivering effective corporate water stewardship.
- Assess water requirement/availability in the supply chain and estimate water stress in regions under the supply chain. (To appreciate the scale that supply chains can assume, consider this: raw material produced in one

Table 9.1. Water accounting tools.

Tool	Description	Scope	Application
FAO/AQUASTAT	It is FAO's global water information system, developed by their Land and Water Division. It is the most commonly used tool for water statistics.	Helps the corporate sector to assess regional water risks.	Country-wise water risk assessment.
GEMI Local Water Tool	It is a free tool for business houses for an evaluation of external impacts, business risks, opportunities and management plans related to water use and discharge. As the name implies, GEMI Local Water Tool is designed for site-specific application.	• Helps companies assess external impacts, business risks, opportunities and to manage water-related issues at specific sites. • Provides a common and consistent visualization platform for internal and external communication. • Provides interconnectivity between global and local water risk assessments and a uniform approach between site assessments. • Provides a central repository of information for the individual user to create reports for internal and external stakeholders.	Evaluation of water-related external impacts, business risks and sufficiency of management plans at specific sites.
IPIECA Global Water Tool (GWT) for Oil & Gas	The IPIECA Global Water Tool for Oil and Gas is a customized version of the the free and easy-to-use World Business Council for Sustainable Development (WBCSD) Global Water Tool; it helps oil and gas companies to water use mapping and risk assessment.	• The GWT for Oil and Gas has a different Excel workbook consisting of eight customized inventory input sheets enabling oil and gas metric and reporting outputs. • Effluent quality parameter (total petroleum hydrocarbons) is included. • Sector-relevant intensity metrics are created. • External datasets employed for mapping are the same for all Global Water Tools. • Mapping icons for the eight main parts of the oil and gas value chain are employed.	The tool will allow users to consider factors such as what percentage of its production volume falls under water-scarce areas, how many refineries are located in water-scarce areas and at what level of risks, and number of sites under water risk.
Open Life Cycle Assessment	Open LCA is a free and professional Life Cycle Assessment (LCA) and Footprint programme with a broad range of functions and available databases, created by Green Delta.		Life cycle assessment of products.

(Continued)

Table 9.1 (Continued)

Tool	Description	Scope	Application
Maplecroft Global Water Security Risk calculator	The Calculator harnesses eight years of proprietary data for 200 political, economic, human rights and environmental risks, covering 198 countries, and provides actionable insight into the global risk and opportunity landscape.	Generates risk index	Identifies 'risk hotspots' and plans mitigation actions.
UNEP Vital Water Graphics	Graphics for visualisation of water risk: An Overview of the State of the World's Fresh and Marine Waters.	2008 update of the 'Vital Water Graphics' is aimed at giving an overview of the state of water resources in the world and providing answers to these important questions.	• Storage, distribution and circulation • Water use and management • A scarce and competitive resource • Water management in urban areas • River's fragmentation • Pollution management • Monitor freshwater biodiversity • Water and health • Pricing water
Water Footprint Network	This is a community for sharing knowledge. This network coordinates efforts to further develop and disseminate knowledge of Water Footprint Assessment applications, methods and tools.	Provides a range of water footprint tools and calculators. Maintains the global water footprint database, 'WaterStat'.	Water footprint assessment in policies, business and in river basins.
WBCSD Global Water Tool	The Global Water Tool (GWT) is a free, publicly available resource for identifying corporate water risks and opportunities and which provides easy access to and analysis of critical data. It includes a workbook (data input, inventory by site, key reporting indicators, metrics calculations), a mapping function to plot sites with datasets, and Google Earth interface for spatial viewing.	Understands water use/needs of operations in relation to local externalities (including staff presence, industrial use and supply chain, water consumption and efficiency) to make informed decisions. Performs a first-level screening through maps or charts capturing key water performance and risk indicators of water consumption, efficiency and intensity.	Can be used for (1) communication with internal and external stakeholders and reporting under corporate disclosure. (2) Global Reporting Initiative, CDP Water, Bloomberg, Dow Jones Sustainability Index, etc.

| WRI Aqueduct WWF-DEG Water Risk Filter | Aqueduct's global water risk-mapping tool helps companies, investors, governments, and other users understand where and how water risks and opportunities are emerging worldwide. | The Aqueduct Water Risk Atlas is an online mapping tool that lets users combine 12 key indicators of water risk to create global overall water risk maps. | Physical risk assessment of
• Baseline water stress
• Inter-annual variability
• Seasonal variability
• Flood occurrence
• Draught severity
• Upstream storage
• Groundwater stress
• Return flow ratio
• Upstream protected land
• Regulatory and reputational risks |
| Water Risk Filter (Mueller et al., 2015) | Developed by World Wide Fund for Nature (WWF) and Deutsche Entwicklungsgesellschaft. | It is applicable to 35 industrial sectors and needs data, related to facility GPS location and the type of industry. Also includes 30-question survey on physical, regulatory, and reputational data. | Global map of facilities overlaid by water-related map layers. Physical, regulatory, and reputational risk at the basin and company level. |

country is processed in another, utilised and manufactured in a third country and finally marketed in the fourth. The water footprint for this product spans across all four countries in the supply chain.)

- Help the decision on how, when and which type of data are required to be generated (1) to achieve a particular goal of the industry and also (2) to cater to the stakeholders' needs. Data-generation tools must be capable of evaluating both water quality and quantity data.
- Harmonise the reporting system (Morrison, Schulte and Schenck, 2010).

9.2.2 Data generation and internal assessment

An industry can prepare its own water accounts independently with the help of the company's captive knowledge and data base. This can be done in two ways: water profiling and water footprint study, each of which merit separate discussion. External consultants can be engaged, if required.

9.3 Water profiling

Water profiling involves water risk disclosure and water accounting report.

The CEO Water Mandate, PWC and IRW have jointly developed detailed guidelines (Brammer and Pavelin, 2008) on water disclosure which include detailed methodology and conceptual framework. Under that guideline, the disclosure is a detailed water profile of the company supported by its current state of performance and compliance. It also encompasses detailed explanation of the implications of the current state of business risk and opportunities. The water accounting report must also link with local sustainability issues and mitigation strategies. In short, a corporate water disclosure report includes:

- Keeping the investor informed of the water performance of the company.
- Preparing (1) inventory of existing water sources and (2) quantifying demand-supply balance of water.
- Preparing water profile of the industry ('water profile' is described below).

9.3.1 Water profile of the basin

Water-related information is becoming more important day by day because of the pressure brought upon industries by the general population. People worldwide are much more environmentally aware these days compared with say, three decades ago; and as a consequence of popular pressure for environmental protection, governments in all countries have made stringent framework laws. National governments require water-related data for three purposes: (1) efficient delivery of water governance, (2) benchmarking industry-specific water consumption to establish globally acceptable standards and (3) information-sharing with interested individuals, researchers, international organisations, and so on.

Multilateral organisations (WWAP, 2012) collect water data from member countries/organisations to set policy goals, guidelines and tools for managing water resources; the UN-based Organisation for Economic Co-operation and Development (OECD) is one of such organisations that sets policy goals for 'decoupling environmental pressures from economic growth' (OECD, 2001). Other organisations, such as the World Resource Institution (WRI), WWF, FAO and so on, are also engaged in developing tools. Corporations reap enormous benefits from these guidelines and tools. These tools operate at the country or regional level and often are not applicable at local or catchment levels.

The water profile of a basin describes the water resource as well as the current status of utilisation by all sectors. To develop a basin-level water profile, a set of data must be generated which is the baseline information. Some salient points are mentioned below:

- Prepare a water resource map showing the surface water sources and aquifer disposition.
- Estimate the total renewable resource of the basin using standard norms and water-mass balance equations. Separate estimations should be done for groundwater and surface water.
- Estimate the current water use in different sectors and identify use-related issues, such as inequitable distribution, wastage, lack of sanitation, water purification issues, and so on.
- Seasonal water quality data collection.
- Procure information on legal issues, if any.
- Compare data generated with pre-existing data.
- Identify the trend of water withdrawal.

A suggestive structure of basin level water profile is given in Table 9.2.

9.3.2 Benefit of a watershed profile

A watershed profile or baseline profile is done by corporations for preparing EIA reports. An established company needs to update the profile for scenario generation, modelling and so on. It is required for planning, extension and enhancement of products or businesses. The factors that facilitate efficient water profiling of a basin are awareness, motivation and capacity of the companies. The World Resource Institute and several companies around the world are working on a collaborative effort, 'Aqueduct', to develop a water risk atlas of the key basins of the world. 'Aqueduct' data is freely available. Water profiles prepared for individual small basins will help in fine-tuning and calibrating the water risk atlas. Figure 9.2 shows an example of a water profile of a small watershed in a hard rock district of India.

Table 9.2 Structure of basin-level water profile.

Data group	Item	Content	Report format
Physical	Physiography	Geographical area of the basin. Map showing topography and drainage pattern.	Data and map
	Land use	Land-use type and percentage.	Description, map and pie chart
	Geology	Geological formations and percentage of land covered by each formation.	Description, map and pie chart
	Major Streams	Whether there are perennial streams from which water can be abstracted.	Description
	Population	Current population and decadal growth rate.	Data table
Meteorological	Precipitation	Monthly rainfall/snowfall.	Data table
	Temperature	Monthly average temperature.	Data table
	Humidity	Monthly relative humidity data.	Data table
Surface water	Lakes and reservoirs	Annual fluctuation of water reserve.	Data table
	Rivers	Monthly flow of river and flow duration curve.	Data table
Groundwater	Aquifer	Depth, extent and type of aquifer.	Data table
	Groundwater	Annual renewable groundwater resource.	Data table
Water withdrawal	Total industrial withdrawal in the basin	Name the sources and quantity of water withdrawal from each source.	Data table
	Domestic water withdrawal	As above.	Data table
	Irrigation water withdrawal	As above.	Data table
Regulatory issues	Governance issues	Whether the basin is declared stressed or notified by the authority.	Statement
	Compliance issues	Description of licensing methods, fees and annual returns and reports	Statement.

9.3.3   Water profile of a company

The water profile is a key feature in the water disclosure document of the company; water footprint assessment and water audit reports are supplementary documents in a corporate water profile. A water profile should describe the company's interaction with water and also with the regulatory system. It also describes the company's activities and impact in a water-stressed area. The CEO Water Mandate has prescribed some guidelines about preparing the water profile of an industry. Based on that recommendation the salient features of a water profile are described here. Companies are required to prepare three types of water profile:

1 Water profile of each production unit.
2 A compiled water profile of all production units which is referred to as 'global profile'; the global profile provides information on percentile performance, regulatory compliances, 'water withdrawn in water-stressed area' (UN Global Compact, 2015) and the percentage of impact on water-stressed conditions globally.
3 Water profile of the value chain.

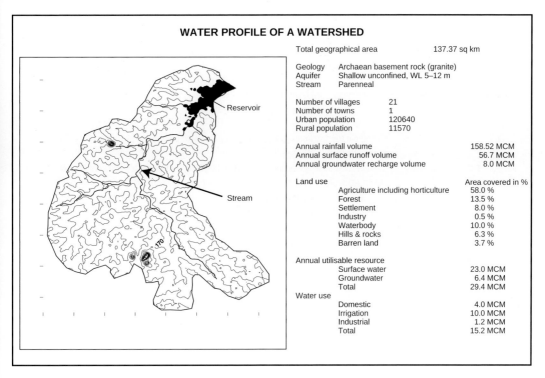

Figure 9.2 Water profile of watershed.

The above three types of water profile are formatted below in Tables 9.3 and 9.4.

A company which has a presence in several locations should prepare its global profile compiling data from every location. In Table 9.4 a sample format of water profile of an industrial unit is given.

In the value chain, the water profile and water consumption in different stages from procurement and sourcing to retail sales can be included if marketing and retail is managed by the same company. See Figure 9.3.

9.3.4   Water balance calculation

The industrial water balance calculation can be simple or complex depending upon the size of the industry: for example, the water balance calculation for a captive power plant can be simple, whereas for a large municipal water supply system or a petrochemical facility, it is highly complex. Basic components of water balance calculation are:

- Amount of daily raw water input;
- Amount of water treated daily;
- Water lost in treatment and purification;
- Amount of water distributed to each unit of the industry;
- Water consumed in each unit;
- Water discharged from each unit;
- Total discharged water treated;

Table 9.3 Items to be included in the water profile for a production unit.

Item	Description
Sourcing	Describe water source, methods of acquisition and abstraction. This includes details of temporary and permanent sources, emergency sources and the types of pumping devices used for each type of sources.
Water treatment	Detailed information on water-treatment plant with flow chart showing treatment system, capacity and amount of sludge generation.
Storage system	Capacity of storage system, number of tanks, location, height and distance from the use point.
Distribution system	Describe the layout of water distribution system in the factory or at the consumer end (in case of water industry) and include a summary of the water audit report. An updated drawing of the layout of pipelines, pumps, manifolds and storage system is required.
Operation and management	The water profile should include a detailed report on the operation and management of water distribution in the factory.
Types of water use	Prepare a detailed map of water use in different production units (which is to be submitted along with water budget showing water used at every point of the production stages).
Quantity and quality assessment	Prepare a document containing quality and quantity of water input and water discharged from the industry. If the factory has adopted a zero discharge policy, the amount and quality of water recycled should be incorporated. A water balance calculation along with a water balance diagram is necessary.

Table 9.4 Global water profile format of a company. Adapted from: UN Global Compact (2015).

Name of the company

Total number of production units

Number of production units in water-stressed area

Reporting period		**From…………..**	**To………….**

Components	**Contents**	**Reporting Unit/Statement**
Water withdrawal	Volume of water withdrawal.	Megalitre (Ml)
Water quality standard	Whether water quality is maintained as per local standards.	Yes/no
Water intensity per unit production	Total water withdrawal/total production.	Kl/unit
Water intensity monetized	Value created per unit of water withdrawal.	USD/Kl
Water consumption	Volume of water consumed.	Megalitre (Ml)
Water discharge	Volume of water discharged.	Megalitre (Ml)
	Destination of discharge (place name and type of location).	Place name(s)
	Statement on quality of water discharged.	Text
WASH service for workers	Number of units or divisions that have full WASH facility.	Number
Regulatory compliance on water withdrawal	Whether annual returns and licence fees are paid for the reporting period.	Yes/no
Regulatory compliance on water discharge	Whether all formalities have been completed and fees are paid for the reporting period.	Yes/no

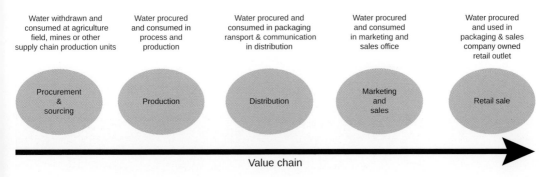

Figure 9.3 Value chain water profile.

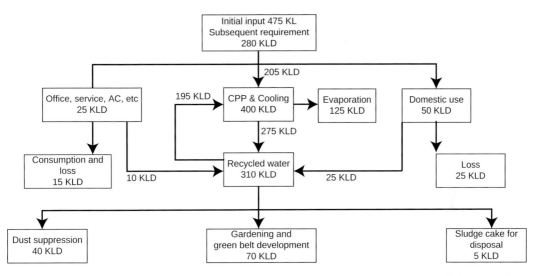

Figure 9.4 Water budget diagram of a 15 Mwh captive power plant (CPP). KLD = kilolitre per day.

- Amount of water discharged into environment; and
- Amount of water recycled.

An example of a water budget diagram of a 10 Mwh captive power plant is shown in Figure 9.4.

It will be noticed in the CPP flow chart above that there is no discharged water; the plant is recycling its total wastewater of 310 kl, that is, the CPP is a zero-discharge unit.

9.3.5   Impact assessment

The impacts of the industry on local water resources vis-à-vis the watershed need to be assessed, on the basis of a standard guideline. The Water Footprint Network (WFN) working group recommends that in a watershed the area of influence should be delineated on the basis of potential ecological and social cumulative impacts. Consequently,

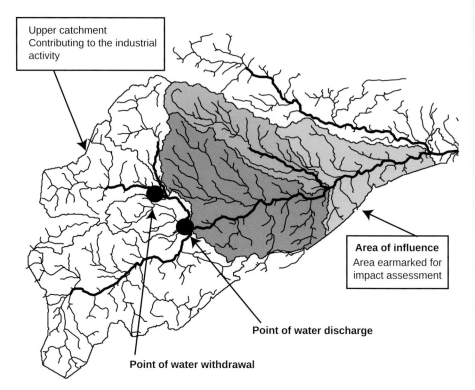

Upper catchment
Contributing to the industrial
activity

Area of influence
Area earmarked for
impact assessment

Point of water discharge

Point of water withdrawal

Figure 9.5 Impact assessment boundaries.

the downstream area from the point of water withdrawal and the point where water is discharged should be marked as area of influence. Two things should be kept in mind:

1 **Boundary delineation:** Delineation of the impact assessment boundary is important. The method of delineating the impact assessment boundary is given in Figure 9.5.
2 **Resource estimation and comparing with baseline data:** The industry must have baseline data for the water resources of the catchment collected before the start of industrial activity. Periodically such data should be updated and compared with the baseline data to estimate the impact.

9.4 Water footprint

With regard to water footprint assessment (WFA), it needs to be appreciated that this is a new concept with new terminology (e.g. blue water, green water, virtual water, etc.). Although water footprint assessment (WFA) is not yet made mandatory in every country, large corporations like the Tata group of industries, PepsiCo and SAB Miller have already taken up WFA as a part of their water sustainability analysis. The reason why large corporate have accepted WFA *suo moto* is because it not only quantifies an impact on the environment but also provides corporations with a quantified basis for the water disclosure document. Industrial water management involves (1) water used in the supply chain (2) water used in industrial process and (3) discharged water. The water footprint details the water consumed by the industry (evaporation and assimilation) and the impact of wastewater discharge into the environment. As a new concept, the water footprint has introduced a number on new terms that are defined hereinafter:

- **Blue water:** Blue water (Hoekstra *et al.* 2011) is the water that can be withdrawn from nature by any device; blue water includes rainwater, surface water and groundwater and is a visible component of the water cycle.
- **Green water:** A large part of the water cycle is captured by soil which is later evaporated and transpired through plants. The amount of water evaporated and transpired from soil is not available for human consumption; this water is called 'green water' (Hoekstra *et al.* 2011) which only plants can use. By definition, green water is that part of precipitation that goes back to the atmosphere through evaporation and transpiration (UNEP, no date). The amount of green water that is available in a watershed depends on the land use and climate. In Figure 9.6 blue and green water distribution of a watershed is explained.
- **Grey water:** Polluted water discharged by industries is called grey water; the term is also applicable to wastewater generated from domestic or agricultural uses. Grey water may be from a point source (industry or town sewage) or a diffused source (agricultural fields).
- **Virtual water:** Virtual water (Hoekstra *et al.* 2011) or embedded water is defined as the water consumed by plants to produce biomass or the water consumed for manufacturing a product or delivering a service. Virtual water of a crop is expressed as the measurable amount of water (green + blue) evaporated and transpired for growing a unit quantity of the crop. For example, the virtual water of rice is 2200 litre/kg. The concept of virtual water is used in estimating the water footprint of individuals, communities

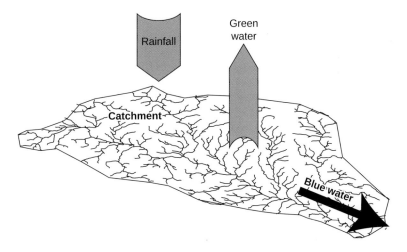

Figure 9.6 Distribution of blue and green water in a catchment. Adapted from Falkenmark (2009).

Table 9.5 Virtual water of some agricultural and industrial products.

Agricultural Product		Industrial product	
Product	**World average (litre/kg)**	**Product**	**World average (litre)**
Rice (paddy)	2,291	Cheese (per kg)	4,914
Rice (husked)	2,975	Leather (per kg)	16,656
Rice (broken)	3,419	1 cotton T-shirt (250 g)	2000
Wheat	1,334	1 sheet of A4-paper (80 g/m²)	10
Maize	909	1 glass of wine (125 ml)	120
Soybeans	1,789	1 pair of shoes (bovine leather)	8000
Sugar cane	175	1 microchip (2 g)	32
Cotton seed	3,644	1 cotton T-shirt (250 g)	2000

Source: Kumar and Jain (2007) and *Virtual water* (2016).

or industries. In Table 9.4 virtual water contents of some agricultural and industrial products are given.

- **Water footprint:** The water footprint (Hoekstra *et al.* 2011) is a consumption-based assessment of water utilised by an individual, community, product, industry, state or country. Any geographical entity can have its own water footprint. The water footprint calculates the virtual and real water consumed by any physical entity (individual, community, country or industry) for a reference period of time. Water footprint assessment (WFA) is a tool to quantify the cost of a product in terms of water used in different stages of the life cycle of the product.

- **Blue water footprint:** The blue water footprint (Hoekstra *et al.* 2011) is the amount of surface or groundwater consumed to produce any goods or to render any service; in this context, 'consumed' refers to the total amount of water, i.e. water evaporated and that incorporated in a product. For example, bottled mineral water has two parts: a plastic container and the water contained; out of these two components, the plastic bottle has virtual water which must be taken into consideration while determining the blue water footprint of manufacturing and marketing of bottled water.

- **Green water footprint:** 'It refers to the total rainwater evapotranspiration (from fields and plantations) plus the water incorporated into the harvested crop or wood' (Hoekstra *et al.* 2011). Green water use refers to the volume of rainwater that evapotranspirates from a crop field during the growing period. Blue water use refers to the volume of irrigation water (withdrawn from surface or groundwater) that is added to a crop field during the growing period. The distinction between green and blue water has been introduced by Malin Falkenmark, a Swedish hydrologist.

- **Grey water footprint:** The grey water footprint (Hoekstra *et al.* 2011) is defined as the volume of fresh water required to dilute the load of pollutants to meet with the existing (Serrano *et al.*, 2016) ambient water quality standards. Calculation of the grey water footprint (for a reference time period) depends on (1) the mean concentration of the reference pollutant in natural water, which is also called 'natural concentration', (2) the ambient concentration of the pollutant in the source water, (3) the total volume of water withdrawn, (4) the total volume of water discharged and (5) the concentration of the reference pollutant in the discharged water.

- **Water footprint of individual**: The water footprint of an individual is the sum total of the blue water consumed by an individual in a year and the total of virtual water content of everything used in daily life (e.g., food, energy, materials and service) by the individual.
- **Water footprint of a community or group of people:** The water footprint of a community or a group of people is the sum total of the water footprints of all individuals in that group.
- **Water footprint of a nation:** The water footprint of a nation (Hoekstra *et al.* 2011) is the measure of the total blue and green water consumed by the nation, including the total national grey water footprint, plus the water footprint imported from other countries (by importing consumer products and other commodities) minus the water footprint exported to other countries (by way of trade of commodities).
- **Water footprint of industry:** This is also termed the 'business water footprint', which is 'a combination of the water that goes into the production and manufacturing of a product or service' (Hoekstra *et al.* 2011) and the water consumed throughout the life cycle of a product. The business water footprint is the total volume of fresh water 'that is used directly and indirectly to run and support a business' (Hoekstra *et al.* 2011), including the virtual water content of raw materials/agricultural products incorporated in the business; that is, the business water footprint is the sum total of blue, green and grey water footprints.

9.4.1 The relevance of WFA to industry

The WFA concept was introduced in 2002 by A.Y. Hoekstra (Hoekstra, 2007) at the International Expert Meeting on Virtual Water Trade, in Delft, the Netherlands. Prior to the introduction of this concept, it was believed that water consumed by an individual relates to such consumptions as drinking water, bath water and so forth, without taking into consideration the virtual water content of the consumer products and food consumed by the individual. Hoekstra introduced the concept of water footprint to illustrate the link between human consumption of goods and water used to manufacture them. Hoekstra introduced a process to evaluate a product in terms of water consumed in manufacturing and marketing. The concept has been discussed in the forum of experts in Delft (2002), and the concepts of water footprint and virtual water have been consequently discussed in the World Water Council in 2003 and in the Fourth World Water Forum in Mexico City in 2006; the concept was recognised as a tangible way of estimating water consumption.

Following the agricultural sector, which is the largest consumer of fresh water, the industrial sector is the second largest (>20% globally). However, the grey water footprint of both sectors is comparable, which implies that water discharged by industries is significantly higher, especially in energy production. Formerly, impact of 'pollution load' on the environment was not properly estimated due to absence of methodology and the unit of measurement. For instance, gross change in the ambient water quality, biological indicators (as discussed in Chapter 6) and its effect on economy

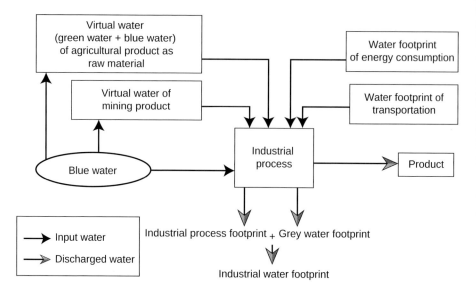

Figure 9.7 Industrial water footprints.

were seen only in qualitative terms. The grey water footprint is a quantifiable parameter now. It is only after the introduction of the concept that industries have started using WFA as a tool to quantify environmental impact. It is now appreciated that the water footprint of an industry is not restricted to the industry premises; it is the sum total of the water footprint of the entire supply chain. See Figure 9.7 for real and virtual water flow of an industry.

WFA is also an indicator of good water stewardship, because it reveals the strength and weakness of the industry in water management; an industry that has a sustainable water footprint is 'well managed'. The water footprint concept can also be extended to the valuation system of product and service, but it must be appreciated that WFA does not produce a fixed number even for a particular product because productivity of water varies from place to place and the cost of raw water is not the same in all countries. Consequently, the value generated in terms of water footprint is variable in space and time (Roson and Sartori, 2015).

9.4.2   Virtual water chain

The virtual water chain is the chain of production and consumption of water-intensive goods. The virtual water chain from primary producer to consumer is shown in Figure 9.8.

9.4.3   Assessment of green water footprint

Soil captures a considerable amount of rainwater which is used by plants (e.g. agricultural crops and forest) for biomass production; this, it will be

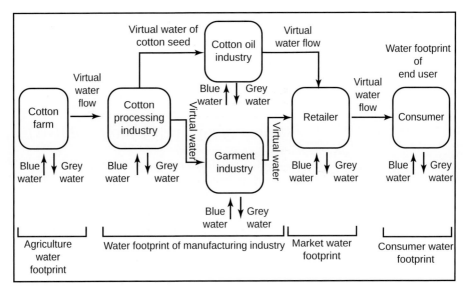

Figure 9.8 Simplified diagram showing virtual water chain and water footprint of a cotton-based industry. Adapted from Hoekstra (2009).

recalled from the definition given earlier in this chapter, is green water. A part of soil water is lost due to evapo-transpiration,[1] and a part is incorporated into plant product (e.g. crop, wood and fodder). Total water lost from soil in these processes is the green water footprint of a plant product. The 'green water footprint' is a measurable parameter which is a spatio-temporal data, that is, the measured value relates to the place and time span of measurement; in the absence of space-time specification, the measured value loses its significance. The green water footprint equation (Hoekstra *et al.* 2011) is:

$$\text{WF}_{\text{proc,green}} = \text{GreenWater}_{\text{Evaporation}} + \text{GreenWater}_{\text{Incorporation}} \qquad (9.1)$$
$$[\text{volume/time}]$$

9.4.4 Assessment of blue water footprint

Unlike green water, which is hidden, blue water is visible, that is, water sourced from surface bodies of water and aquifers. The blue water footprint is the volume of water consumed in production processes.

As defined by Hoekstra *et al.* (2011), the blue water footprint of industries is the total water consumed in various processes. The components of blue water footprint are (1) water that evaporates during the industrial processes (2) water incorporated in the product and (3) water lost to return flow. The water footprint has both temporal and spatial dimensions.

In an industrial process, the equation (Hoekstra *et al.* 2011) for the blue water footprint is:

$$\text{WF}_{\text{proc,blue}} = \text{Blue}_{\text{WaterEvaporation}} + \text{Blue}_{\text{WaterIncorporation}} + \text{Lost}_{\text{Returnflow}} \qquad (9.2)$$
$$[\text{volume / time}]$$

Where:

$WF_{proc,blue}$ = Blue water footprint of process
$Blue_{WaterEvaporation}$ = Blue water lost due to evaporation
$Blue_{WaterIncorporation}$ = Blue water incorporated into the product
$Lost_{Returnflow}$ = Return flow or unrecovered or non-recycled wastewater.

9.4.5 Assessment of grey water footprint (GWF)

The grey water footprint (GWF) depends on the volume of wastewater discharged from an industrial unit and is measured by the amount of water required to dilute pollutants to bring the wastewater quality to ambient water quality standards. GWF estimation as sourced from Hoekstra *et al.* (2011) involves:

- The amount of water withdrawn from the environment (symbol and unit: '*Abstr*' in volume/time);
- Amount of grey water discharged into the environment (*Effl*, in volume/time);
- Natural concentration of the 'quality parameter' (c_{nat}, in mass/volume); quality parameter, c_{nat}, is the reference parameter against which grey water footprint is estimated.
- Actual concentration of the reference pollutant (quality parameter) in intake water (c_{act}, in mass/volume);
- Maximum acceptable concentration of the quality parameter in water (c_{max}), also called the ambient water quality standard for that pollutant;
- Concentration of the quality parameter in discharged water (c_{effl});
- Pollutant load (L), which is the total amount of pollutant in discharged water during the assessment period (mass/time);
- Runoff of the flowing body of water (R, in volume/time);
- Critical load (L_{crit}, in mass/time), a parameter that defines the maximum capacity of the flowing body of water to assimilate the pollutant. According to Hoekstra *et al.* (2011) the concept of the 'critical load' is similar to the 'total maximum daily load' (TMDL) developed by the US Environmental Protection Agency.

Critical load is expressed as (Hoekstra *et al.*, 2011):

$$L_{crit} = R \times (c_{max} - c_{nat}) \quad [\text{mass/time}] \tag{9.3}$$

Grey water footprint, thus, is expressed as (Hoekstra *et al.*, 2011):

$$WF_{proc;grey} = \frac{L}{c_{max} - c_{nat}} = \frac{Effl \times c_{effl} - Abstr \times c_{act}}{C_{max} - C_{nat}} [\text{mass/time}] \tag{9.4}$$

Thermal pollution, also a parameter to estimate the grey water footprint, can be estimated in the same way, where (Hoekstra *et al.*, 2011):

$$WF_{thermal} = \frac{T_{effl} - T_{act}}{T_{max} - T_{nat}} \times Effl \tag{9.5}$$

BOX 9.1

Grey Water Footprint Calculation (Point Source)

Let us assume that a reference pollutant (referred to as a 'quality parameter') is discharged through effluent. The reference time is one year.

- The amount of water withdrawn from the environment $(Abstr) = 25000$ m³/year
- Amount of grey water discharged into the environment $(Effl) = 15000$ m³/year
- Natural concentration of the quality parameter $(c_{nat}) = 0.05$ mg/l
- Actual concentration of the quality parameter in intake water $(c_{act}) = 0.06$ mg/l
- Maximum acceptable concentration of the quality parameter in water $(c_{max}) = 0.10$ mg/l
- Concentration of the same quality parameter in discharged water. $(c_{effl}) = 0.5$ mg/l
- Pollutant load $(L) = Effl \times C_{effl} - Abstr \times C_{act} = 15000 \times 0.5 \times 1000 - 25000 \times 0.06 \times 1000 = 6000000$ miligram $= 6000$ grammes

$$WF_{proc;grey} = L/(C_{max} - C_{nat})$$
$$= (6000000 \text{ mg}/0.05 \text{mg})/l/\text{year}$$
$$= 120000 \text{ m}^3/\text{year}$$

Where T_{eff} is the temperature (⁰C) of the effluent, T_{act} is the temperature (⁰C) of the intake water, T_{max} is the maximum permissible temperature (⁰C) of water in the environment and T_{nat} is the temperature (⁰C) of water unaffected by human intervention.

The maximum acceptable temperature increase $(T_{max} - T_{nat})$ depends on the type of water and local conditions. If no local guideline is available, the European Union recommends a default value of $(T_{max} - T_{nat})$ as 3⁰C (EU., 2006).

An example of grey water footprint calculation is given in Box 9.1.

9.4.6 Assessment of business water footprint (BWF)

The industrial water footprint, also termed business water footprint (BWF), can have several subsets. An industry may select one or all subsets for WFA depending on the objectives of water management. For example, a cement industry that runs on a captive power plant (CPP) has several subsets, for instance, (1) a cement grinding unit (2) CPP, (3) township, and so on, of which the cement industry may chose to estimate BWF for the CPP only.

Parameters BWF measurement as sourced from Hoekstra *et al.* (2011) are:

- **Water footprint of a business:** It is the measure of water consumed by a business in its value chain. The concept of value chain can be used for branding a product by its water footprint, which gives a competitive edge to the product and can be used to promote marketing.
- **Water footprint of a business sector:** A business can opt for assessing the WF of a particular sector or all of its sectors separately, so that it can compare performance of different sectors in terms of water footprint.

- **Water footprint of a process step:** A business can assess the WF of a particular industrial process (say, cement grinding) in order to compare it with the WF of an alternative technology.
- **Water footprint of a product:** Each product has a discrete water footprint per unit of production. The company can compare between products (or compare the same product produced in different locations) in terms of water consumption.
- **Water footprint of a consumer:** The water footprint of a consumer is a combination of the water footprint of a consumer product and the direct water use by the consumer; this has direct market implications. For instance, a consumer can reduce his or her WF by substituting a consumer product that has a large water footprint by a different type of product that has a smaller water footprint (Hoekstra *et al.* 2011).

The water footprint of an industry (BWF) is the sum total of the water consumed by the industry, directly or indirectly, to run or support the business (Gerbens-Leenes and Hoekstra, 2008). There are two parts in BWF: (1) the operational water footprint or direct water use and (2) the water footprint in supply chain.

1 **Operational water use:** Operational water use includes the water used by an industry for running its production process, that is, water used for production, cooling, heating, drinking, washing, office use, toilets, landscape development and maintenance (this list is only suggestive, and there are many more operations in a business process).
2 **Water footprint in supply chain:** An industry procures its raw materials from various other industries, mines and agricultural fields/agro-industries, which constitute the supply chain of the industry. Water use in the supply chain consists of the virtual water content of raw materials and the water consumed in the process of delivering (including transport) for processing.

The prerequisites for water footprint assessment as described by Gerbens-Leenes and Hoekstra (2008) are:

- **Defined entity:** The entity (e.g. an industry or a country) must be a geographically delineated area; be it a unit of a production system or the whole of the establishment. The physical boundary thus delineated includes the water source, supply chain and water discharging sites.
- **Distinguishable inventory boundary:** The entity for which the water footprint is assessed should be distinguishable from other entities in the business environment. There must be enough information to develop a working knowledge of the use of water within the industry itself and its supply chain.
- **Well known inputs and outputs:** The input-output of material and water can be identified and quantified.

Gerbens-Leenes and Hoekstra (2008) have formulated a method of business water footprint calculation, which is done in six steps. These are as follows.

Delineation of inventory boundary

A defined business unit is a physically defined entity with specific annual input and output, that is, it receives raw materials and/or products which are either sourced internally, externally or both. The schematic diagram of business units, their water footprints and flow of products is shown in Figure 9.9.

An industry may have several such business units or a number of products. While estimating the water footprint of a business, each unit should be separately assessed.

Operational water footprint per business unit

According to Gerbens-Leenes and Hoekstra (2008), the operational water footprint is the sum total of blue, green and grey water footprints of the unit. These three components should be distinguishable and should be measured separately. The blue water footprint is the volume of freshwater that is consumed to produce the goods in the particular unit, especially the evaporative loss; it can be derived from the total blue water input by subtracting the volume of wastewater released. The grey water footprint of a production unit is calculated from the wastewater released and the amount of intake water. The green water footprint is the amount of water lost to evapo-transpiration from the business unit.

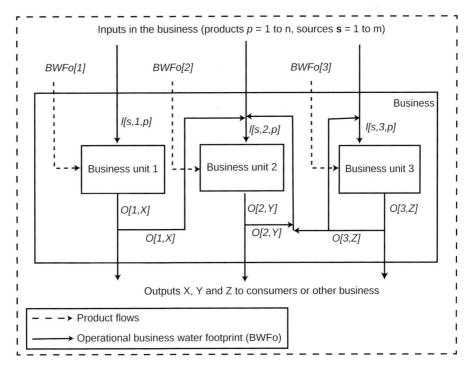

Figure 9.9 Schematic diagram of business units, their water footprints and flow of products. Printed with permission from Professor Anjen Y. Hoekstra.

This step takes into consideration the green, blue and grey water footprints to calculate the operational water footprint per business unit per year. The equations given below are sourced from Gerbens-Leenes and Hoekstra (2008).

$$\text{Business } WFP_o = Operation\,WFP_{green} + Operation\,WFP_{blue} + Operation\,WFP_{grey} \tag{9.6}$$

in which:

Business WFP_o = the operational water footprint of business unit (m³/year)
Operation WFP_{green} = the green operational water footprint of business unit (m³/year)
Operation WFP_{blue} = the blue operational water footprint of business unit (m³/year)
Operation FP_{grey} = the grey operational water footprint of business unit (m³/year)

Generally, the green water footprint (GWF) of a business unit is very small and may be absent unless the unit has trees/garden/green belt within the unit. It may, however, have a large green water footprint in its supply chain if it deals with agricultural products. The process of GWF assessment is highly complicated as it involves long time data-generation from agricultural fields and plantation crops.[2]

Supply-chain water footprint per business unit

The data source of supply chain water footprint is available from the suppliers. Gerbens-Leenes and Hoekstra (2008) have suggested, if it is assumed that in a supply chain there are 'i' types of different products coming from 'x' different sources, then the water footprint of supply chain is:

$$\text{'}WF_{bus,sup} = \sum_x \left(\sum_i \left(WF_{prod}[x,i] \times I[x,i] \right) \right) [\text{volume/time}] \tag{9.7}$$

$WF_{bus,sup}$ = supply-chain water footprint of the business unit
$WF_{prod}[x,i]$ = the water footprint of input product i from source x
$I[x,i]$ = the volume of input product i from source x into the business unit'

Blue water footprint per business unit

The blue water footprint of a business unit is expressed as the sum total of the WFs of all input products in the business unit added to the water footprint of the operation of the business unit. As adapted from Gerbens-Leenes and Hoekstra (2008), the equation for total water footprint of a business unit is:

$$WFP \text{ of a business unit} = WFP \text{ of operation of the business unit} + WFP \text{ of supply chain} \tag{9.8}$$

Product water footprint per business unit

The water footprint of a product is defined as the total volume of fresh water that is used directly or indirectly in production. To get a finished product, several steps are involved, and the water consumption of each step in the production chain is to be estimated. While estimating the water footprint of a product, one has to conceptualise the sequential processes involved in the production system, starting from the production of raw material to marketing. For example, the steps in the production of a cotton shirt are cotton growing, harvesting, ginning, carding, knitting, bleaching, dying, printing, stitching and finishing. Another example is the cement industry, where raw materials come from both mines and industries in the supply chain.

There are two ways of estimating the water footprint of a product:

1 **The chain-summation approach:** The chain-summation approach is the simplest when there is only one product from a business unit (e.g. cement production from slag). The production may involve several steps in the process, but all are attributed to a single product. The water footprint of the product is equal to the sum of water footprints of all steps divided by the production quantity (Pegasys, 2011).

2 **The stepwise accumulative approach:** this approach is a generic way of calculating the water footprint of a finished product, which is the sum total of water footprints of all input products. Suppose, a business unit has seven input products for one output product, in that case, the water footprint of the output product is simply the sum total of water footprints of seven input products (adding the process water footprint). In another case, where we have one input product for a number of output products, the water footprint of the input product is distributed proportionately to each of its products.

The formulae for calculating (1) chain-summation approach and (2) stepwise accumulative approach are (Gerbens-Leenes and Hoekstra, 2008):

1 **Chain summation approach:**

$$PWF[u,p] = BWF[u] / O[u,p] \qquad (9.9)$$

2 **Stepwise accumulative approach:**
The equation is sourced from Gerbens-Leenes and Hoekstra (2008)

$$\text{'}PWF[u,p] = \left\{ \left(E[u,p] / E_t[u] \right) \times BWF_u \right\} / O[u,p] \qquad (9.10)$$

Where

$PWF[u,p]$ = the water footprint of output product p from business unit u (m³/unit of product).

$O[u,p]$ = the annual volume of output product p from business unit u (units/year).

$E[u,p]$ = the economic value of output product p of business unit u (euro/year)

$E_t[u]$ = the economic value of the total output of business unit u (euro/year)'.

Water footprint of the total business

The water footprint of the total business is the summation of all components discussed above (caution should be taken to avoid double counting of virtual water flows between various business units). The total water footprint of a business (BWF) is thus expressed by the equation (Gerbens-Leenes and Hoekstra, 2008):

$$\text{`}BWF = \sum_{u=1}^{x}\left(BWF[u]\right) - \sum_{u=1}^{x}\left(PWF[u,p]\right) \times O^{*}[u,p] \qquad (9.11)$$

$O^{*}[u,p]$ stands for the annual output of product 'p' coming from 'u' number of production units.'

In Figure 9.10, the method of water footprint estimation by the stepwise accumulation method of a hypothetical industry having 5 supply chains, 4 processing units and 2 products is shown in a 'Sankey' diagram. Units 1 and 2 are the supply units to Unit 3 and 4. The industry consumes blue water to the tune of 357 CM. The total grey water production of the factory from all four units is 162 CM which amounts to a grey water footprint (GWF) of 2162 CM. The total business water footprint (BWF) is 2652.

A case study on the water footprint of Stora Enso's Skoghall Mill of Sweden is given in Box 9.2.

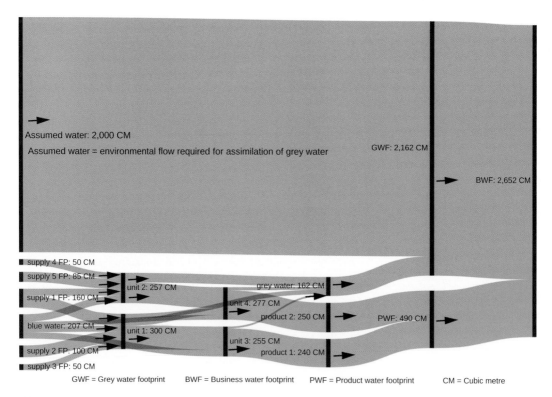

Figure 9.10 Sankey diagram showing the daily flow of the supply chain water footprint, blue water consumption and resultant business water footprint (BWF) of an industry having more than one supply chain and more than one product.

Case Study of Water Footprint of Stora Enso

The Stora Enso's Skoghall Mill of Sweden reported their water footprint to the Alliance for Beverage Cartons (ACE) and the WWF in 2011 (Stora Enso, 2011). In this study, they have separately assessed the water footprints of the supply chain, process, purchased electricity and biofuel.

The mill has the following supply chain water footprint components:

- Wood from Scandinavian forest: green WF only;
- Purchased pulp: blue and green WF;
- Chemicals (potato starch): blue green and grey WF;
- Purchased electricity: blue, green and grey WF; and
- Biofuel: green water WF.

The water footprint of the Stora Enso's Skoghall Mill for process and production has been estimated for its blue water footprint only. The mill does not have any green WF, and grey WF was not estimated due to lack of data. It is estimated that blue WF of the mill is 1.9 m³/MT.

The study reveals that the total water footprint of the Mill is 2194. This is elaborated in Figure 9.11.

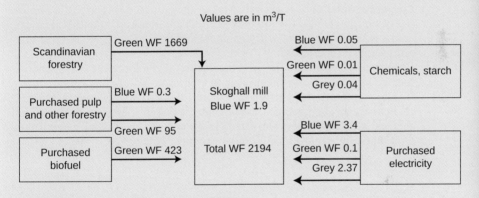

Figure 9.11 Water footprint of Skoghall Mill.

9.4.7 Life cycle–based assessment

Life cycle–based assessment is the assessment of water consumption of a product from procurement of raw material to marketing or 'from cradle to grave'. The stages in the life cycle of a given product are:

1 Production and processing of raw materials;
2 Packaging of raw materials;
3 Transport of raw materials;
4 Product manufacturing;
5 Packaging of finished product;

6 Transport to regional distribution centres;

7 Use or consumption;

8 Waste management (packaging and product leftovers); and

9 Production of auxiliary materials and energy carriers.

It needs to be appreciated that out of the nine stages mentioned above, the last three are passed on through the consumers. The production units are required by regulatory laws to keep a documented account of materials consumed in the supply chain; as such, data for steps 1 to 6 are available with the production units. Collection of data for the last three steps involves the man on the street and is therefore, difficult to obtain. Generally data collection from consumers is carried out through research programmes.

9.4.8   *Application of water footprint assessment*

Major applications of water footprint assessment in a corporate environment have been discussed in the Technical Report of UNEP and Water Footprint Network (UNEP, 2011). According to the UNEP report, the most important applications of business water footprint are:

- **Corporate water policy:** An industry can develop its water policy to reduce the water footprint in the finished product and its supply chain. Life cycle assessment of a product will enable an industry to design products with reduced water footprint.
- **Corporate water accounting and disclosure:** Water footprint is an essential tool in corporate water accounting. It reinforces accounting by providing insight into methods and practices and identifying gaps and challenges. Inclusion of water footprint assessment (WFA) in a disclosure document thus empowers investors and other stakeholders to benchmark water use in a particular process of production or service delivery.

9.4.9   *Benefits of WFA*

The most important benefits of water footprint assessment are:

- WFA is an integrated approach that enables companies to evaluate the impact of its products on freshwater resources.
- Water footprint is a consumption-based assessment; industries can evaluate the consumptive use of water and undertake planned reduction of water consumption.
- WFA helps industries in 'sustainability assessment' and identification of social and environmental impacts, and business risks.
- Grey water footprint assessment helps an industry to quantify the environmental impact, based on which environment management plan (EMP) can be formulated. It needs emphasising that EMP reports are vital for business certification.
- Corporations can include the WFA report in its sustainability report. At the moment, the WFA is not a mandatory requirement because it is a relatively

new concept, but considering the rate at which big business is beginning to implement it, it could well become mandatory sooner rather than later. Pending WFA becoming a regulatory requirement, big business houses are doing it *suo moto* for their own good; it provides a scientific basis for formulating sustainable business plans.

- WFA promotes transparency. An industry can enhance its brand value by showcasing its water conservation effects reflected in the WFA. Company can show the virtual water content of its products on the product label.

9.4.10 Water footprint assessment as a framework for corporate water sustainability

As mentioned earlier, a water footprint sustainability assessment provides a scientific basis for long-term business sustainability. In plain words, water footprint assessment is crucial to determine the carrying capacity of a catchment; without this information, industries often unwittingly create water-stressed conditions. There are many methods of water footprint assessment; the company must be well enough informed to be able to select the method that applies to its own industry.

Formerly, the environmental impact of industries was *incomplete* because it was based on water withdrawal only; that is, based only on blue water; *the grey and green water footprint was ignored.* In water footprint assessment, the combined impact of green, blue and grey water embedded in a product or service is quantified; if the grey water footprint is higher than the other two footprints, it indicates that the industry has an inefficient effluent management system. It is important to point out that this quantification procedure is process-/product-specific and is carried out separately at every stage from fetching raw material to marketing. This stage-wise assessment enables industries to locate the actual sites of impact and formulate location-specific impact management plans.

Sustainability is a function of three areas: environmental, social and economic. The spatial boundary of a WFA is either at local level, river basin level or a macro level, extending beyond the river basin. Under each of the three boundaries, the environmental, social and economic components of sustainability are evaluated. Sustainability assessment of water footprint addresses all three colours of water and assesses whether water consumption is sustainable in the following way (Sala *et al.* 2003):

- **Environmental sustainability:** at a local level, e.g. in a factory premises or in an agricultural field the assessment evaluates (1) whether the green water footprint of production exceeds the green water requirement for natural vegetation, (2) whether the environmental flow of blue water is impaired and (3) whether the grey water footprint has violated the ambient water quality standard.
- **Basin level:** the assessment evaluates whether the green and blue water footprints have adversely impacted the natural flow of the stream or affected

the environmental flow requirement of the lower catchment. At the macro level, the assessment will evaluate global impact of WF. Green water footprint sustainability depends on the availability of green water in the catchment for evapo-transpiration from productive lands other than land covered by natural vegetation. The green water footprint is sustainable if the available green water for ET is greater than the actual ET from the production system. A grey water footprint is deemed sustainable if the grey water footprint is within the limit of the assimilation capacity of the environment, the water quality standard is not violated and discharged water is socially acceptable.

- **Social sustainability:** the WFA will evaluate whether the water footprint has affected the blue water availability of community and local water users—people living in the water basin—or it has any visible impact on the global availability of freshwater.
- **Economic sustainability:** Economic sustainability is assessed by exploring the scopes and opportunities of water conservation through WFA. The assessment will estimate the economic viability of reducing WF by adopting water conservation measures without any adverse impact on production and market. The assessment also evaluates the impact of water allocations on the water uses within the basin. At the macro level, the economic impact of virtual water trade is a criterion to evaluate economic sustainability.

When the green, blue or grey water footprint in a catchment does not fulfil one of the criteria of environmental, social or economic sustainability, the water footprint cannot be considered to be geographically sustainable.

The major elements in a business water strategy are good water stewardship (Zhang, Hoekstra, and Mathews, 2013), value chain water management and water disclosure and reporting. These are the basic requirements for meeting water challenges and developing water policies. Sustainability reporting is a major task for business, required for legal compliance and for informing stakeholders. Water footprint assessment is a useful tool in sustainability reporting and response formulation.

WFA assessment provides answers to major questions in the sustainability assessment analysis. It also provides measurements that quantify the indicators answered in the questions. Major environmental questions that are integral parts of a sustainability report (Step2 Sustainability, 2014) are:

- What is the environmental impact related to water utilisation?
- What opportunities are there that can help in conserving water?
- What are the environmental risks created by any product of the company or the performance of a production unit?
- What are the performances of water treatment facilities, ETP and waste water disposal system?
- Is there any over exploitation of water resource, either in the agriculture field, mining or at any stage of supply chain?
- Does it include enough information to help strategic planning, water governance and water risk assessment?

9.4.11 International standards of water footprint assessment

Water use and management is a key consideration for any organisation in the light of growing demand for resources and increasing water scarcity. Water footprint assessment, therefore, should be improved to develop a highly accurate and feasible assessment (Gu *et al.*, 2014) that can be accepted globally. ISO 14046 is the new standard for water footprint assessment that will provide this consistency and give water footprint results credibility (BSI, 2017).

ISO 14046:2014 is designed to prepare step-by-step assessment and reporting of the water footprints of products, processes or organisations. It also claims to provide transparency, consistency, reproducibility and credibility for assessing a water footprint conducted according to the International Standard.

This assessment is also based on other background documents released by ISO such as 'life cycle assessment' (ISO, 2014). The ISO 14046:2014 methodology is modular, that is, it can be assessed under discrete modules for different life cycle stages and can be assimilated later. By providing water quality inputs, it can also identify potential environmental impacts in different geographical and temporal dimensions.

Building technical capacity: Water footprint analysis is a complex technical process, but building the technical capacity for assessing water footprint in developing countries will bring about competitive market opportunities (ISO-WWW, 2014).

9.4.12 Case studies

A large range of companies have successfully applied WFA in the sustainability statement: Some examples are:

- **SABMiller and WWF-UK:** SABMiller and WWF-UK have jointly conducted WFA in the value chain of SABMiller and have identified a number of sustainability issues like a wide difference of WF in beer production between their Czech units (45 litre/litre) and their South African counterpart (155 litre/litre). They used WFA as indicator of direct and indirect water use by consumers and producers (SABMiller and WWF, 2009) and guiding factors in their water strategy formulation.
- **UPM-Kymmene:** UPM-Keymmene, a leading paper manufacturer, assessed that 99% of their water footprint is attributed to forest products in the supply chain. The WF of one A4 size wood free (pulp made from plywood) uncoated paper 'is 13 litres and for wood free coated (WFC) 20 litres, consisting of 60% green WF, 39% grey WF and 1% blue WF' (UPM-Kymmene, 2011).
- **The Coca-Cola Company:** Coca-Cola is using water footprint assessment tools to measure the total volume of water consumed and polluted to produce a product. It also uses the tool for assessing sustainability of that water use. Coca-Cola is a partner of the Water Footprint Network (The Coca-Cola Company, 2012) and focused studies on the blue, green and grey water footprints

of sugar beets, orange juice and Coca-Cola. Coca-Cola Europe also published a report on its assessment of the water footprint of sugar use in Europe.

9.5   Industrial response to WF assessment

Water footprint assessment gives a business insight into how water is being used in the different units of production and in the supply chain. Industries can assess the sustainability of the footprint. Industries can use the assessment for:

- **Using water in a better way:** Managing loss at the field level and in the production system and thus reducing both green and blue water footprints;
- **Using water smarter:** Selecting best cropping and irrigation systems and best practices in the industrial manufacturing units;
- **Economic opportunities:** The economic sustainability assessment will lead to exploring investment opportunities in more economical water use systems and practices.
- **Development possibilities:** Industries can act to achieve a water footprint which is more socially sustainable, i.e. it does not affect the access of common people to fresh water.
- **Discharge:** The grey water footprint indicates actual impact of wastewater discharge into the environment. Industries can manage effluent treatment in a better way.
- **Trade:** Selecting products for import and export which have the most sustainable water footprint.

9.6   Water disclosure document

A water disclosure document is intended to appraise the investors and stakeholders about to water-related performance of the company, especially in water-stressed areas. The document should include:

1 Corporate water policy, strategies, and performance targets that drive performance improvements and address risks and negative impacts;
2 Water profile of the basin/watershed. Report of water monitoring and changes in basin conditions;
3 Water profile of the industrial unit or the company as the case may be;
4 Water audit report and the industrial response in water conservation;
5 Water footprint assessment and sustainability study report;
6 Water-related business risks and opportunities and negative impacts of the industry;
7 Report on the implementation of water strategies and policies throughout the company and across the company's value chain; and
8 Report on strategies of disclosure to local stakeholders and engaging them in water stewardship and IWRM activities.

9.7 Benefits of water disclosure

Water disclosure is crucial for business and has a number of valuable functions in safeguarding business.

1 **Enhancing business awareness and understanding:** Under water-stressed conditions, business has to deal with various issues such as sustainable water sourcing, managing discharge and managing internal water use. Water accounting and disclosure generate overall awareness and understanding of the risks of businesses under different scenario and climate change conditions. This can help in meeting physical water risks.

2 **Providing relevant information for investors and other key stakeholders:** Industries can provide valuable information on the water stress, regulatory risks and social risks related to water to their investors. Investors, on the basis of reports, provide valuable inputs in business and can take informed decisions to direct capital towards solutions.

3 **Raising public awareness:** Water accounting also generates documents to share with the public at large on how the industry is dealing with local water resources. Watersheds require collective management, and water management, being a politically sensitive issue, can provide better water-related information and help businesses and other stakeholders to identify long-term trends and potential hotspots, which can in turn be tackled individually or collaboratively. This is helpful in reducing reputational risk in business.

4 **Better regulatory compliance:** Water accounting and water profiling can reduce regulatory risks by disclosing water assessment and accounting reports to the authorities. These documents will help authorities to become more informed about the performance of the industry, and in the event of litigation these documents can defend the case of the industry in a court of law.

9.8 Conclusion

Recently some corporations have already started reporting water risks to investors and stakeholders, and many of them are proactively (Daniel and Sojamo, 2012) engaged in water stewardship practices. This communicates to the investors that the industry is trying to adopt a sustainable water strategy. Daniel and Sojamo (2012) find that there is a lack of sufficient research in quantifying corporate initiatives in water accounting and disclosure but observed that many companies apply various tools for assessing their water risks.

Water disclosure is one of the processes in the pathway of sustainability that helps in strategic planning for adopting programmes under water stewardship. Disclosure is practically a company's survival strategy during water crisis. This also establishes a dialogue (Morrison *et al.*, 2012) with the investors regarding future water-related programmes and may enhance credibility of the business with other stakeholders. Industries like Unilever have made it possible to publish proven records of their success.

Brammer and Pavelin, (2008) have discussed the present state of corporate environmental disclosure and mentioned that the drivers behind such disclosures are legal, competitive, strategic and financial. In their view, the present state of reporting is highly sophisticated and addresses both social and environmental policies.

According to the study of Brammer and Pavelin (2008), empirical studies have revealed that voluntary environmental disclosure has been practised by many companies for a long time, but the quality of reporting is not uniform across companies and countries. This depends on a variety of characteristics of industries; the reasons behind such differences are the methods of data collection, the persons involved in the process and the methods of statistical analysis. Grey *et al.* (1995) have observed that there are two types of reporting; one is very sketchy and unfocussed while the other type 'involves a very detailed textual analysis of disclosures', but those are very few in number.

Corporate accounting and disclosure should be streamlined by uniform definitions, units of calculation and methods of assessment and reporting. The methods discussed in this chapter, such as using international tools of corporate water disclosure, water profiling through baseline data generation and water footprint assessment followed by sustainability assessment are deemed beneficial to industries.

Notes

1 Evatranspiration is defined as the water lost through plant transpiration and evaporation from soil. For further study on evapotranspiration the reader may consult the published literature, available at http://www.fao.org/docrep/x0490e/x0490e04.htm

2 Recent research on the water footprint has developed a database of the GWF of different crops and agricultural products across the globe, which is available from: http://waterfootprint.org/en/resources/interactive-tools/water-footprint-assessment-tool/.

Bibliography

Brammer, S. and Pavelin, S. (2008) 'Factors influencing the quality of corporate environmental disclosure', *Business Strategy and the Environment*, 17(2): 120–136. doi: 10.1002/bse.506.

BSI (2017) *ISO 14046 Water footprint – Principles, requirements and guidelines.* Available at https://www.bsigroup.com/en-IN/ISO-14046-Water-footprint/ (accessed: 5 January 2017).

The Coca-Cola Company (2012) *Water stewardship: The Coca-Cola sustainability project.* Available at http://www.coca-colacompany.com/sustainabilityreport/world/water-stewardship.html#section-replenishing-the-water-we-use (accessed: 4 April 2016).

Daniel, M.A. and Sojamo, S. (2012) 'From risks to shared value? Corporate strategies in building a global water accounting and disclosure regime', *Water Alternatives*, 5(3).

Deloitte (2014) *Pl reputation risk survey EN.* Available at http://www2.deloitte.com/content/dam/Deloitte/pl/Documents/Reports/pl_Reputation_Risk_survey_EN.pdf (accessed: 30 April 2016).

European Environment Agency (no date) Available at http://www.eea.europa.eu/themes/water/wise-help-centre/glossary-definitions/water-stress (accessed: 13 July 2016).

Falkenmark, M. (2009) *Food and environmental sustainability: The water perspective.* Available at http://chaire-landoltetcie.epfl.ch/files/content/sites/chaire-landoltetcie/files/Documents/Food_and_environmental_sustainability_-_The_water_perspective_-_M._Falkenmark.pdf (accessed: 18 July 2016).

Gerbens-Leenes, P.W. and Hoekstra, A. (2008) *Business water footprint accounting: a tool to assess how production of goods and services impacts on freshwater resources worldwide.* Delft: UNESCO-IHE Institute for Water Education.

Grey, R., Javad, M., Power, D.M. and Sinclair, C.D. (2001) 'Social and environmental disclosure and corporate characteristics: A research note and extension', *Journal of business finance accounting*, 28(3-4): 327–356. doi: 10.1111/1468-5957.00376.

Gu, Y., Xu, J., Wang, H. and Li, F. (2014) 'Industrial water footprint assessment: Methodologies in need of improvement', *Environmental science & technology*, 48(12): 6531–6532. doi: 10.1021/es502162w.

Hoekstra, A.Y. (2007) *Human appropriation of natural capital: comparing ecological footprint and water footprint analysis value of water.* Available at http://waterfootprint.org/media/downloads/Report23-Hoekstra-2007.pdf (accessed: 1 June 2016).

Hoekstra (2009) *A comprehensive introduction to water footprints.* Available at http://www.pseau.org/outils/ouvrages/waterfootprint_comprehensive_introduction_to_water_footprints_en.ppt (accessed: 12 June 2016).

Hoekstra, A.Y., Chapagain, A.K., Aldaya, M.M. and Mekonnen, M.M. (2011) *The water footprint assessment manual.* Available at http://waterfootprint.org/media/downloads/TheWaterFootprintAssessmentManual_2.pdf (accessed: 12 January 2017).

ISO - WWW (2014) *Using ISO 14046.* Available at http://www.iso.org/iso/iso14046_briefing_note.pdf (accessed: 7 June 2016).

ISO (2014) *ISO 14046-2014. Environmental management – Water footprint – Principles, requirements and guidelines.*

Kumar, V. and Jain, S. (2007) 'Status of virtual water trade from India', *Current Science*, 93(8).

Morrison, J., Schulte, P., Koopman, L., Allan, N., Norton, M. and De Souza, K. (2012) *Corporate water disclosure guidelines toward a common approach to reporting water issues.* Available at https://www.pwc.com/us/en/corporate-sustainability-climate-change/assets/pwc-corporate-water-disclosure-gudelines.pdf (accessed: 29 March 2016).

Morrison, J., Schulte, P. and Schenck, R. (2010) *The CEO water mandate: Corporate water accounting, An analysis of methods and tools for measuring water use and its impacts.* Available at http://pacinst.org/app/uploads/2013/02/corp_water_accounting_exec_sum3.pdf (accessed: 21 June 2016).

Mueller, S.A., Carlile, A., Bras, B., Niemann, T.A., Rokosz, S.M., McKenzie, H.L., Chul Kim, H. and Wallington, T.J. (2015) 'Requirements for water assessment tools: An automotive industry perspective', *Water resources and industry*, 9: 30–44. doi: 10.1016/j.wri.2014.12.001.

OECD (2011) *Indicators to measure decoupling of environmental pressure from economic growth: Executive summary.* Available at http://www.oecd.org/environment/indicators-modelling-outlooks/1933638.pdf (accessed: 1 April 2016).

Pegasys (2011) *Water footprint and competitive advantage and trade in the Nile Basin countries.* Available at http://nileis.nilebasin.org/system/files/Nile%20Basin%20Water%20Footprint%20Virtual%20Water%20Training%20Report.pdf (accessed 25 March 2017).

Reig, P., Shiao, T. and Gasser, F. (2013) *Aqueduct water risk framework.* Available at http://www.wri.org/sites/default/files/aqueduct_water_risk_framework.pdf (accessed: 30 April 2016).

Roson, R. and Sartori, M. (2015) 'A decomposition and comparison analysis of international water footprint time series', Working Paper Series, IEFE: The Centre for Research on Energy and Environmental Economics and Policy at Bocconi University (77).

SABMiller and WWF (2009) *Water footprinting identifying & addressing water risks in the value chain.* Available at http://awsassets.panda.org/downloads/sabmiller_water_footprinting_report_final_.pdf (accessed: 4 July 2016).

Sala, S., Bianchi, A., Bligny, J.C., Bouraoui, F., Castellani, V., De Camillis, C., Mubareka, S., Vandecasteele, I. and Wolf, M.A. (2003) *Water footprint in the context of sustainability assessment,* JRS Scientific and Policy Report of the European Union. Luxembourg, ISBN 978-92-79-28279-9.

Serrano, M.T.C., Mariscal, M.I.R., Causapé, M.C.C., Martínez, A.A., Elcorobarrutia, F.R., Cabello, M.J.C. and Alfonso, A.M.T. (2016) 'Effect of mineral and organic fertilization on grey water footprint in a fertirrigated crop under semiarid conditions', *Geophysical Research Abstracts,* 18.

Step2 Sustainability (2014) *ISO publishes water footprint standard.* Available at http://www.step2sustainability.eu/news.asp?id=18 (accessed: 8 July 2016).

Stora Enso (2011) *Case study on the water footprint of Stora Enso's Skoghall Mill report to the Alliance for Beverage Cartons and the Environment (ACE) and WWF.* Available at http://www.ace.be/uploads/Modules/Publications/case_study_on_the_water_footprint.pdf (accessed: 14 October 2016).

UN Global Compact (2015) *Company water profile.* Available at http://ceowatermandate.org/disclosure/develop/company-water-profile/(accessed: 1 April 2016).

UNEP (2011) *Water footprint and corporate water accounting for resource efficiency.* Available at http://waterfootprint.org/media/downloads/UNEP-2011.pdf (accessed: 21 June 2016).

UNEP (no date) *Sourcebook of alternative technologies for freshwater augmentation in some countries in Asia.* Available at http://www.unep.or.jp/ietc/Publications/TechPublications/TechPub-8e/conservation.asp (accessed: 23 March 2016).

UPM-Kymmene (2011) *UPM 2011 1.* Available at http://waterfootprint.org/media/downloads/UPM-2011_1.pdf (accessed: 4 July 2016).

Virtual water (2016) in *Wikipedia.* Available at https://en.wikipedia.org/wiki/Virtual_water (accessed: 18 July 2016).

Water Footprint Network (no date) *Glossary.* Available at http://waterfootprint.org/en/water-footprint/glossary/#BWF (accessed: 7 June 2016).

WWAP (World Water Assessment Programme) (2012) *The United Nations world water development report 4: managing water under uncertainty and risk.* Available at http://www.unesco.org/new/fileadmin/MULTIMEDIA/HQ/SC/pdf/WWDR4%20Volume%201-Managing%20Water%20under%20Uncertainty%20and%20Risk.pdf (accessed: 1 April 2016).

Zhang, G.P., Hoekstra, A.Y. and Mathews, R.E. (2013) 'Water footprint assessment (WFA) for better water governance and sustainable development', *Water resources and industry,* 1-2, pp. 1–6. doi: 10.1016/j.wri.2013.06.004.

10

Detection of Water Loss and Methods of Water Conservation in Industries

10.1 Overview

Water conservation is a set of scientific, economic, political and social activities that ensures a sustainable supply of water to meet the demands of humankind as well as the demands of ecosystems. Except for the small amount of desalinated seawater, science and technology cannot augment renewable freshwater resources. The renewal activity is built naturally within the water cycle. Water conservation is therefore a global issue that brings all water-using sectors into convergence towards maintaining the balance of natural water cycle.

Since the cost for procuring, purification and redistribution of water is on the rise, water conservation is now step towards economic benefit. It has been observed that a few simple conservation methods go a long way to benefit the value chain, within a year or two of initiation.

In all water conservation initiatives, it is important to assess or determine to what extent the conservation makes sense (Connecticut Department of Environmental Protection, no date). Water conservation can start with reducing water consumption and may be extended further by deploying sophisticated technologies for effectiveness. The capital expenditure made in deploying sophisticated technologies not only generates fixed assets (like land, treatment plants, etc.) but also effectively increases the profit margin by reducing the cost per unit consumption of water. Water conservation has two aspects: (1) quantitative and (2) qualitative. As we have discussed in previous chapters, industrial activities can have impacts upon the resources in both aspects. Water conservation activities in industries are a channelizing method in water stewardship initiative or corporate social responsibility. Whatever may be the domain of conservation activities,

Industrial Water Resource Management: Challenges and Opportunities for Corporate Water Stewardship, First Edition. Pradip K. Sengupta.
© 2018 John Wiley & Sons Ltd. Published 2018 by John Wiley & Sons Ltd.

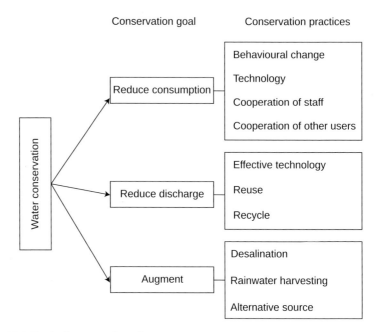

Figure 10.1 Basic frameworks of water conservation.

it includes an engagement of stakeholders, involvement of technology and assessment of impact. See Figure 10.1.

Water conservation has a number of goals that ensure availability of water, provided that water withdrawal never exceeds the regenerating capacity of an ecosystem; this idea is the crux of preserving an ecosystem and keeping ecological services in good health. Because water and energy are intimately related, water conservation amounts to energy conservation, and vice versa. In other words, industries need to appreciate that conservation of one simultaneously saves the costs of the other. Conservation in the strict sense covers all methods that substantially reduce water withdrawal and deliver a good quality of discharge as return flow. It is not just about optimisation of water use by industries; water conservation also reduces costs to society in two ways (1) by cutting costs of the water supply, (2) by ensuring adequate water supply.

10.2 Getting started: Develop a water conservation strategy

In the Third Millennium, the corporate sector has stepped into a period of worldwide environmental awareness. For their own well-being, therefore, industries have adopted better water conservation strategies to avoid falling afoul of the times. Water conservation in the industrial environment includes (1) detection of overuse and loss, (2) quantification of loss, (3) conserving water through various methods and (4) augmenting resources through rainwater harvesting/recycling/reuse.

10.3   Detection of overuse

Since the Industrial Revolution in the 18th century, development of water infrastructures by small, medium, large and heavy industries has continued unrestricted until the 1980s, a period which witnessed a sharp rise in environmental awareness. Following the rise of global awareness, the supply-oriented approach to water has always been the main focus of water management, which is in sharp contrast to unrestricted water use, or to be more precise, water abuse, that lasted for more than two centuries.

Water benchmarking and water auditing are the two processes of estimating water overuse and water loss respectively. Water-sensitised infrastructure development planning had never considered actual demand and conservative use of water; this phenomenon is most prominent in water-intensive irrigation sectors and construction of hydroelectric projects. Abusive water praxis is hard to detect until benchmarking of water use is conducted. It is critical for the corporate sector to realise that benchmarking is *the* new praxis.

10.3.1   Benchmarking

Benchmarking is a tool for comparing performance over time and across water utilities (Berg and Padowski, 2007). A benchmark value represents the quantity of normative water use, calculated from relative performance, baseline performance and trends. Benchmarking allows the company to plan its future water use without imposing any stress on water by improving the very productivity of water. Efficient benchmarking should be both diagnostic and remedial, and is therefore a comprehensive exercise. A single benchmarking model or method may not be sufficient. Berg (2006) has given three major reasons for benchmarking: (1) improving service quality, (2) expanding networks and (3) optimizing operations. Following the guidelines from Sydney Water (2016) benchmarking can help businesses to:

- Measure performance;
- Compare performance against other similar businesses;
- Identify inefficiencies that become noticeable only after benchmarking;
- Determine realistic operating targets; and
- Organise better planning for changes in business operations.

Example: A water use benchmarking study (Nelson, 2013) under the water stewardship programme was conducted by the Beverage Industry Environmental Roundtable for 1,600 locations representing 17 different beverage companies from 2009 to 2012. It was found that companies improved their 'water use efficiency' in production per litre of beer from 3.06 litres in 2009 to 2.76 litres in 2011, giving a quantified picture of water optimisation.

An industry or a consortium of industries may carry out benchmarking surveys for different industries (say, steel, beverage, mining, etc.) and evaluate the water use in each division of the industry for comparison. A hypothetical example of benchmarking analysis is given in Table 10.1.

Table 10.1 A simple benchmarking analysis.

Parameter	Unit	Industry A	Industry B	Industry C	Industry D
Total area	m²	18000	15460	20350	9800
Green belt area	m²	7000	4500	6000	2300
Annual production	MT	90000	65000	120000	50000
Number of employees	person	90	72	89	45
Water use					
Domestic	m³	1900	2600	3200	1500
Process	m³	1825	1500	3000	1500
Cooling	m³	1460	2300	2300	2000
Green belt	m³	10950	3650	4380	1460
Total use	m³	16135	10050	12880	6460
Analysis					
Water use per m² of green belt	m³m⁻²	1.56	0.81	0.73	0.63
Total water use per tonne of production	m³/T	0.18	0.15	0.11	0.13

The analysis shows that water consumption per m² in greenbelt development is 1.56 m³ in industry A, whereas the same in industry D is much lower at 0.63 m³; performance of industry D then becomes the benchmark for greenbelt development. Likewise, industry C becomes a benchmark for industry A, B and D to follow in respect of total water use.

Through benchmarking (and passing through other accounting systems like water audit, water footprint assessment and associated life cycle assessment), a company can identify the possible locations of overuse of water and can adopt mechanisms for reducing water use and take steps for conserving water.

10.4   Water audit

Water loss occurs due to leakage in the water distribution system, as already discussed. Another type of loss, which is difficult to detect and therefore control, is water theft and unauthorised usage. In combination, detection of these two is collectively referred to as water auditing. Reduction of the 'unaccounted for water' in an industry is a measure of the efficiency of the water supply system; simply put, water theft, which is otherwise hard to detect, is revealed in a thorough inventory in the audit process.

10.4.1   Fundamentals of water audit

The first step to achieve water sustainability in industries is to quantify water use. Regardless of the type of usage, industries need to appreciate that every use is measurable, and an account of use is derivable from a systematic audit. The process of measuring water from the point of abstraction

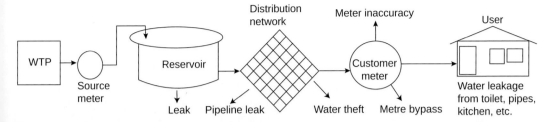

Figure 10.2 Different stages of water flow path and possible losses. Adapted from US-EPA Region 9, (2012).

to the final discharge is called a water audit. In many countries, water audits are encouraged by the government and regulatory authorities, who also release guidelines for conducting such audits.

A water audit is a process by which the amount of water loss in an industry can be quantified. The loss may be due to leakage and other reasons such as theft or unauthorised or illegal withdrawals from the systems. Effectively, water audit is a fundamental method of profiling distribution, use and disposal in the system. The Figure 10.2 shows different stages of water flow path and possible losses.

From the point of view of water auditing, industries may be divided into two categories:

1 Water industries which procure water from their own sources or purchase and distribute it to the consumers (e.g. municipal water supply which is privately owned in some countries or a government department).
2 Industries that procure water for their own use (processes, cooling, office use and domestic use in townships).

Large industries with a large township generally have separate water establishments for the township and factory, and often consider conducting separate water audits for township and process facility. A good example of this is the Digboi Refinery of Oil India Limited. This petrochemical industry abstracts water from its own sources (both surface and groundwater) and distributes a large part of the water to the neighbouring township, including its own establishments.

10.4.2   Benefits of water audit

The Central Water Commission (CWC), Government of India (2005) has published a guideline for water audits citing specific reasons for conducting them. The benefits of water audits according to the CWC guidelines are (a) reduced water losses, (b) improved financial performance, (c) improved reliability of supply system, (d) enhanced knowledge of the distribution system, (e) efficient use of existing supplies, (f) safeguarding public health and property, (g) improved public relations, (h) reduced legal liability, and (i) reduced disruption.

CWC guidelines are designed for all types of water users across the spectrum. In the context of the present book, in addition to the general guidelines, specific guidelines for the corporate sector need to be enunciated, which are as follows (Tanner and Braver, 2001):

- Coordination between water planning and other aspects of the concerned industrial facility, including planning and management divisions of the facility.
- Knowledge about the amount of water use at the management level.
- Knowledge about the 'water cost' determined with proper tools for water costing.
- Knowledge about cost-effectiveness of the water conservation projects.
- Generation of data and reports for regulatory compliance.
- Water disclosure to the investors.

10.4.3   Scopes and objectives of water audit

The goal of water audit, as has already been stated, is the conservation of water and cost reduction. Listed below are activities that lend a cutting edge to the conservation and cost-reduction plans:

- Systematic documentation of water input (i.e. groundwater or surface water).
- Mapping the distribution network
- Estimate efficiency of pumps, purification system, pipelines etc.
- Identify and quantify water loss to recommend plumbing retrofit (preceded by cost estimation of retrofitting and the payback period) and other water-saving initiatives.
- Identify water conservation opportunities.
- Evaluate existing water conservation systems to identify areas of improvement.

10.4.4   Human resource requirements for water audit

A water audit is an expert job which is often counter-intuitive. It requires trained personnel. The corporate sector therefore, has to have a clearly defined training programme formulated as a part of company policy. If expertise is available in-house, then the company by itself can conduct the required training programme; if not, the same may be outsourced. What the water auditor needs to keep in focus is to ensure a measurable improvement in water efficiency and financial gain.

In the event where water auditing is outsourced, there are guidelines available for selection of water auditors. The International Standard Organisation (ISO) has developed a standard (ISO, 1996) for environmental auditing in the ISO 14012:1996(E) Document which categorically defines the quality and qualification of environmental auditors. Similar guidelines need to be drawn up for the selection of water auditors. The Australian Union has developed a training guideline for water auditors in which knowledge requirements are outlined:

- Hydrology in general;
- Knowledge relating to hydraulic systems, design concepts, measurements and calculations;

- Knowledge of industrial operations;
- Knowledge of plumbing, pumps, water meters and installation methods used in hydraulic systems, including terminology, definitions, codes and specifications;
- Knowledge of plumbing codes, building codes etc. of the country;
- Knowledge of water quality standards and requirements; and
- Knowledge of water regulations and laws.

10.4.5 Corporate process in water audit

There are four major stages for successful water audit, as illustrated in Figure 10.3.

Requirements for initiating audit

The corporate manager or the decision maker of the industry is required to initiate the audit process. Management should understand the purpose and outcome of the audit, there being two types: (1) internal audit and (2) external audit. Internal audit suffices for management requirements and is done in-house, for which the company is required to ensure their staff are trained in water audit by a reputable organisation. A civil engineer with knowledge

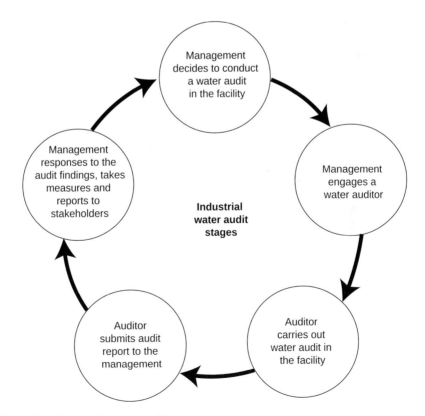

Figure 10.3 Stages of water audit.

of hydraulics, water distribution systems and plumbing is ideally suited for training in water audit. In contrast, external audit is a *statutory requirement* for regulatory compliance and also towards water disclosure.

Engaging an auditor

There should be evidence to show that water auditors engaged by the company are qualified both in terms of theoretical knowledge and work experience. As in the selection of candidates in any sector, evaluation of a potential auditor must go through the following: interviews, written and/or oral assessment, presentations and work experience (if and when required). Professional certifications and qualifications facilitate the process of engaging an auditor.

Conducting an audit

A water audit leads to a consolidated water conservation plan (TIDE, 2010). It should be channelised through several steps that will lead to a decision-support system. A simple flow chart of a water audit is shown in Figure 10.4.

Audit report

This report is the vital document containing the auditor's observations on water procurement, use and loss. Auditors will report the locations and causes of water loss and possible remedial measures and financial implications.

Response

Management will examine the audit report, take appropriate water conservation measures as found in the audit report and inform the stakeholders about actions taken in response to the audit report.

The water audit process is not universal (the audit process is complex and lengthy for large industries in comparison with medium/small-scale industries).

10.4.6   Water audit processes

The New Hampshire Department of Environmental Service (2013) released a document containing the different steps in water audit. The major steps are (a) a 'walk-through' exercise to itemise data required for audit, check availability of required data, identify areas that need data generation; (b) following itemisation, collection of water sourcing data from company records; (c) quantification of water delivered to different units in the facility; (d) measurement of water flow; (e) detection of leakages and quantification of loss through leakage in different units; (f) quantification of consumption and (metered, unmetered, authorised and unauthorised) uses, (g) preparing an audit report containing suggestive measures for loss control.

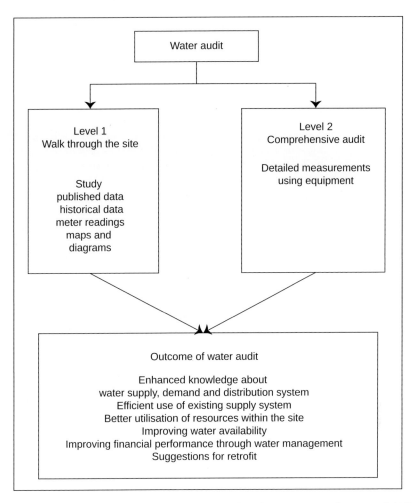

Figure 10.4 Components of water audit. Printed with permission from TIDE (TIDE, 2010).

Walk through exercise

A 'walk-through' exercise to itemise data required for audit, check availability of required data, identify areas that need data generation.

Conduct a meeting with management to discuss audit requirements, access to information, documentation, followed by a walk-through exercise. The auditor's first job is to verify whether water use is conservative or wistful, explore ways for optimising water use, and quantify water loss through leaks, pilferage and unauthorised uses. In the absence of such data, the audit team will have to generate it and prepare an inventory. Where companies have advanced data generation systems (on daily input of water, uses in different sections, flow meter logs, pumping logs, pressure gauge readings etc.), the audit becomes easier.

In the next step, auditors are required to inspect water services, meters, pumps, reservoirs, tanks and water fixtures throughout the building/centre.

Several data tables are required for sector-wise inventory. For household audit in industrial townships, questionnaire-based information is collected and inventoried. To quantify leakage, separate forms are required.

Collection of detailed information about the industry involves:

- Name of the industry and main products;
- Names of different units and how the different units are linked (for example, in thermal power plants there are 'procurement and storage' units, boiler, cooling towers, offices, staff quarters, etc.)
- Information about permanent and temporary staff/occupants in each unit and their schedules and the number of daily visitors.
- Working schedule, i.e. working hours, daily shifts, and so on.
- Water vision and policy of the company. This will guide the audit process to identify the focus area.

Audit data needs to be collected in a prescribed format. A suggested pre-audit format is shown in Table 10.2 (the table is suggestive, and more information may need to be obtained).

Collection of water sourcing data from company records

The auditor will inspect all records pertaining to procurement, delivery and use of water by the company. The company may have separate records on different production units, townships and supply systems. The auditor will identify the source and quantify water input from all sources, such as:

- Own sources of raw water.
- Raw water purchased from outside sources;
- Whether purified water is procured or there is an in-house purification facility. If purified water is supplied from outside, then quantify it.
- Quantify of total available water after purification;
- Gather detailed information about cooling towers, DM plants, RO plants, and so on.

Quantification of water delivered to different units in the facility

The auditor will gather information about water distribution and storage systems, and inspect all systems to verify the data supplied by the company. The processes involved are:

- Gathering information about pump capacity, pump running hours, capacity of sumps and tanks, length and diameter of pipelines, number of faucets, showers, flush, joints, meters (flow and pressure), manifolds and valves. The auditor will cross-check the data if any anomaly is observed.
- Gather information about schedules of water supply to different units. Get information about flow and pressure head at different points in the distribution system.

Table 10.2 Basic facility header sheet.

Item	Input data
Name of the industry.	
Type of industry (manufacturing, extractive, water supply, etc.).	
Contact person with designation.	
Address.	
Phone and E mail.	
Age of facility.	
Number of occupants/employees.	
Average daily visitors and nature of their water use.	
Total area of industry.	
Land use details (built-up area, factory, office, store, yard, etc.).	
Total air-conditioned area, with details of air conditioning systems.	
Different water uses in the facility (e.g. production, cooling, domestic, etc.).	
Water supply sources in details (e.g. utility, borewells, private tankers etc.).	
Details of water storage system (e.g. tank capacity, height from ground, etc.).	
Average cost of water per m³ (a) from own sources, (b) from external sources.	
Location and number of water main metres and sub-meters (indicate type, size and manufacture of each metre).	
Details of water metre readings (log).	
Describe water distribution system from source to last point of distribution, including pipe length, diameter and materials.	
Total number of toilets, water connections, flush, faucets, washing machine, water coolers etc.	
Do you have a single line drawing (SLD) of your water distribution system?	
Details of fire fighting facility.	
Details of dust-suppression facility.	
Details of gardening, and green belt irrigation system.	
Records of expenditures on maintenance of water-using equipment.	
Service contracts	
Do you have any regular leak detection and repairing programme? If yes, state details.	
Details of existing water conservation measures, already implemented (if any).	
Is there a water balance done for the facility? If yes, how often is it done?	
Has a water audit been conducted for the facility? If yes, provide the audit report.	
Any other relevant information?	

Adapted from Cilantro (2013).

- Prepare maps and floor plans of distribution system, showing pumps, pipelines, tanks, plumbing details and delivery points. A single line drawing of the distribution system (e.g. as shown in Figure 10.5) is to be developed.
- Collect water and sewer bills, meter readings, logs etc.
- Collect data about water use of fixtures, hoses, rinse tanks, cooling towers, recycling ponds and other water-using equipment and structures.
- Consult user manuals to gather specifications on appliances, fixtures, pumps, and so on and to verify the same against actual performance.

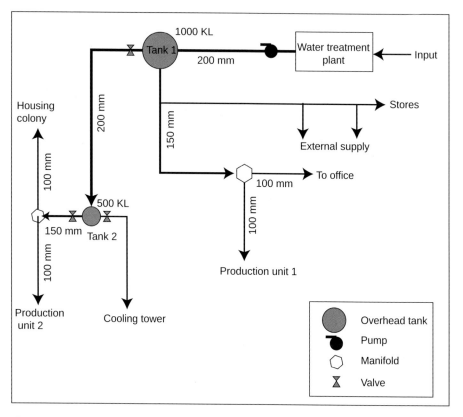

Figure 10.5 A single line drawing of water distribution system.

Measurement of water flow

It is also necessary to quantify water flow from water meter readings. Water meters are generally attached at the starting point of a water input system. There are several types of flow meters with various capacities, specifications and water flow units (e.g., gallon/minute, litre/second etc.). A water meter has a dial that has an arm (like the arm of a clock) to read the water flow rate and a counter, which is called a register or 'totaliser', that reads the total water passed through the pipe since calibration. Collect water meter reading logs from the company office and check the integrity of the readings. The method of calculating the daily flow involves a recording of water meter readings at a fixed time of the day. A typical water meter is shown in Figure 10.6.

There are numerous instances where water flow meters are not installed or are dysfunctional (i.e., data on water flow is absent/incomplete/inadequate). In such instances, the auditor is required to use an 'acoustic flow meter' or 'ultrasonic flow metre' to generate the data. An ultrasonic flow metre has coupled sensors (transducers) which are attached to the water pipe to generate data on the rate of flow and the velocity of water flowing through the pipe. An Ultrasonic flow meter operates on the basic principle of Doppler Effect. An ultrasonic signal is generated and that is reflected by suspended particles or gas bubbles in the flowing liquid. The change

Figure 10.6 Water flow meter By Andrevruas (Own work). CC BY 3.0 via Wikimedia Commons (Water metering, 2016)

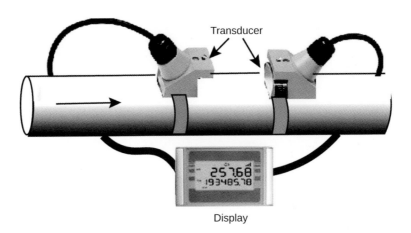

Figure 10.7 Ultrasonic flow meter.

in frequency of the reflected wave is the key phenomenon that is utilised for measuring fluid velocity. The average rate of flow can be estimated from multiple readings at the same point. The total flow in a day is then estimated from the average rate of flow (i.e. total flow in a day = average rate of flow in litres per second × 24 × 60 × 60 litres per day). See Figure 10.7.

Inspecting water services plans

Factory premises typically have numerous supply points that need to be catalogued for evaluation of water use. There are standardised formats for preparation of such catalogues. For example, the format to catalogue water used in toilets is shown in Table 10.3.

Table 10.3 Data collection form for toilet inventory.

			Information on toilets			
Location	Type of toilet men/ladies	ID	No. of faucets	No. of flushes	No. of showers	No. of urinals
		T1				
		T2				
		T3				
		T4				
		T5				
		T6				

Table 10.4 Leakage data collection form.

Sl No.	Location	Type of leaking device (pipe, joint, flush, faucet, etc.)	Leak (Y/N)	Leak rate (lpm)	Daily leak (L)
1					
2					
3					
4					
5					
6					
7					
8					
9					
10					

Detection of leakages and loss

It is impossible to visualise a factory that does not have leaking pipes, joints, faucets, flush and so on; these leakages and losses must be catalogued. Leak detection is a methodical task, which can sometimes even be hazardous (e.g. leak detection in underground pipelines and inaccessible locations). Different types of leak detection methods are given in Box 10.1; leak detection is followed by cataloguing. Table 10.4 shows a format for cataloguing leakage points and quantifying water loss.

Leak Detection Methods

Leak Detection Surveys: Industries should conduct periodical leak detection surveys by trained personnel. The areas where leaks are detected must be pin pointed.

There are several devices, like listening rods, aquaphones and geophones, which are used for leak detection. These instruments capture the sound of the leaking pipe and precisely mark the leaking point. Two major sensor based techniques are:

1 **Tracer gas technique:** In this technique, nontoxic light gases like hydrogen or helium are injected into a segment of the underground water pipe. These gases escape where there is a leak in the pipe and come to the surface of the ground. The presence of the gas is detected by a sensor.
2 **Thermography:** Thermography is based on the principle that where there is a leak in the underground pipe the thermal signature of the surrounding soil will change. This thermal change is detected by a sensor.

BOX 10.1

Quantification of consumption and uses

In this audit step, water consumption and use are quantified; although 'consumption' and 'use' sound similar, it needs clarifying in this context that these two terms are two different parameters in audit procedure. For instance, water lost to evaporation, for instance, in a cooling tower, is considered 'consumption' because it is not a part of 'return flow'; likewise, the water left in linens after washing that is subsequently lost in the drying process is also consumption. Water 'use', on the other hand includes the total water delivered to the industry, that is, the sum total of consumption, loss and return flow. The water audit steps involve:

- Calculation of water consumption for each facility (e.g. cooling tower, toilets, canteens, etc.), where Consumption = Water in – Water out (waste) – Leakage
- Calculation of water inflow, where Total inflow = Wastewater outflow + Consumption + Leakage.
- Compare measured water consumption of devices against specifications in operating manual.
- Tabulate the amount of 'lost' water for each device.
- Water 'lost' entails leakage, theft and unauthorised use. The amount lost to theft and unauthorised uses also need to be reflected in the audit report.
- The audit report must contain recommendations for repairing leakages.

Calculation of the cost of water procurement involves: cost of procurement of raw water and cost of purification when water is sourced from outside. If, on the other hand, water comes from a captive source (e.g. tube wells installed by the company within its premises), the cost will include power consumption in pumping plus quantification of loss in conveyance, cost of purification and cost of maintenance. From this exercise, the cost of water per 1000 cubic metres can be arrived at. A simple example is given in Table 10.5.

Table 10.5 Cost of water calculation.

Input m³	Water bill USD	Purification cost USD	Loss of raw water at purification plant m³	Cost of water USD	Cost per m³
1000	100	120	10	(1000–10) x 100/1000} + 120 = 219	USD 0.219

A calculation of all costs involved in water conservation measures includes repairing or replacement of old pipes, tanks, pumps and so on, retrofitting fixtures and plumbing (notably water-saving toilets), installation of sensors, and finally a calculation of the total annual water savings after such conservation measures.

Other things that need to be incorporated in the audit report are historical water demand, future expansion, an increase or decrease in the number of employees/occupants, planned water-conservation practices, retrofits and upgrades, and the impact of seasonal changes.

Preparing an audit report

All information gathered through the audit procedure is analysed and compiled in a report along with recommendations for water-saving methods and practices. Although there is no prescribed format for water audit reports, the report must essentially contain the following chapters:

- The cover page of an audit report can be so designed that it conveys useful information; for instance, name of the site, name of the client, location and period of audit coverage can be displayed.
- An executive summary that briefly outlines (a) audit methods, (b) key findings and recommendations.
- Then comes the main body of the report containing chapters on (a) introduction (b) methodology (c) water sourcing and use, (d) water loss, (e) evaluation of existing water conservation practices, (f) concluding chapter with recommendations and finally (g) the appendix containing data charts, specification of standards, instruments used and all extraneous information relevant to the audit.

10.4.7 Water audit software

The American Water Works Association (AWWA) is a leading organisation in the field of water audit. They have standardised the audit process and prepared a mass balance table which is very convenient to prepare and consult. The AWWA software (freeware) is comprehensive and covers *all* aspects of the water industry, including both revenue and non-revenue water components. The components of AWWA software that need special mention are (a) reporting worksheet, (b) water balance (c) grading matrix and (d) water-loss-control planning guide. This mass balance table is shown in Table 10.6.

Table 10.6 Water balance table of water audit.

Volume from Own Sources (Corrected for Known Errors)	System Input Volume	Water Exported (Corrected for Known Errors)		Billed Water Exported		Revenue Water
		Water Supplied	Authorised Consumption	Billed Authorised Consumption	Billed Metered Consumption	Revenue Water
					Billed unmetered consumption	
				Unbilled Authorised Consumption	Unbilled Metered Consumption	Non-revenue water
					Unbilled Unmetered Consumption	
			Water Losses	Apparent Losses	Systematic Data Handling Errors	
					Customer Metering Inaccuracies	
					Unauthorised Consumption	
Water Imported (Corrected for Known Errors)				Real Losses	Leakage On Transmission and Distribution Mains	
					Leakage and Overflows at Utility's Storage Tanks	
					Leakage on Service Connections up to the Point of Customer Metering	

NOTE: all data in volume for the period of reference, typically one year.

Source: AWWA Free Water Audit Software, Copyright © American Water Works Association. Printed with permission from AWWA (2016).

The above table contains terms that require definitions. The water balance terms and definitions are given below (adapted from AWWA Free Water Audit Software).

System Input Volume (C) The total system input volume of water, that is volume from own source (A) + water imported (B) +/− metre error adjustment.

Tools for estimating system input volume:

Two data input tables can be used for this purpose: (1) Water supplied log sheet (Table 10.7) and (2) master meter tracking form (Table 10.8).

Volume from own source (A) The volume of water abstracted by the facility from groundwater or surface water sources through own infrastructural arrangements.

Water imported (B) Water imported, i.e. water purchased, from water utility or water authority or from privately owned water sources.

Table 10.7 Water supplied log sheet.

Month			Date metre was installed	
Year			Date metre was tested	
Facility name				
Metre number			Model	
		Metre reading log		
Date	Time	Metre Reading	Quantity (L or m³ or gallons)	Remarks

Adapted from EFCN.

Table 10.8 Metre testing form.

Metre location		Manufacturer	
Model		Serial No.	
Date	Tested by	Testing Report	

Adapted from EFCN.

Water Supplied (D)   The annual volume of treated water delivered to the retail water distribution system. This equals system input minus the water exported.

Authorised consumption   Authorised consumption is the amount of water supplied to consumers. A company may or may not realise the cost of water from its consumers. Authorised consumption has two components: (1) Billed authorised consumption and (2) Unbilled authorised consumption. The methods of estimating authorised consumption are given in Table 10.9.

Table 10.9 Method of estimating authorised consumption.

Item		Method of estimation
Billed authorised consumption	Billed metered consumption	Data available from direct metre reading of individual consumers. Daily consumption = metre reading of the day – metre reading of previous day.
	Billed unmetered consumption	Total consumption data is collected from supply source. Estimation also done through door-to-door surveys.
Unbilled authorised consumption	Unbilled metered consumption	Data is collected from consumer metres periodically. An overall estimation is done.
	Unbilled unmetered consumption	A gross or overall estimation is done. Alternately a door-to-door survey is conducted to collect data on water consumption directly from the consumers.

Adapted from AWWA (2016).

Table 10.10 Methods of estimation of water loss.

Item		Method of estimation
Apparent losses	Systematic data handling errors	Assessment of data-handling errors conducted internally and then audited.
	Customer metering inaccuracies	Old customer meters often give inaccurate readings and under-register the consumption of water. Data on this can be obtained from sample checking and estimating the quantum of error.
	Unauthorised consumption	Unauthorised consumption is classified as water theft. It may arise if water is pilfered from fire hydrants, metre bypass and illegal line tap. It is difficult to estimate the quantity accurately. Some auditors fix a percentage of error (say, 0.25%). This can also be evaluated from sample checking and monitoring.
Real losses	Leakage on transmission and distribution mains	Can be estimated separately through leak detection surveys and must be entered separately. Instruments and appropriate methods for leak detection in underground pipes should be adopted. Survey logs should be maintained.
	Leakage and overflows at utility's storage tanks	Actual leak-detection survey to be conducted and estimated and entered in survey log.
	Leakage on service connections up to the point of customer metering	Actual leak detection survey to be conducted and estimated and entered in survey log.

Adapted from AWWA (2016).

Water Exported (E) The amount of water that is not a part of its own revenue water but is sold to the neighbouring water service providers. Data on this head can be directly obtained from the company.

Water losses Water loss arises for two reasons: (1) *apparent loss*, which is a function of inaccuracies in the system, like faulty meters, faulty billing; also including unauthorised consumption, theft of water, etc. and (2) *real loss*, i.e. visible losses, caused by overflow, leaks and breaks in pipelines, reservoirs and other storage systems. Methods of estimation of water loss are given in Table 10.10.

Table 10.11 Leak Detection Survey Log.

Water system name		ID (if any)
Leak Detection Survey Log		
Date		Survey time
Description of the location of survey		
Distance surveyed		Method of leak detection
Brief description of each leak detected		
Detection point	Description	Quantified loss per day
Remarks		
Signature		

Example:

Adapted from EFCN.

For efficient leak detection a leak detection survey log should be maintained. The log table, as adapted from EFCN is given in Table 10.11

A water audit performed in a hypothetical company 'ABC Water Inc.' and hypothetical data have been entered in the AWWA software for analysis. Figures 10.8 is a screenshots of the data-entry table and the water mass-balance table.

From the point of view of business, the AWWA software considers 'delivered water' in two parts: (a) revenue water, which includes the volume of water sold to the customers and reflected in their bills and (b) non-revenue water, which is lost or accounted as unbilled authorised use of water. It needs mentioning that the components of AWWA software on 'revenue' and 'non-revenue' water are fully applicable to water industries and a water distribution system, thereby making the AWWA water audit method versatile.

(Note: The term 'water industries' incorporations water utilities and all agencies that trade in water.)

10.4.8 Industrial response to water audit report

The water audit report contains details about how an industry is using water, including evaluation of unauthorised use and water lost through

AWWA WLCC Free Water Audit Software: Water Balance					Water Audit Report For:	Report Yr:
Copyright © 2010, American Water Works Association. All Rights Reserved. WAS v4.2					ABC Water	2015
	Water Exported 12,056.000				Billed Water Exported	
—	—	Authorized Consumption 18,569.813	Billed Authorized Consumption 17,525.000	Billed Metered Consumption (inc. water exported) 11,500.000		Revenue Water 17,525.000
Own Sources	—			Billed Unmetered Consumption 6,025.000		
(Adjusted for known errors)	—		Unbilled Authorized Consumption 1,044.813	Unbilled Metered Consumption 800.000		Non-Revenue Water (NRW)
1,245.000	—			Unbilled Unmetered Consumption 244.813		
—	Water Supplied 19,585.000	Water Losses 1,015.188	Apparent Losses 299.993	Unauthorized Consumption 48.963		2,060.000
—				Customer Metering Inaccuracies 251.020		
—				Systematic Data Handling Errors 0.010		
Water Imported			Real Losses 715.195	Leakage on Transmission and/or Distribution Mains **Not broken down**		
30,396.000				Leakage and Overflows at Utility's Storage Tanks **Not broken down**		

Figure 10.8 Screen shot of water balance table of ABC Water Inc. Source: AWWA Free Water Audit Software, Copyright © American Water Works Association. Printed with permission from AWWA (2016).

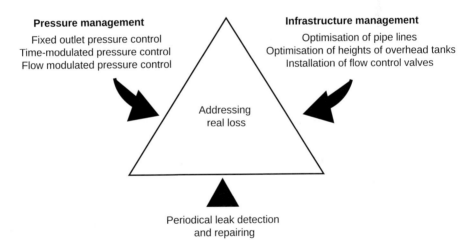

Figure 10.9 Proactive measures to control real loss.

leakage. The audit report also contains suggestions for improvements through various methods. On the basis of the audit report, industries are required to take the following measures:

- Repairing leaks;
- Installing methods of reuse and recycling of water;
- Scheduling flow of water or reducing flow of water to minimise water abuse;

- Retrofitting with water-efficient devices, such as water-saving flush toilets, urinals and aerated faucets;
- Vigilance system to control water loss due to unauthorised consumption;
- Installing water metering systems in places indicated in the audit report.
- Educating employees to change water-use behaviour and creating a 'workplace culture' that focuses on fostering a sense of pride among employees on the company policy on water conservation;
- In addition to a creation of 'workplace culture', industries will serve their own cause more effectively by also launching water awareness programmes with local community.

10.4.9   Real loss management

Babić and Djukić (2012) analysed IWWA methods in the light of case studies in Belgrade suburban areas. They have drawn from IWA and suggested that real loss cannot be completely eliminated. The unavoidable annual real loss can be estimated as a factor directly proportionate to the length of pipelines and average pressure in the distribution network.

Real loss can be reduced if certain proactive measures are taken by the industry. See Figure 10.9. The measures are:

- Periodical leak detection and repairing
- Infrastructure management
  - Reduction of unnecessary length of pipelines
  - Managing heights of overhead tanks
  - Installation of flow control valves
- Pressure management
  - Fixed outlet pressure control with pressure reducing valve (PRV)
  - Time-modulated pressure-control reduction in pressure during off-peak periods
  - Flow-modulated pressure control using electronic control system that controls pressure based on the flow.

10.5   Methods of water conservation

The goals of water conservation are to preserve water for future generations and to keep the ecosystem healthy. The methods are as follows:

10.5.1   Water use management

Water use management is one of the vital issues in all industries and can be achieved through a combination of changing behaviour (Danielsson and Spuhler, 2008) and adopting water-saving technologies in the process. Efficient water use has a positive impact on the industrial economy and may enhance profits. Against the backdrop of global water crisis, industries are under pressure to adopt optimised water use through certain best practices.

Water consumption and cost of water processing can be minimised through optimisation of water use. It can lower water withdrawal from local sources, reduce pumping and processing costs and release less wastewater. Reduction in use may be achieved through (1) increasing the productivity of water, (2) reducing water use in thermal-energy production, (3) reducing pollution load and (4) reducing processing cost. The major interventions that can result in efficient water use management are as follows.

10.5.2 Demand management

Demand management involves activities for manipulation (Beecher, 2016) of the level and timing of water usage. This is true for manufacturing industries, mining and water service providers. A part of demand management is the load management that enables the user to reduce the load on facilities. The whole process of demand management should be viewed from the economic standpoint, and reduction of demand will result in more profit generation. A demand management plan may involve a wide range of measures including:

- **Water demand profiling:** Water demand profiling is an important component of demand management. In an industrial project, a water demand profile (IPIECA Water Working Group, 2014) should be prepared at every stage and include every component. Water utilities may prepare their water demand profile on a daily or monthly basis to increase the efficiency of water supply and manage cost-reflective tariffs. For other industries profiling may be done month-wise for a period starting from the construction stage to the life of the project. See Figure 10.10.
- **Cost-reflective tariff:** Water utilities can adopt water-conserving rate structures to reduce per capita water use and maintain revenues. Cost-reflective tariffs help water utility managers to reduce water demand during peak hours. If the peak hour tariff is fixed at a high rate, people will to move their water consumption from peak periods to off-peak periods.
- **Regulate the efficiency of water-using appliances:** Water utilities can regulate the efficiency of water-using appliances like washing machine, dish washers etc. of their end users. These measures will reduce water use at the demand side.

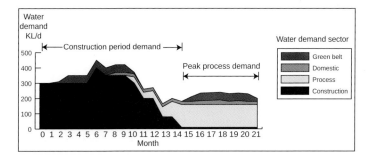

Figure 10.10 Water demand profile.

10.5.3   Changing the water use behaviour

The primary driver for any change is the change in behaviour, and it is not an exception in the industrial sector. An initiative for behavioural change may include government, society and even the industry itself. To bring about the desired change, some social action programmes like awareness generation, demand creation and certain activities like action research may be helpful. The components of behavioural change programme for water use management are:

- **Inculcate the importance of water scarcity issues among employees:** Employees should be exposed to regular orientation courses, workshops, group meetings and site visits to develop an environmentally informed understanding of water use in different sectors of the industry.
- **Engaging employees in in-house conservation:** Industries should motivate employees to adopt water conservation practices in the workplace.
- **Educate employees in the identification of problems and searching for solutions:** Employees should be engaged in the process of problem identification and encouraged to develop innovative ideas in water conservation, wherever possible.

10.5.4   Water use assessment

Water use assessment is the second step in water use management. It can be done through:

- Identification of peak periods of consumption through analysis of bills.
- Calculation of average water consumption in different departments and sections of the industry.
- Water use mapping and database development.

Industries should conduct water use assessment at regular intervals of one to five years and compare current usage with previous assessments. It may be conducted separately or along with the water audit and water footprint study. This is discussed in Chapter 9 of this book on water accounting and disclosure.

Water use may also be assessed in terms of money spent for acquisition or withdrawal of fresh water. Industries can arrange for data-logging of consumption and identify areas of avoidable excess, if any, and any anomalies in use. This may also help in comparing water consumption in various units. Water use assessment is a statutory requirement under the water acts of some countries. In that case, there are return forms that have to be submitted to the appropriate competent authority.

10.5.5   Reduced consumption and water loss

As discussed, benchmarking can compare the water consumption of different industries and can set up a standard of water use. Industries should adopt the benchmark and reduce consumption through retrofitting equipment and improving the water use behaviour through the workplace culture.

Water audits can detect water loss and suggest the ways to control it. In an industrial establishment, the main cause of leakage is equipment malfunctioning. Common locations of malfunctioning are (a) leaking flow valves and faucets, and (b) dysfunctional float valves that cause wastage by spilling, especially in toilet cisterns. Regular surveillance is crucial to detection and prevention of water loss.

10.5.6   Reuse and recycle

Reuse and recycling technologies are becoming increasingly popular in water conservation. They reduce both the discharge of wastewater into the environment and withdrawal of water from sources. A wide range of proven technologies (conventional and advanced) is available. Water recycling systems can be tailored to meet specific demands. There is a fine distinction between 'reuse' and 'recycle': In recycling systems, wastewater is purified and commonly used for industrial processes; it is also used for groundwater recharge. Water is said to be 'reused' when 'greywater' is directly used for other industrial purposes. Some of the methods of direct reuse are:

- **Irrigation:** Discharged water can be used for irrigating gardens and for green belt development.
- **Dust suppression:** Wastewater can be used for dust suppression through bag filters or other methods.
- **Fire fighting:** Wastewater, if quality permits, can be stored for emergency fire fighting.
- **Washing.**
- **pH adjustment.**
- **Cooling water:** Boiler blow-down water to be used as cooling-tower make-up water

The US-EPA has released a guideline on how wastewater can be reused at different stages of purification. See Table 10.12

Industries recycle wastewater in various methods and for different purposes. The most popular applications are:

10.5.7   Zero liquid discharge plants

The principle of 'zero discharge' as the name implies, is recycling of all industrial wastewater within the industry; that is, there is no wastewater exiting from the factory premises. Different industries have different water standards; thus any given zero discharge facility is designed to meet a specified grade. There is no rule of thumb for designing universally applicable zero discharge systems, except for the fact that all zero discharge systems produce sludge from the water purification system, which is collected in evaporation tanks. The sludge contains solid waste/salts which are periodically cleaned and disposed of. Some solid wastes have beneficial uses: for instance, nitrate-rich solid wastes from petrochemical industries are used as fertilisers.

Table 10.12 Suggested water recycling treatment and uses.

Increasing levels of treatment, increasing acceptable levels of human exposure		
Primary Treatment: Sedimentation	**Secondary Treatment:** Biological Oxidation, Disinfection	**Tertiary/Advanced Treatment:** Chemical Coagulation, Filtration, Disinfection
No uses recommended at this level	• Surface irrigation of orchards and vineyards • Non-food crop irrigation • Restricted landscape impoundments • Groundwater recharge of non-potable aquifer** • Wetlands, wildlife habitat, stream augmentation** • Industrial cooling processes**	• Landscape and golf course irrigation • Toilet flushing • Vehicle washing • Food crop irrigation • Unrestricted recreational impoundment Indirect potable reuse: Groundwater recharge of potable aquifer and surface water reservoir augmentation**

* Suggested uses are based on Guidelines for Water Reuse, developed by US-EPA.

** Recommended level of treatment is site-specific.

Source US-EPA (2008).

10.6 Water saving in agriculture industries

Agriculture-based industries, such as food products, beverages, paper, biodiesel, and so on are major consumers of water. The majority of water used by agro industries is used in irrigation. There are many water conservation technologies for agro industries, some of which are mentioned below:

10.6.1 Soil moisture sensors

Over-watering of soil by irrigation is not only wasteful but also detrimental to the crop. Sensors measure soil moisture based, on which the volume of water and the area to be irrigated can be calculated.

10.6.2 Rain sensors

Sprinklers used in irrigation operate on fixed schedules programmed into the irrigation equipment. Rain sensors need be applied to automated irrigation systems to prevent unnecessary wastage of water during and after rains.

10.6.3   Drip/micro–irrigation

There was a time when irrigating fields by pumping water left virtually the entire field awash, which was enormously wasteful. Micro-irrigation or drip systems have revolutionised water conservation in agriculture by delivering only the required amount of water directly to the plant roots; thus minimizing losses to wind, runoff, evaporation, or overspray. Drip irrigations systems can save more than 50% of water compared with conventional pumped irrigation. See Figure 10.11a.

10.6.4   Sprinkler heads

Sprinkler systems are used to spray water to irrigate crops. There are two types of spray systems: (1) mist spray and (2) rotary spray. In mist spray systems, water is sprayed as a stream of water droplets. In rotary spray systems, the spray head rotates on an axis and irrigates a circular field around it. Rotary spray heads deliver a thicker stream of water than the mist spray heads. Water lost due to evaporation and blowing wind is comparatively lower in a rotary spray head. See Figure 10.11b.

10.6.5   Centre pivot irrigation

Centre pivot irrigation systems are popular in agriculture industries, especially in plantations, sugar crop production, and so on. In the conventional system of sprinkling water in the air, a good amount of water is lost through evaporation. This system has now been improved and made more efficient. This efficiency comes from putting sprinkler heads, or nozzles on hose drops, as

Figure 10.11 Water-saving irrigation systems: (a) micro irrigation, (b) sprinkler, (c) centre pivot irrigation. Source: USDA (no date).

pictured in Figure 10.11c, to minimise water drift and evaporation (McDonald, 2015). This system is also energy efficient. Soil sensors can be employed to monitor soil moisture levels that help in controlling excess irrigation.

10.7   Rainwater harvesting

10.7.1   Introduction

Rainwater harvesting (RWH) is not a new concept; it has been practiced for several centuries in low rainfall areas and other places where fresh water is insufficient. There is historical evidence of RWH from almost all countries. For instance, in India, RWH dates back to Indus Valley Civilisation. RWH is becoming more and more popular with the water crisis we have been witnessing in recent times. Individual houses, communities, schools, hospitals, government buildings and corporations are all adopting RWH to achieve water security. Rainwater harvesting is a globally recognised best practice of industries. In areas where the groundwater level is severely depleted and where natural groundwater recharge is severely affected by construction on the land surface, RWH is most commonly practiced as an alternative for depleted natural water resources. Although the RWH technologies are expected to be applied universally, they are most needed in water-scarce countries. The technology is easily implementable, and the cost can be easily recovered within a short period; RWH also has a long-term positive impact on local water resources because, it reduces load on groundwater and surface water. The currently growing trend of adopting RWH is a direct consequence of growing public awareness worldwide. Some of the major RWH options are:

- Rooftop rainwater harvesting for storage use;
- Rooftop rainwater harvesting for groundwater recharge; and
- Surface runoff harvesting for storage (percolation tank) and enhancing groundwater recharge.

Rainwater harvesting has two purposes: (1) using captured water *in situ* and storing it for consumption and (2) recharging groundwater. Both systems are beneficial to the environment but succeed only if meteorological, hydrogeological and social conditions are factored into the design. Industries can opt for an installation of RWH systems in public places under CSR programmes.

Is it really necessary to adopt rainwater harvesting system everywhere? The answer is, probably not. But there are definitely some situations in which rainwater harvesting is strongly advisable to resolve water-related issues.

Natural 'allocation' of rainwater involves: (1) evaporation, (2) transpiration, (3) runoff, (4) storage in soil moisture, and (5) surface storage and storage in aquifer. The natural processes mentioned in (4) and (5) is collectively defined as the 'water retention capacity' of the area. RWH is basically aimed at capturing 'runoff' to increase water retention capacity; it can be readily appreciated therefore that RWH is best suited to hilly regions where runoff is high. This natural allocation process is impaired by human intervention. For example, if a large building is constructed on open ground, the area for rainwater percolation is reduced, directly affecting groundwater and soil moisture storage. As a consequence, the enhanced runoff from the constructed area has to accounted for in the RWH design to compensate for the reduced percolation.

Arid and semi-arid regions are defined in terms of the aridity index. An aridity index (AI) is defined as: $AI=P/t$, where P=annual total precipitation; t=annual mean temperature. In semi-arid areas, AI is between 15 and 23 whereas in arid areas the AI is between 8 and 15 (Leech, 2014).

RWHs designed for humid and arid regions differ radically because of low precipitation and high evaporations rates in arid climates. In confined aquifer regions, where *in situ* groundwater recharge is not possible and water is over exploited, artificial recharge of groundwater through RWH is an essential technological option.

10.7.2 Regulations and guidelines

Industries can take up RWH schemes in their factories or outside depending on the (1) water sustainability policy of the company, (2) the water-stress condition of the region, (3) regulatory requirements, and (4) as a component of CSR programmes. Though RWH is becoming popular in water-stressed nations, there is a little focus on its importance in the regulatory guidelines in many countries. A literature survey has revealed the following status of inclusion of RWH in the water framework of different countries.

- **Africa:** While discussing RWH policy in 10 countries of Sub-Saharan Africa, Hartung and Patschull (no date) state that the national water laws of Botswana, Ethiopia, Kenya, Lesotho, Namibia, South Africa, Tanzania, Uganda and Zambia have not mentioned RWH in anywhere in the national water framework. Only South Africa has mentioned the possibility of RWH in domestic establishments.
- **USA:** In the United States, Texas offers a sales tax exemption on the purchase of rainwater harvesting equipment. New Mexico has established Water Harvesting Income Tax Credits to incentivise individuals and businesses to collect rainwater for future uses, (Meehan and DuBois, 2016). In many states of the USA, RWH installation requires prior approval from concerned water authorities.
- **UK:** In the United Kingdom, RWH is included in the Code of Sustainable Homes, and new constructions are required to have RWH installation for building plan approval. There are official guidelines and standards on RWH in the UK. Tax exemption is also provided.

- **India:** India is one of the pioneering countries advocating for RWH. In India, the state laws also direct industries to declare their RWH plan in the EIA report before installation or expansion of a plant.
- **South Africa:** In South Africa, domestic RWH is allowed without any licence or permit. In the National Water Law, 1998, South Africa has no provision for abstracting flowing water without licence. It implies that RWH for business establishments will require licences (Allen, 2012).
- **Australia:** RWH is a traditional system in Australia, and there are institutions for technical guidance. The National Water Quality Management Strategy (NWQMS) by NRMMC has released risk-based guidelines for storm-water reuses in the country.
- **China:** RWH has traditionally been popular in China, and the government now encourages it, though it is not categorically included in governance documents.

10.7.3   Why industries should take up RWH

The need to capture, store, filter and reuse rainfall (Stormsaver, 2016) has become the responsibility of industries, not only as regulatory requirements but also to ensure alternative sources of water. RWH is also essential for the shift towards sustainable development. The cost-saving capacity of RWH is now being appreciated by many industries across the globe. The benefits of RWH are:

- Water saving and reducing the water bill.
- A good technology for storm-water management solutions.
- The technology is very simple and can be easily adopted; maintenance cost is very low.
- RWH can supply water to industries even during water cuts; for example, in some industries in Maharashtra, India, manufacturing industries with RWH installations are continuing with their production despite frequent water cuts by the government (The Times of India, 2016).
- In RWH, water collection requires no power.
- Dissolved salt content in rainwater is low; consequently, scale formation in the RWH pipelines is low.
- Cost of purification of water drawn from RWH installations is low.
- RWH reduces pressure on groundwater use; it can supply a substantial amount of freshwater for industrial use.
- RWH reduces overland runoff, thus reducing soil erosion by storm runoff.

There are three main drivers of RWH:

1 **Cost saving:** Cost saving is one of the main drivers. It reduces consumption of treated municipal water, hence reduces the utility bill. But to what extent the cost is saved and how the investment is paid back depends on the designed capacity of the RWH installation.
2 **Corporate policy:** Manufacturing plants includes RWH programmes because it is a part of the corporate policy of most countries.
3 **Community engagement:** Industries may find opportunities in entering partnerships with community based organisations through project funding (Davies and Roy, 2004) for community RWH schemes.

10.7.4   Components of RWH

The components of RWH depend on the purpose and the technology used. The essential components of RWH are (1) catchment, (2) conveyance system, (3) filtration system, (4) recharging facility and (5) ponding or storage system. If rainwater is harvested for groundwater recharge, the storage system is replaced by a filtration and recharging shaft. There may be a combined storage and recharging arrangement.

RWH Catchment

For the purpose of RWH, a catchment is defined as an open place from which rainwater moves towards a single outlet point and then can be collected; the catchment must have a defined boundary so that there is no water loss due to an overflow across the boundary. The boundary may be constructed (i.e. parapet, dyke and wall) or natural. Different types of catchments suitable for RWH are shown in Table 10.13.

Methods of rainwater collection, storage and utilisation depend mostly on the type and surface material of catchment, quantum of annual rainfall, intended use and demand.

Conveyance system

Conveyance systems take harvested water from the roof to the storage tank and are composed of several components.

- **Gutter:** The gutter is the lower part of the catchment where rainwater runoff accumulates. The cross section of a gutter may be semi-circular, triangular or rectangular. The downspout is attached to one end of the

Table 10.13 Different types of catchment suitable for RWH.

Catchment type	Description
Roof catchment	Rooftops are the first choice for catchment; water quality from different roof catchments is a function of the type of roof material, climatic conditions and surrounding environment.
Pavements or any other paved land surface	Examples of such type of catchments are metalled roads, tiled walkways in gardens, parking lots and so on.
Open barren land	If there are open barren lands within the project area (especially in water-stressed conditions) where percolation is much more than runoff; the runoff water cannot be dismissed as not being viable for RWH. It is important for industries to realise that even the nominal runoff from open barren land are to be harvested.
Rocky outcrop	Rocky outcrops are solid rocks with high runoff (and very low percolation) which are ideally suited for RWH.
Land covered with vegetation	The vegetation covered catchment retains water in the root zone and minimises runoff, but following heavy rains they do have flowing water as sheet flows and small ephemeral rivulets which can be harvested.

gutter through which rainwater comes down to the ground level. The downspout is also referred to as the down pipe or roof drain pipe. The gutter sizes should be proportional to the rainwater runoff from a '100-year 24 hour' storm event.

- **Down pipe/downspout:** Rainwater that is channelled through the gutter flows downward through downspouts. Downspouts are usually vertical pipes and are designed with 50 cm^2 of downspout area for every 100 m^2 of roof area. There should be a leaf screen at the junction of the gutter and the down pipe to protect the pipe from clogging. The leaf screen itself can get clogged and needs periodic cleaning.
- **First flush:** During dry months, dust and debris accumulate on the rooftop. The first rain with significant runoff washes dust and debris into the down-pipe and ultimately to the storage tank. To avoid debris flow through the conduits a first flush system is attached. Figure 10.12
- **Conduits:** Conduits are pipes that carry water from catchment to storage tanks or the recharging system. Conduits can be of many kinds of materials, (e.g. PVC pipes, GI pipes, ceramic pipes, etc.) or materials that are locally available. Table 10.14 gives an idea about the diameter of pipe required for draining rainwater based on rainfall intensity and roof area.

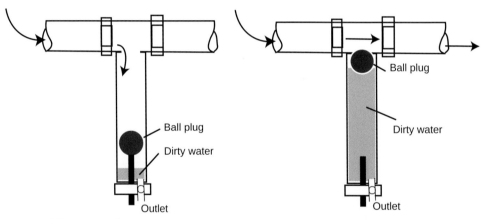

(a) During the first rain dirts are collected in the vertical pipe through the hole at the mouth of the vertical pipe.

(b) The dirty water is accumulated in the vertical pipe and the ball rises and closes the mouth of the vertical pipe. Dirty water is released through bottom outlet.

Figure 10.12 First flush separator.

Table 10.14 Sizing of rainwater pipe for roof drainage

Roof area in m^2			500				1000	
Rainfall intensity in mm/h	10	20	50	100	10	20	50	100
Diameter of down pipe in mm	90	100	150	225	100	150	225	300

Filtration system

Despite dust being removed by the first flush system, there are suspended particles in the harvested rainwater. To eliminate such particles, a filtration system is attached before the water reaches the storage tank. There are several types of filters but pop-up filters and dual-intensity filters are the most popular in India. The working principles of the filters are shown in Figure 10.13. These filters, depending on models, are suitable for catchments from 200 to 500 m².

For large industries which store large volumes of harvested rainwater, or use it for artificial recharging, sand and gravel filters are used (Figure 10.14). Harvested rainwater passes through the filter chamber with layers of sand, gravel and charcoal to remove suspended impurities. The capacity of such filters depends on the area of the roof catchment and

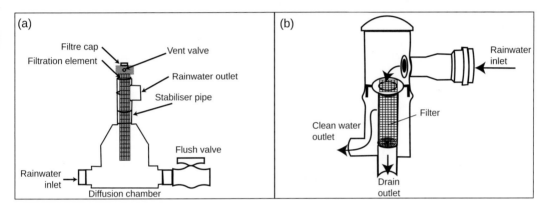

Figure 10.13 Filter for rooftop RWH, (a) pop-up filter and (b) dual-intensity on line filter.

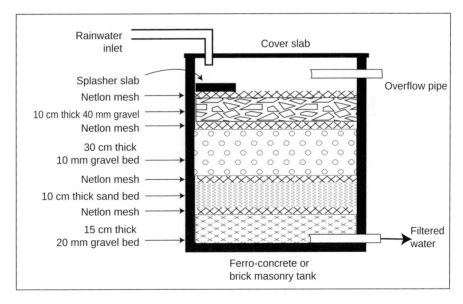

Figure 10.14 Cross section of sand and gravel filter.

the amount of rainfall. In general, the filter capacity should be 0.5 to 0.8 litres per m² of the roof area. Sand and gravel filters are of two types: ground filters and underground filters, depending on the purpose and position of the storage tank.

Storage device

The rainwater storage tank collects the filtered rainwater and keeps it for future use. The storage tank is set up above the ground, on a platform. It may also be an underground sump in some cases when there is a shortage of space; for example, locating the sump under a parking lot. The storage structure must be well protected and secured from external disturbances. A storage system is shown in Figure 10.15.

- **Capacity of storage tank:** The capacity of the storage tank is designed according to rainfall intensity, duration of the dry period and water demand. If harvested rainwater is utilised daily and the long-time storage is not required, then a tank size can be estimated based on the following calculations.

Figure 10.15 Rainwater tank at Ceres Environment Park. By Pengo assumed (based on copyright claims), Published under GFDL, CC-BY SA-3.0 or CC BY 2.5, via Wikimedia Commons.

Average daily rainfall = P mm

Roof area = A m^2

Runoff coefficient = C

Then the tank size should be equal to the daily harvesting potential = A × P × C × 0.001 m^3. However, the actual size of the tank depends on the management decision of whether the water will be used in rationing method or rapid depletion method.

- **The rationing method (RM):** In RM method, only a small amount of water is released for daily use and the maximum amount of water is stored. In RM therefore, rationing method large tank is required.
- **The rapid depletion method (RDM):** There is no restriction on the use of harvested rainwater in RDM. During monsoonal rains, RDM is considered to be the only source of water. Conventional water sources are used only if RDM supply dries up. By removing pressure on conventional sources of water, RDM has the effect of conserving other freshwater resources.
- **Types of tanks:** The tanks are either surface tanks or underground tanks. Some examples of surface tanks are:
  - **Rain Barrel:** This is a barrel-shaped cylindrical tank that holds 1000 litres and can be placed close to the building. Rain barrels are the most common types of tanks where RWH is practiced. Figure 10.16.a.
  - **Rain Silo:** Silos are larger than rain barrels; standard sizes vary from 2,500 to 15,000 litres. These are industrial-quality tanks with significantly thicker walls for long-term durability. There are other types of rain silos made of corrugated steel, available in 2 to 30 metre dia silos with capacities from 5000 to more than 1 million litres. The steel shell is heavily galvanized for long-term durability. See Figure 10.16b.
  - **Rain Box:** These are small PVC tanks with 200 to 1000 litres capacity and are suitable for small projects. Water is typically drawn through a fitting at the bottom of each rain box and flows by gravity for use in gardens or recharging ponds. A ball valve and hose adapter are standard; an external overflow is available as an option. See Figure 10.16c.
  - **Concrete or masonry tanks:** Concrete and masonry tanks were commonplace before the advent of plastic tanks. Currently, concrete or masonry tanks are used in civil construction sites or for sumps. A typical masonry surface tank is shown in Figure 10.17.
  - **Underground Tanks:** Construction of underground tank is typically considered where there is a shortage of space; the surface of the underground

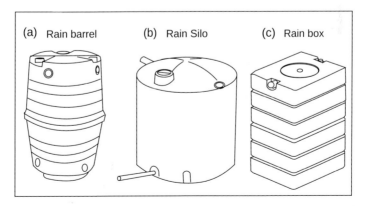

Figure 10.16 Different types of RWH storage tanks.

Figure 10.17 A masonry rainwater harvesting tank, Rajasthan, India. Taanka (2016).

tank then becomes available for a variety of purposes, like car parking, cycle sheds, gardens, and so on.

○ **Storage ponds:** Storage ponds are surface tanks, meant for storing harvested rainwater, and can be excavated or may be remodelled from an existing tank. In large industrial establishments where RWH potential is high, such types of tanks are useful, but a good amount of water is lost through evaporation.

10.7.5 Rainwater harvesting potential

Liaw and Chiang (2014) proposed three categories of rain RWH potential: (1) theoretical, (2) available and (3) environmentally supportable rainwater potential. Theoretical estimation of RWH potential is a function of long-term average rainfall of an area, catchment area and runoff coefficient. However, since it is not possible to harvest the total rainfall, the available RWH potential is estimated based on the storage capacity. Environmentally supportable rainwater potential is estimated considering the required environmental flow. To determine how much rainwater within the industrial facility should be released for environmental flow, it is necessary to have a clear understanding of the surrounding ecology. Regulatory authorities often set guidelines for offsetting environmental flow that is taken during a given RWH project.

Estimating the theoretical RWH potential

There are three variables for this estimation: (1) mean annual precipitation, (2) area of the catchment and (3) runoff coefficient. The runoff coefficient is defined as a dimensionless coefficient relating the amount of runoff to the amount of precipitation received. If the catchment is impermeable and has high slope, the coefficient value is higher, and if the catchment is a permeable formation or covered with vegetation that retains a good amount of water, the coefficient is lower. A commonly used table of runoff coefficients for different types of catchment is given in Table 10.15.

Runoff calculation: The purpose of runoff calculation is to estimate the amount of rainfall available for use. The simple equation of runoff calculation is:

$$Q = A \times C \times P$$

Table 10.15 Recommended runoff coefficient table.

Description of area		Runoff coefficients (C)
Lawns	Sandy soil	0.10 to 0.20
	Clay soil, flat, 2%	0.17 to 0.35
Unimproved Areas (forest)		0.15
Business	Downtown areas	0.95
	Neighbourhood areas	0.70
Residential	Urban	0.50 to 0.70
	Suburban	0.40 to 0.70
Industrial	Light Areas	0.70
	Heavy Areas	0.80
Parks, cemeteries		0.25
Playgrounds		0.35
Railroad yard areas		0.40
Streets	Asphalt and concrete	0.95
	Brick	0.85
Drives, walks, and roofs		0.95
Gravel areas		0.50
Graded or no plant cover	Sandy soil	0.30 to 0.40
	Clay soil	0.50 to 0.60
Open water and wetland		0.00
Land	Barren land, grass land, cultivated crop land	0.05
Roof	Metal	0.95
	Asphalt and concrete	0.90
	Tar and gravel and asbestos	0.80 - 0.90
	Thatched	0.02

Adapted from Cross (2013); Brainard (2016); and Guinnett County, USA (no date).

where Q is the amount of harvestable water, A is the area of catchment, C is the runoff coefficient of the catchment and P is precipitation.

For example: if the catchment area is 100 m² and C=0.8 then for every 100 mm of rainfall, the runoff potential is $100 \times 0.80 \times 0.1 = 8000$ litres.

10.7.6   Artificial recharge of groundwater

In this system, harvested rainwater is discharged into a recharging system instead of a storage tank. It is most suitable in areas where natural recharge of groundwater is very slow or impacted due to land-use change. The schematic of an artificial recharge system (filter and recharge shaft) is shown in Figure 10.18.

Rainwater recharge is done both from rooftop rainwater harvesting and from surface runoff harvesting. Different technologies have adopted different perspectives. Some of the options are:

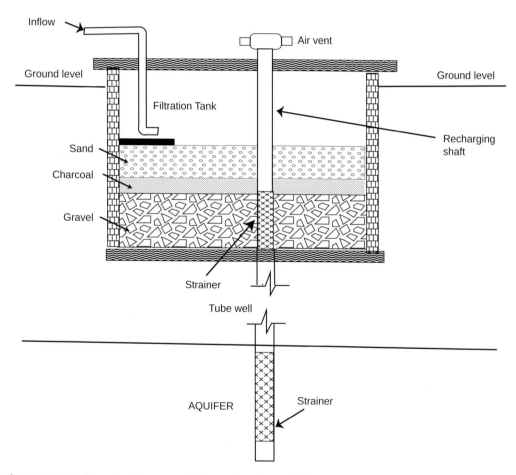

Figure 10.18 Schematic diagram of filter and recharge shaft.

Figure 10.19 Artificial recharge from rooftop RWH. Photo courtesy Arijit Das.

Groundwater recharge from rooftop catchment

Groundwater can be recharged directly from the rooftop after passing the water through the siltation chamber and the 'gravity sand filter' and directing it to the aquifer through a recharge shaft. A recharge shaft is dug manually or drilled by the reverse/direct rotary method. Diameters of recharge shafts vary from 50 to 300 mm depending upon the availability of water. It is constructed where the groundwater is depleted or being 'mined'. Groundwater is said to be 'mined' when the withdrawal rate exceeds the natural recharging capacity, causing a severely depressed aquifer level. Figure 10.19 shows a rooftop RWH and artificial recharge system.

Estimation of recharging capacity of a recharging well (recharge shaft)

Designing a recharge well and estimating the number of wells required in a RWH project area demand a calculation based on Darcy's Law. The flow in the recharge well is the reverse of a pumping well. When water is admitted into a well, a cone of recharge is formed surrounding the recharging well, which is just the reverse of a cone of depression. For estimating the recharge capacity of the well, it is assumed that (1) rainwater is admitted to the well at the ground level and (2) the effective depth of recharging well is the length of the water column in the well.

Other artificial recharging structures

In addition to the above methods, there are other types of recharge structures; in all such recharge methods it must be ensured that the water that is to be fed into the structure is filtered for the removal of suspended load. Common groundwater recharge techniques are:

- **Recharge pit:** Recharge pits are used for shallow aquifers. These pits are generally 1.5m × 1.5m in area and 2m in depth, with masonry side lining. Filtered rainwater is directly drained into the pit through an inlet-pipe. During heavy rainfall, the water level in the pit rises, which subsequently descends to the groundwater level. The pit floor is carpeted by gravel beds, which act as both filter and recharge site. See Figure 10.20a
- **Dug well:** A dug well doubles up as a recharge structure in many instances of RWH. Here, the recharge water is guided through an inlet pipe to the dug well. The discharge end of the inlet pipe is immersed. See Figure 10.20b
- **Abandoned/running hand pump:** An abandoned/running hand pump can be used for recharge. It is suitable for small buildings having a roof area up to 150 m². Water is diverted from the rooftop to the hand pump through 50 to 100 mm diameter pipe.
- **Recharge trench:** It is constructed where there is a permeable stratum of adequate thickness available at shallow depth (≤3m). The trench is 0.5 to 1.0 m wide, 1 to 1.5 m deep and 10 to 15 m long and filled with pebbles and boulders. See Figure 10.20c.
- **Gravity head recharge well:** In these wells, water goes down to the aquifer by the action of gravity. Where the aquifer is deep and the water level is depleted, abandoned bore holes can be used as recharge wells. The rooftop rainwater is channelled into the shaft, and recharge takes place under gravity flow conditions.

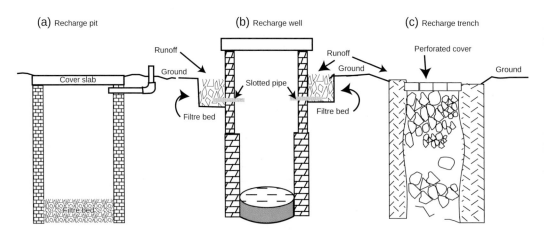

Figure 10.20 Artificial recharge structures: (a) recharge pit, (b) recharge well and (c) recharge trench.

10.7.7   Surface runoff harvesting

When rain falls on the land, part of it is arrested by vegetation, part evaporates, and part of it percolates into the aquifer; the rest of the water generates the surface runoff. The amount of runoff and the technology to be adopted depend on several ground conditions and factors. The efficiency of RWH structures depends on an evaluation of the following factors:

- **Climate:** The main climatic considerations are:
  - Monthly, annual and mean monthly rainfall.
  - Peak monthly rainfall and return period.
  - Ambient temperature; monthly maximum-minimum and mean monthly average.
  - Monthly rate of evapo-transpiration.
- **Terrain Condition:** Arguably the most important consideration in surface runoff harvesting is the terrain condition. The following terrain factors are to be studied before the RWH planning:
  - A physical inventory of the current socio-economic situation.
  - Inventory of existing sources of water.
  - Land condition and land availability.
  - Hydrological and hydrogeological factors.

Climate data are to be acquired from the local Meteorological Department.

Structures for surface runoff harvesting

There are several types of technologies for harvesting surface runoff, some of which have existed for generations as traditional practices in countries like Kenya, South Africa and India. In these traditional methods, there are instances of remarkable adaptation and innovation born of cultural wisdom. Structures for surface runoff harvesting are 'percolation tanks' and 'rock catchment harvesting':

- **Percolation tanks:** Percolation tanks are artificially created surface bodies of water. The part of the land under submergence in these tanks requires having adequate permeability to facilitate percolation into the aquifer. Surface runoff and rooftop water can be diverted to the percolation tank. Although percolation tanks are constructed even today, the practice goes back centuries. See Figure 10.21.
- **Deep groundwater recharge from surface runoff:** Surface runoff within factory premises can be utilised for groundwater recharge to a great extent if the integrity of water quality is maintained. This is also an option to manage storm runoff in peak storm conditions to avoid water-logging. In this system, surface runoff is collected in a pond for initial retention storage. An artificial recharging structure is constructed within the pond or near the boundary of the pond. It consists of a filtration tank for filtering the pond water and a recharge shaft to recharge groundwater. The structure has three parts: (1) a horizontal conduit pipe delivering pond water into the filtration chamber, (2) a sand and gravel filter and (3) a recharging shaft.
- **Harvesting from rock catchment:** Rock-catchment harvesting systems are practised in rocky, uneven terrain. Runoff from natural rock surfaces is collected in large ponds. This practice is quite successful in Kenya; in

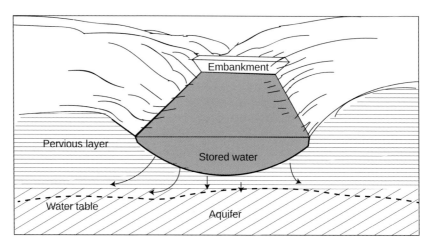

Figure 10.21 Percolation tank.

Figure 10.22 Rock cut temple of Himachal Pradesh serves as a rock catchment RWH system in India. Photo Courtesy Arunabha Das.

India, it is practised traditionally and is often found associated with temples located on hills where local people undertake the RWH facility on their own initiative. The advantage of rock catchment harvesting is that it can collect up to 90% of the local rainfall. Figure 10.22 shows a rock catchment RWH system in India.

10.7.8   Issues in RWH

Common issues in a RWH system are:

- **Water quality issues:** A RWH system should be well designed from the hygienic point of view. Although rain is contaminated by atmospheric pollutants, it is relatively free from impurities compared to harvested water. Harvested rainwater gets contaminated during conveyance and storage, which is a major health risk; dust, leaves, insects and animal excreta are the main sources of contamination. Improper storing, open drains for conveyance, leaks in tanks and pipes are avenues through which particulate contaminants and pathogens pollute harvested rainwater. To get rid of the risk of contamination, the catchment should be washed frequently, and the tanks and pipelines should be inspected to avoid cracks and leaks.
- **Mosquito issue:** Mosquitoes are vectors of malaria parasites, dengue virus and several other life-threatening pathogens. Stored rainwater can be a breeding ground of mosquitoes. Prevention of mosquito breeding requires 'insect proofing' of all tank openings (inlets and overflow outlets).

10.7.9   Maintenance of RWH system

The maintenance requirements for a rooftop RWH system are simple and inexpensive:

- Rooftop should always be kept clean.
- Inlets of downspouts should be protected with a mesh to prevent entry of leaves, etc.; this mesh needs periodic cleaning.
- The filter should be cleaned periodically and replaced when necessary. Sand filter beds should be backwashed so that pore spaces are not clogged.
- Storage tank manholes should be properly fastened to prevent contamination. They should also prevent the entry of sunlight to avoid growth of algae.
- Outlets of overflow pipes should be covered with fine mesh to prevent mosquitoes from entering the tank or filter chamber.

10.7.10   Constraints in adopting a rainwater harvesting system

The School of Engineering, University of Warwick (2002) conducted a survey in Sri Lanka, Ethiopia and Uganda on people's views on adopting RWH in the domestic sector. The main issues addressed by the Warwick survey report are rainwater quality, socio-economic issues, governance and technology. The public opinion that emerged from the survey is that rainwater is not potable. Though the water policy of the government in these countries is pro-poor, there are no specific guidelines regarding RWH. Indeed, the fact that domestic RWH is not economical in Uganda is expressed even in publications (Thomas and Kiggundu, 2004). In contrast, RWH systems are gaining popularity in China, Indonesia, India, and in several other developing countries. It is reasonably certain that RWH will gain in popularity and with improved technology will be accepted as an alternative freshwater source worldwide. Unlike public opinion from civil society, public opinion in the corporate sector is fully in favour of RWH. Participation of the corporate sector will therefore, enhance the social acceptance of RWH. Pending worldwide acceptance, some technological and physical issues regarding RWH system are discussed below:

- **Lack of proper guidance:** Although rainwater harvesting is a very simple technology, it nevertheless needs location-specific guidance or guidelines. For example, designing conveyance, storage or filtration systems in RWH may turn out to be highly cost intensive if installed without the guidance of experts. Enlisting expertise not only reduces installation and maintenance costs but also increases the efficacy of RWH.
- **Absence of proper space for storing water:** Storage space for harvested water is a major constraint in industrial establishments. Construction of most industrial premises and their townships predate the idea of RWH as a viable practice; consequently, there was no provision for RWH storage in the original planning. To continue business viability, the absence of RWH storage space has to be compensated for by innovatively designed constructions.
- **Absence of suitable aquifer to recharge:** Use of RWH for groundwater recharging requires design compatibility with local hydrogeological conditions. For example, sometimes it is difficult to have a suitable aquifer to recharge, even though RWH installations are in the proper place. Likewise, recharging can become difficult and expensive if the aquifer is very deep.
- **Regulatory bindings:** Although regulations for RWH are meant to optimise/compensate water withdrawal, sometimes the regulatory guidelines are inconsistent with local hydrogeological conditions. In such situations, industries should not be charged with noncompliance.

10.7.11   Promotion and further development of rainwater utilisation

Rainwater utilisation is becoming popular and is also being accepted by industries. For promotion of the idea, the following steps should be undertaken.

A Systematic approach

Rainwater utilisation should be incorporated into municipal ordinances and regulations alongside water conservation and wastewater reclamation. It is not advisable that RWH installations be carried out with the help of common sense only. Technical consults along with the use of standardised material are vital to effective use of RWH.

Implementation policy

It does not suffice simply to bring the industries under the regulatory framework for RWH. Implementation policies should be formulated to educate society about RWH, which make the local government a pivotal agency for popularising RWH in civil society. The education process must begin with familiarising civil society with the natural hydrological cycle. Consideration should be given to subsidising facilities for rainwater utilisation.

Technology Development and training

Encouraging technology and human resources development to support rainwater utilisation is important. It is also important to promote development of efficient and affordable devices to conserve water, facilities to use rainwater and devices to enhance the underground seepage of rainwater. Together with this, there is a need to train people on technologies and devices used in RWH.

Networking

To promote rainwater harvesting and utilisation as an environmentally sound approach for sustainable urban water management, a network should be established involving government administrators, citizens, architects, plumbers and representatives from all sectors of the national economy, especially from the manufacturing sectors. It is essential to encourage regional exchanges amongst public servants, citizens and industry representatives involved in rainwater storage, seepage and use, as well as the conservation and reclamation of water.

10.7.12 Example of an industrial RWH

A hypothetical industry in India proposed the construction of a cement grinding plant, along with a captive power plant, covering a total land area of 26.39 hectares of barren land. The location is a seasonally water-stressed area where groundwater is the only industrial water source. A study of the groundwater resource revealed that it is not sustainable to meet its daily water demand of 400 m³/day. To augment local water resources and to comply with the regulatory requirements, the industry decided to take up an RWH scheme. To estimate the RWH potential and its optimum use, the following information has been collected:

1 Annual and monthly average rainfall.
2 Average monthly evaporation.
3 The total land use plan proposed for the 26.39 hectares of barren land, given the area to be covered by each specific land use, e.g. roof area, road, car park, green belt, garden etc.

The study reveals that:

- Annual monsoon rainfall = 1200 mm
- Total roof area available as catchment = 33,968 m²
- Total land area available for storage system = 3900 m².
- Annual evaporation rate = 1520 mm

Table 10.16 RWH potential estimation in an industry.

Item		Quantity (m³)
Input	Rooftop RWH (m³)	35,956
	Precipitation over the tanks/reservoirs (m³)	4587
	Total RWH (m³)	40,543
Loss	Evaporation loss from reservoirs/tanks (@ 1.52 m/year over an area of 3900 m²) (m³)	5928
	Net water available from RWH (m³)	34,615
Consumptive use	Planned utilisation for dust separation, green belt development, gardening, etc. throughout the year (m³)	18,000
	Planned utilisation for industrial process from June to September (m³)	16,615
	Total RWH water available for planned utilisation (m³)	34,615

- Total storage area $= 11,700$ m³.
- Runoff coefficient of roofs $= 0.9$
- Total rooftop RWH $= 35,956$ m³
- Total annual industrial water requirement $= 400 \times 365 = 146,000$ m³

The RWH and utilisation plan developed from the study is shown in Table 10.16.

Out of the total annual requirements of 146,000 m³, the industry generates 34,615 m³ through RWH, thus cutting down their groundwater withdrawal by 23%.

10.8 Conclusion

The World Water Council has released a document on world water vision which states that industrial water demand and use has increased to such an extent that it has surpassed all previous records and much exceeded nature's capacity to replenish the withdrawal. Water use and management in mining and in the extraction of petroleum and natural gas are considered to be environmental challenges as they cause high withdrawal, pollution and diversion of natural flow. Management is also an issue in the thermo-electric power sector, which is highly water intensive, and the gap between withdrawal and use is extremely high. Through adoption of some of the best practices, industries can manage their water use more efficiently. This chapter attempts to deliver a basic idea on different types and sectors of industrial water use and emphasise the key processes of water use.

The major aspects of water conservation are behavioural changes of the consumers and using appropriate technology for conservation. Industries need to conserve water for their own benefit and can do so through arresting wasteful use and leakage. They can augment their net resources through technological adaptations like rainwater harvesting, reuse and recycling of wastewater.

Bibliography

Allen, J.E. (2012) *Tank sizing from rainfall records for rainwater harvesting under constant demand.* Available at https://ujdigispace.uj.ac.za/bitstream/handle/10210/8319/Allen.pdf?sequence=1 (accessed: 3 May 2016).

AWWA (2016) *Free Water Audit Software.* Available at https://www.awwa.org/resources-tools/water-knowledge/water-loss-control.aspx (accessed: 12 September 2016).

Babić, B. and Djukić, A. (2012) 'Estimation of water balance and water losses in water utilities: Experiences from the Belgrade waterworks', *Djukić Water Research and Management*, 1(4): 27–34.

Berg, S.V. (2006) *Conflict resolution: Benchmarking water utility performance.* Available at http://warrington.ufl.edu/centers/purc/purcdocs/papers/0641_Berg_Conflict_Resolution_Benchmarking.pdf (accessed: 24 March 2016).

Berg, S. and Padowski, J.C. (2007) *Overview of water utility benchmarking methodologies: From indicators to incentives.* Available at http://warrington.ufl.edu/centers/purc/purcdocs/papers/0712_Berg_Overview_of_Water.pdf (accessed: 24 March 2016).

Brainard, S. (2016) *Rainwater harvesting course 6 1.* Available at http://slideplayer.com/slide/1519628/(accessed: 24 May 2016).

Cilantro (2013) *Sustainable solutions for factories.* Available at http://www.cilantrotech.com/energy-efficiency-solutions/factories/#Water_form_pop (accessed: 1 May 2016).

Connecticut Department of Environmental Protection (no date) *Funding provided by US Environmental Protection Agency Clean Water Act Program.* Available at http://infohouse.p2ric.org/ref/01/00524.pdf (accessed: 24 March 2016).

Cross, J. (2013) *Handling stormwater runoff, implications for the city of Portland.* Available at https://ds.lclark.edu/sge/2013/05/01/handling-stormwater-runoff-implications-for-the-city-of-portland/(accessed: 24 May 2016).

Danielsson, M. and Spuhler, D. (2008) *Reduce water consumption in industry.* Available at http://www.sswm.info/category/implementation-tools/water-use/hardware/optimisation-water-use-industry/reduce-water-consum (accessed: 22 March 2016).

Davies, R. and Roy, B. (2004) *Water and community development: Rainwater harvesting and groundwater recharge.* Available at http://www.sswm.info/sites/default/files/reference_attachments/DAVIES%202004%20Water%20and%20Community%20Development.pdf (accessed: 3 May 2016).

Government of India Ministry of Water Resources (2005) *General guidelines for water audit & water conservation.* Available at http://www.cwc.gov.in/main/downloads/Water%20Audit%20&%20Water%20Conservation%20Final.pdf (accessed: 7 October 2016).

Guinnett County US (no date) *Recommended runoff coefficient values table.* Available at https://www.gwinnettcounty.com/content/LocalUser/pnd/stormwater_design_guide/WebHelp/stormwater_design_guide/Hydrology/Rational_Method/Recommended_Runoff_Coefficient_Values_Table.htm (accessed: 24 May 2016).

Hartung, H. and Patschull, C. (no date) *Legal aspects of rainwater harvesting in Sub-Saharan Africa.* Available at http://www.eng.warwick.ac.uk/ircsa/pdf/10th/5_01.pdf (accessed: 15 April 2016).

IPIECA Water Working Group (2014) *Identifying and assessing water sources.* Available at http://www.ipieca.org/resources/good-practice/identifying-and-assessing-water-sources/(accessed: 22 March 2017).

ISO (1996) *ISO 14012.*

Leech, G. (2014) *Classification of arid & semi-arid areas: A case study in western Australia*. Available at http://press-files.anu.edu.au/downloads/press/n1673/pdf/guy_leech.pdf (accessed: 9 October 2016).

Liaw, C.-H.and Chiang, Y.-C. (2014) 'Framework for assessing the rainwater harvesting potential of residential buildings at a national level as an alternative water resource for domestic water supply in Taiwan', *Water*, 6(10): 3224–3246.doi: 10.3390/w6103224.

McDonald, K. (2015) *Big picture agriculture*. Available at http://bigpictureagriculture.blogspot.in/2015/10/thirty-five-water-conservation-methods_20.html (accessed: 3 July 2016).

Meehan, K. and DuBois, G. (2016) *State rainwater harvesting laws and legislation*. Available at http://www.ncsl.org/research/environment-and-natural-resources/rainwater-harvesting.aspx (accessed: 15 April 2016).

Nelson, L. (2013) *Water use benchmarking in the beverage industry trends and observations 2012*. Available at http://www.bottledwater.org/files/BIER%20Water%20Use%20Benchmarking%20Report%202012.pdf (accessed: 29 March 2016).

New Hampshire Department of Environmental Service (2013) *Water efficiency: Business or industry water use and conservation audit*. Available at http://des.nh.gov/organization/commissioner/pip/factsheets/dwgb/documents/dwgb-26-16.pdf (accessed: 8 April 2016).

School of Engineering, University of Warwick (2002) *Very low cost DRWH in the humid tropics: constraints and problems*. Available at http://www2.warwick.ac.uk/fac/sci/eng/research/civil/dtu/pubs/reviewed/rwh/dfid/r2.pdf (accessed: 6 April 2016).

Stormsaver (2016) *Industrial rainwater harvesting*. Available at http://www.stormsaver.com/Industrial-Rainwater-Harvesting (accessed: 15 April 2016).

Sydney Water (2016) *Benchmarks for water use*. Available at https://www.sydneywater.com.au/SW/your-business/managing-your-water-use/benchmarks-for-water-use/index.htm (accessed: 24 March 2016).

Taanka (2016) in *Wikipedia*. Available at https://en.wikipedia.org/wiki/Taanka (accessed: 7 October 2016).

Tanner, S. and Braver, D. (2001) *Implementing water conservation goals at federal facilities—Lessons learned*. Available at http://www.nrel.gov/docs/fy02osti/31303.pdf (accessed: 7 October 2016).

Thomas, T. and Kiggundu, N. (2004) 'Constraints to domestic roofwater harvesting uptake in Uganda: An assessment', Available at http://wedc.lboro.ac.uk/resources/conference/30/Thomas.pdf (accessed: 6 April 2016).

TIDE (2010) *Water Audits for Industries, Farms and residences*. Available at http://dsttara.in/UploadedFiles/CSG_ActivitiesDocuments/13.pdf (accessed: 25 April 2016).

The Times of India (2016) *Rain water harvesting boon for industries during water cuts*. Available at http://timesofindia.indiatimes.com/city/mumbai/Rain-water-harvesting-boon-for-industries-during-water-cuts/articleshow/48784426.cms (accessed: 15 April 2016).

USDA (no date) *NRCS photo gallery image detail*. Available at https://photogallery.sc.egov.usda.gov/netpub/server.np?find&catalog=catalog&template=detail.np&field=itemid&op=matches&value=730364&site=PhotoGallery (accessed: 3 July 2016).

US-EPA (2008) *Water recycling and reuse*. Available at https://www3.epa.gov/region9/water/recycling/(accessed: 2 July 2016).

US EPA Region 9 (2012) *Using water audits to understand water loss: A joint presentation of the USEPA office of groundwater and drinking water and the American water works association*. Available at https://www3.epa.gov/region9/waterinfrastructure/docs/water-audits_presentation_01-2012.pdf (accessed: 27 August 2016).

Water metering (2016) in *Wikipedia*. Available at https://en.wikipedia.org/wiki/Water_metering#/media/File:Hidr%C3%B4metro.JPG (accessed: 7 October 2016).

11

Corporate Social Responsibility: Way Ahead in Water and Human Rights

11.1 Introduction

Rousseau, the 18th-century French philosopher postulated that there should be a social contract between business and society (Byerly, 2013). What is implicit in Rousseau's prescription is that big business should *suo moto* undertake public welfare activities instead of using its financial might solely to maximise profit. It needs appreciating in this context that the 'contract' prescribed by Rousseau dates back to the earliest stages of Industrial Revolution in Europe. Two centuries have gone by, and yet the idea of a social contract between profit-makers and people has not found its feet. Indeed, far from entering into a contract, the corporate world prefers to operate from ivory towers. Although Rousseau's prescription is lost in time, the idea has resurfaced after two hundred years; in modern-day parlance, it is called corporate social responsibility (CSR) which is now making the rounds in the corridors of legislative power. Arguably the most fundamental area where CSR is needed is in protecting people's right to water. Access to fresh water is a basic human right. That right is violated when corporate activities prey upon people's access to water. The subject matter of this book—corporate water stewardship (CWS)—does not acquire any degree of meaningfulness unless it is read in conjunction with CSR.

According to Bichta (2005), CSR has evolved over the years and has been associated with corporate philanthropy. After two centuries of air and water pollution, civil society movements have eventually succeeded in compelling the corporate-government combine to recognise its responsibility in providing a clean environment to people. For example, during the late eighties and early nineties, chemical industries in Canada, the US and the UK started programmes like 'responsible care' in which industries worked in close association with society for effective environmental management; the government,

Industrial Water Resource Management: Challenges and Opportunities for Corporate Water Stewardship, First Edition. Pradip K. Sengupta.
© 2018 John Wiley & Sons Ltd. Published 2018 by John Wiley & Sons Ltd.

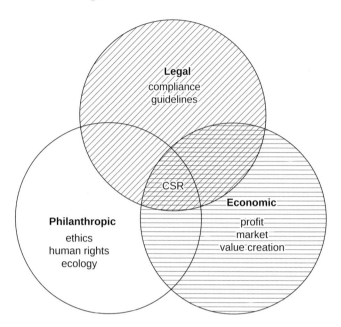

Figure 11.1 Conceptual framework of CSR.

for its part, reinforced the clean environment movement with legislation. During this period, the term *corporate citizenship* evolved to safeguard the interest of stakeholders. Three concepts have become associated with business since the late nineties: (1) corporate citizenship; (2) corporate sustainability, and (3) equal promotion of the economic, social and environmental interests of a business. And thus, corporate social responsibility is now seen as profitability plus compliance plus philanthropy. See figure 11.1.

11.2 Public policy on CSR

In the past decade, governments have become increasingly proactive in promoting corporate social responsibility (CSR) through their public policies (Ascoli and Benzaken, 2009). In many countries like India, Brazil, Mexico, China, UK and South Africa, government is being involved in the activities of the private sector to promote CSR investment in social development. Government ministries, departments and institutions are becoming instrumental in ethical practices in business to address human rights, child labour, workplace security, water and sanitation and environment.

Legislating CSR is fraught with complexities. If the CSR laws are perceived as being corporate friendly and indifferent to the interests of the general population then:

1 Government will suffer massive loss of face if the legislation is quashed by the judiciary;
2 Confrontation between the legislature and judiciary;
3 Political upheaval of constitutional amendment; and
4 Popular uprising.

Legislating CSR is, therefore, a slow process in multiparty democracies, and draft bills are under review in many countries. Pending issuance of government guidelines, companies often develop their own CSR strategies. These strategies are sometimes found to be deficient in philanthropic content (designed as they are to step around taxation laws) and face strong criticism from the civil society, which can lead to media storm—a situation that the corporate sector desperately wants to avoid. To avoid such reputational threats, the corporate sector has to be constantly on guard regarding the quality of soil, air and water. Even though CSR has not quite found legally stable feet, the mere fact that legislators are working at it has put vested interest groups on red alert.

The United States, India, countries under the European Commission and some other industrialised countries have CSR policies and legal statements for the corporations. CSR in the United States is a part of responsible business conduct (RBC), and the RBC team 'provides guidance, promotion and support for responsible business practices, engaging the private sector, labour groups, non-governmental organisations, and other governments' (US Department of State, 2011). 'EU's widely-disseminated definition stresses that CSR is voluntary, goes beyond what the law requires, and is an integral part of the business' (Insead Knowledge, 2013). Fifteen European Union countries have actively engaged in CSR regulation and public policy development. Canada adopted CSR in 2007. Prime Minister Harper encouraged Canadian mining companies to meet Canada's newly developed CSR standards. The CSR policy of the United States and the European Union are highlighted in Box 11.1

Highlights of CSR policy of US and European Commission

US businesses freely acknowledge their ethical and social obligations. They accept the idea that businesses bear economic, legal, ethical, and discretionary responsibilities. Williams (2010) defines discretionary responsibility as the presumption that a company will voluntarily serve society. Such service reaches beyond economic, legal, and even ethical responsibilities. Most businesses support the fight against hunger. The US Sentencing Commission's Guidelines for Organizations in the United States helps facilitate such goals. Almost all businesses are covered. This includes labor unions, partnerships, unincorporated organizations and associations, incorporated organizations, nonprofits, pension funds, trusts, and joint stock companies.

In July 2001, the European Commission presented a Green Paper. This laid the foundation for CSR in Europe. The Green Paper, 'Promoting a European Framework for CSR' (Commission of the European Communities, 2002), states that corporate social responsibility is a 'concept whereby companies integrate social and environmental concerns, their business operations, and their interaction with their stockholders on a voluntary basis'.

BOX 11.1

Source: Forte (2013) printed under CC BY licence.

A few case studies on CSR Laws from selected countries are listed below:

In India CSR is addressed in the Company's Act, 2013. Any company having a net worth of INR 5000 million (USD 1.0 = INR 67 approximately) or more or a turnover of INR 10,000 million or more or a net profit of INR 50 million or more has to spend at least 2% of last three years' average net profits on CSR activities as specified in Schedule VII of the Companies Act, 2013. The Government of India also released specific guidelines on CSR activities to be taken up by industries in India.

The South African government regulates the usage of CSR to a large extend (Postma, 2012). The Department of Trade and Industry (DTI), the vital public policy actor, implements CSR policy through Companies Act No. 71 of 2008. The South African constitution admits a 'Bill of Rights' which is aligned with CSR, and the courts of law admit intersection between CSR and the Constitution.

In Kenya, on the other hand, CSR-related legislation is only faintly defined in the Kenya National Environment Action Plan and draft guidelines by the Kenya Bureau of Standards (KEBS). But, in Kenya, the strong ancient value and indigenous concept of mutual assistance and joint efforts are important factors in defining CSR (UK Esseys, 2015).

For many years, Indonesia has been practising corporate social responsibility (CSR) on a voluntary basis. Recently, it has been included in the Company Liability Act No 40/2007 where two types of CSR implementation are prescribed: (1) voluntary, for companies not based on natural resources and (2) mandatory, for natural resource–based companies. Large companies functioning in Indonesia are voluntarily complying with international standards of CSR framework, such as 'ISO 26000 on Social Responsibility', 'OECD Guidelines for Multinational Enterprises', and so on.

11.3 CSR policy of corporations

Holme and Watts (2000) define CSR as the human face of business. A survey conducted by the World Business Council for Sustainable Development (WBCSD) towards a comprehensive definition of CSR received varied response from the members of the council. By analysing all those responses, the council concluded that:

- A CSR policy must make a positive contribution to the wellbeing of the society.
- The business community can demonstrate its human face through CSR.
- CSR provides opportunities to establish dialogue with stakeholders and establish partnerships with the government, inter-governmental organisations (IGO), NGOs and the civil society.
- CSR can explore local customs, cultures and ethnic diversity, and can deliver social services while maintaining high and consistent global standards and policies.
- CSR can generate action programmes for sustainable development.

Although a comprehensive and unanimous definition remains elusive, these points broadly define the objectives of CSR. While discussing water management and its relation to CSR, we need to elaborate the above points in the context of water sustainability.

CSR applications in different countries are being used as tools for mitigating local problems. Companies incentivise local people to participate in sustainable business plans through CSR activities. As mentioned earlier, pending issuance of government guidelines, CSR initiatives are designed and implemented by the company itself, but voluntary application of CSR can fall foul of civil society unless it is realised that CSR should be observed as local actions for global change.

11.4   Addressing water in CSR

Addressing water in CSR is addressing human rights in the water sector. Through CSR, a corporation can address and mitigate global issues via local community actions. The most important CSR related issues in water sector are as follows.

11.4.1   Water security

Water security has become a global issue. Water security is under threat for many reasons: rising population, urbanisation, increasing water pollution, over-use of groundwater, water-related disasters and climate change. CSR activities can be instrumental in improving water security of a region, where investments from the government side is not adequate.

11.4.2   Drinking water and sanitation

Providing fresh water to citizens is the responsibility of the government, but a piped water supply has not reached the rural regions in many developing countries. According to the WHO, one-third of the world population is still deprived of access to fresh water (WHO, 2010). CSR can significantly improve freshwater supply in such cases (Lambooy, 2011). If the industries are not familiar with the technicalities of water infrastructure they can seek assistance from governments or international agencies for technical guidance or grants. There are matching grant funds from donor countries and government provisions for Private-Public Partnership (PPP) programmes ideally suited to such CSR initiatives; for instance, water infrastructure and WASH for the community can benefit immensely from PPP programmes. These are welcomed by society and therefore benefit business sustainability. The Government of India (Ministry of Drinking Water and Sanitation, no date) has released a guideline for the corporate sector on CSR activities under water and sanitation programme.

The Government of India has specific directives on CSR activities in the water sector. The main objectives of taking up CSR activities in the rural drinking water sector are:

- Provisioning of affordable safe drinking water in adequate quantities within minimum distances to all areas and at all times.
- Initiating conservation, recharge and sustainability measures with regard to drinking water in the rural areas.
- Special attention are required for weaker sections of the society.[1] One of the major examples of water and sanitation in CSR has been executed by the Gas Authority of India Limited (GAIL), a public sector company in India (GAIL, no date). According to reports published in the website of GAIL, the rural poor have improved access to safe drinking water through CSR activities in Madhya Pradesh, Uttar Pradesh, Uttarakhand and Haryana.

11.4.3   Ecological development

Industries can engage in ecological development in the areas of their operations. 'Ecological development' refers to both social and environmental responsibility that addresses issues of human rights and conservation of natural resources. According to Fenwick (2007), organisations involved in CSR activities are seeking to develop practices and policies that are ecologically sustainable and socially responsible. The components of ecological responsibility are:

- Minimising negative impacts over the environment;
- Promoting involvement of stakeholders in clean environment programmes;
- Preventing pollution of air, water and soil;
- Saving energy and raw materials;
- Protecting biodiversity; and
- Protecting wetlands and bodies of water.

A good example of such an activity is showcased by Rand Water, a water company in South Africa, which is engaged in the KZN Donga Rehabilitation Project. This project is aimed at preserving sensitive catchment areas through reclamation and rehabilitation of dongas (gullies in the veld) in the Tugela and Mnweni catchments. 'The project activities include mapping, plotting and recording of dongas, construction of gabions and planting of grass species. The project constructed a total of 54 gabions structures including planting of 400 square meters of Vetiva grass.' (Rand Water, no date).

11.5   CSR management framework

Like all other aspects of big business, CSR must be well managed. In a business environment, all CSR activities are implemented through a structured management system. There are several CSR models tailored for small to large-scale business. A CSR management framework is shown in Figure 11.2.

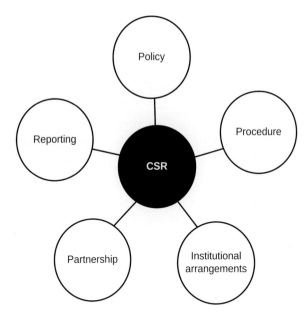

Figure 11.2 CSR management framework.

The pillars on which a good CSR rests are as follows:

11.5.1 Policy

A corporate policy on CSR should establish core values as a starting point with a code of conduct that includes (1) management's commitment, (2) engagement policy and (3) respect for international and national guidelines. The policy should also include fund raising and funding opportunities.

11.5.2 Procedure

A CSR procedure includes an identification of risks and opportunities in the company's initiatives for implementation. It also covers the scope of monitoring, evaluation and impact assessment. The performance should attain due diligence, extending activities to supplier and distribution chains and collaboration with other actors. CSR projects/programs can be implemented through (1) NGO or trusts created by concerned corporate houses for CSR activities; (2) collaboration with other companies; (3) collaboration with institutions, NGOs, government, semi-government, autonomous bodies, trusts and so forth. The chosen entities are required to have an established track record of undertaking such projects/programs.

A good CSR initiative calls for a good management. Depending on the volume of investment and size of CSR project areas, managers are specially trained for CSR projects; in general, a CSR Manager reports directly to the CEO. In large companies, CSR committees are formed to review implementation of CSR projects. The CSR committee submits periodical reports to the board of directors.

11.5.3   Institutional arrangement

While implementing CSR programmes, a company has to engage in partnership with other organisations at various stages of the project implementation. This can be done in two ways:

1 Engaging organisations (institutes, NGOs, consulting firms, etc.) by funding their projects or implementing projects through them. For example, GAIL India Ltd. finances projects that are compatible with the company's CSR programmes which are being run by a second-party organisation, provided such organisations have a proven track record on infrastructure, expertise and experience. There are multinational companies who implement their CSR projects on water issues and ecological development through organisations like the World Wide Fund for Nature (WWF), Green Peace and Circle of Blue.

2 Forming institutions of their own: Companies can form their own organisations in the form of registered societies and invest CSR funds through those organisations. For example Ambuja Cement (Ambuja Cement, 2014) in India has two institutions that look after their CSR activities. These are (1) the Ambuja Cement Foundation, the CSR arm of the company; and (2) the Ambuja Vidya Niketan Trust that look after education.

11.5.4   Partnership and stakeholders' engagement

Like other aspects of water management and stewardship, stakeholders' engagement is essential. Stakeholders' engagement can take many forms across the full spectrum of engagement levels. They may be engaged indirectly or directly. Indirect engagement is:

- Listening to and understanding stakeholders;
- Distributing information, promoting ideas and education;
- Crowd sourcing information;
- Discussing solutions interactively;
- Consulting and collecting quantitative and qualitative submissions;
- Visualizing and co-designing options;
- Co-creating and collaborating to deliver solutions; and
- Building long-term online stakeholder communities as a source of reputational capital.

Indirect engagement can be beneficial because it builds a relationship of trust with stakeholders and stakeholders feel that they are an essential component of the CSR initiatives.

Direct engagement is engaging stakeholders as promoters, motivators and implementing agencies. For example, local people are appointed as labourers in excavating a body of water, thus generating work. Local NGOs may be engaged as propagators of messages on water conservation and water literacy or for implementation of developmental projects. Often large NGOs and research organisations can be engaged for organising awareness camps and seminars. The company may engage organisations through funding their social/community development projects.

11.5.5 Reporting

In general CSR in large companies is performed through CSR departments, who are answerable to the board of directors. A general guideline is that the quarterly reports on expenditure and activities are presented by the CSR committee at the board/committee meetings. The board of directors will review the implementation of CSR periodically. If required by the law, such CSR reports are also presented to the government. The annual reports of companies include a section on CSR activities to keep stakeholders informed of CSR initiatives undertaken by company. Business initiatives to support sustainable water management include voluntary codes and assessment tools. However, reporting requirements vary, and legislation does not specify the technical tools of measurement. In most businesses, water management and disclosure are still weak, even for businesses operating in areas of high water stress.

11.6 CSR initiatives in the water sector

CSR initiatives in the water sector cover both human rights and social development. While delivering CSR on water, a company can showcase best practices in environmental conservation. A company can take up several result-oriented CSR programmes in the water sector. Some specific projects are mentioned here:

- **Water Harvesting:** A company can construct check dams, bunds, drainwater recovery, rooftop systems and percolation ponds under its CSR programme in water-stressed areas. See Figure 11.3.
- **Sustainable irrigation systems:** Under CSR programmes, a company can promote water-saving irrigation practices like low-cost drip irrigation.
- **Watershed management programmes:** Watershed management programmes under CSR are not uncommon in some countries of South Asia; in India, large companies like ITC, Nestle India Ltd, Amul, and so on are actively involved. To date, ITC's watershed development programme has generated over 4.7 million person days of work.
- **Providing safe water to the community:** As already been stated, CSR funds can be utilised in drinking water projects in remote and stressed areas. BASF's Sozialstiftung (Social Foundation), a social foundation of the BASF chemical company in co-operation with the United Nations Environmental Programme (UNEP) has embarked on a drinking water project at the Maloti Drakensberg Mountains, a central water source for South Africa. Cargill is providing safe drinking water to four thousand villagers in South Sumatra (Indonesia) from the water-treatment facilities built at a cost of one million USD in Palem. They also provide piped water to 400 families.

Figure 11.3 A water harvesting tank at Ankleswar industrial area, Gujarat, India.

11.7 International standards and guidelines

The European Union released a Green Paper in 2001 for developing an overall European framework, in partnership with the main CSR actors, aiming at promoting transparency, coherence and best practice methods. This document aims to set standards and introduces CSR actors to sustainability tools (IISD, 2013). The International Standard Organisation (ISO) released ISO 26000 to help organisations effectively assess and address all aspects of CSR (American Society for Quality, no date). The scope of ISO 26000 includes the following:

- Provide practical guidance on CSR and developing step-by-step CSR policy.
- Assist in stakeholders' engagement and enhancing credibility of reports.
- Promote common terminology in the field of CSR activities.
- Promote public awareness of CSR activities.
- Assist in preparing financial reports and physical progress reports.
- Assessment of CSR activities in the context of social and cultural background of the country of operation.

This ISO standard is not intended to reduce government's authority to address the social responsibility of organisations, neither is it a process of certification. It is intended to provide guidelines to industries to operate efficiently in the CSR sector (Mea, 2016).

CSR can also be viewed as a global governance mechanism. Understanding CSR from this perspective is useful for realising why civil society and government are interested in CSR. The international guideline documents for CSR activities are:

1 OECD's Guidelines for Multinational Enterprises,
2 UN Global Compact,
3 UN Guiding Principles on Business and Human Rights,
4 International Labour Organisation's conventions, and
5 United Nations 2030 Agenda for Sustainable Development.

Companies are expected to use these guidelines as a starting point for their efforts. The guidelines of OECD, UN Global Compact and United Nations are briefly described in Box 11.2.

Three International Guidelines on CSR

- **OECD's Guidelines for Multinational Enterprises (MNE)**
  MNE Guidelines provide a comprehensive framework for responsible business conduct (OECD, 2011). These guidelines are for avoiding adverse impact on society, environment and the business itself. The OECD guideline covers specific business issues like (i) general policies, (ii) disclosure, (iii) human rights, (iv) employment and industrial relations, (v) combating bribery, bribe, (vi) combating bribery, bribe solicitations and extortion, (vii) consumer interests, (viii) science and technology and (ix) competition.
- **The Ten Principles of UN Global Compact for Business** (UN Global Compact, 2016)
  The ten principles of the UN Global Compact covering areas of human rights, labour, environment and corruption are:
  1 Businesses should support protection of internationally proclaimed human rights.
  2 Ensure business is not complicit in human rights abuses.
  3 Businesses should recognise the right to collective bargaining through associations.
  4 Elimination of all forms of forced and compulsory labour.
  5 Effective abolition of child labour.
  6 Elimination of discrimination in recruitment procedures and workplaces.
  7 Businesses should support a precautionary approach to environmental challenges.
  8 Undertake initiatives to promote greater environmental responsibility.
  9 Encourage development and diffusion of environmentally friendly technologies.
  10 Businesses should work against corruption.
- **UN Guiding Principles on Business and Human Rights**
  These Guiding Principles have three pillars: (1) Responsibility of the States to protect and fulfil human rights; likewise (2) corporate responsibility is to respect human rights with due diligence; and (3) the state and the government is responsible to provide judicial, administrative, and legislative remedies when human right is breached. (United Nations, 2012).

BOX 11.2

11.8   Case studies

A few case studies are showcased here:

11.8.1   Coca-Cola

The Coca-Cola Foundation is an institution under the CSR initiative of Coca-Cola. In 2009, the Replenish Africa Initiative (RAIN) an organisation under the Coca-Cola Africa Foundation (TCCAF) had pledged 35 million USD in addition to its initial commitment of 30 million USD to support Pan-African sustainable safe water access and sanitation programmes. This funding will help to improve the lives of more than 6 million Africans through sustainable access to safe drinking water, sanitation and hygiene (WASH) by 2020.

The Coca-Cola Foundation also funded Cambodian Women for Peace and Development with an amount of 50,000 USD for their Community Clean Water Supply and Sanitation Project 2015 aimed to extend support for water access and hygiene education efforts in Cambodia (The Coca-Cola Foundation, 2016).

11.8.2   Nike

In addition to continuing to work on improving conditions for workers, Nike has updated its materials development and management to address the impact of climate change on the supply chain and reduce its environmental impact. One area of concentration is water conservation.

11.8.3   Swiss Re Group

The Swiss Re Group, a leading wholesale insurance provider, identified climate change as a major sustainability challenge. They give awards for best practices in watershed development. Swiss Re's International 'ReSource' Award for Sustainable Watershed Management is one of the best examples that demonstrate corporate responsibility; the award provides USD 150,000 as grants to several projects in Solomon Islands, Afghanistan, China and Guatemala.

11.8.4   Molson Coors

This beer maker recently partnered with Circle of Blue, which describes itself as an 'international network of leading journalists, scientists and communications design experts'. Molson Coors also belongs to the Beverage Industry Environmental Roundtable, a corporate attempt 'to define a common framework for [environmental] stewardship'—without any pesky regulatory agency or independent watchdog groups present. Molson Coors also signed into the Global Compact programme of the UN.

11.8.5   *Levi Strauss & Co*

Levi Strauss & Co. has partnered with water.org and has committed USD 250,000 to fund programs that collectively will provide at least 200 million litres to safe water projects underway in South Asia, Latin America and Africa.

11.9   Future of CSR

How a company practices CSR ultimately reflects its values and its relationship to the society it inhabits and upon which it depends. CSR is here to stay, and every company needs a cohesive, integrated CSR strategy that plays on its core strengths and institutional capacities. By strategically managing its CSR initiatives, every corporation can maximise its benefits to society and the environment, create societal value and fulfil the expectations of its stakeholders. The future of corporate social responsibility rests on several principles as advocated by Wayne Visser (Visser, 2010):

- **Innovations:** Businesses should direct their creativity toward solving social and environmental problems. An innovative example is Cirque du Soleil, an entertainment company that 'installed a seven layer filtration system for their O show in Las Vegas in 2009 (The Centre for Sustainable Practice in Arts, 2009). They also are involved in the One Drop Foundation as part of their CSR initiative; they use a ground-breaking method of using theatre and arts to educate people in Central America about watershed management and water conservation practices' (Vijayara, 2011).
- **Compatibility:** Social problems and issues are properly addressed when the solutions are compatible with local conditions. For example, water harvesting is something simple that every company can do, and it has become a major CSR focus.
- **Scalability:** As defined in Investopedia (2007), 'scalability is a characteristic of a system, model or function that describes its capability to cope and perform under an increased or expanding workload.' A scalable system will always perform better in conditions of increased workload. In the water sector, scalability is best tested in the event of climate change or extreme climate conditions. Scalability can be enhanced and reinforced by adopting appropriate technology. For example, in watershed management projects, an application of geographical information systems (GISs) enhances the capability of decision-making processes.
- **Responsiveness:** Business should be responsive, i.e. reacting quickly to emergent situations. This involves responding with emotion to people and events.
- **Global localisation:** This means that companies should understand both local and global issues, and act locally with a focus on mitigation of global

problems. It is best exemplified by watershed management projects that locally augment flow of water in a stream and have cascading effects, improving the overall environment of the larger basin, to which the locally augmented stream belongs.

11.10   Conclusion

Cubism, the painting style made famous by Pablo Picasso, is the art of first studying an object from different angles and later assembling different perspectives on canvas for the viewer. So, what would be the cubist view of a glass of water placed on a table? Water is where life began. Water sustains life on earth. There are industrial giants like Pepsi and Coca-Cola for whom water is an invaluable resource that helps them make billions of dollars. There is an irate mob somewhere because factory effluents are polluting their source of water. On the one hand, there are poor people from rural areas who are contemplating suicide because of the scarcity of rain, and they cannot afford pumped water for irrigation; on the other, rich urban people are careless enough to leave a full bathtub unplugged and so on. Perspectives on water are endless, but the defining element is perhaps its divisiveness. Water, by its quality and available amounts, separates the rich from the poor. And, the division is so sharp that these two classes effectively live in two different worlds. The reader will appreciate that the cubist view of a glass of water sitting innocently on a table will be a very complex picture indeed. It is this complexity of the water problem that has attracted world attention and therefore, also the attention of the United Nations, and has given rise to such modern concepts as Integrated Water Resources Management (IWRM), Corporate Social Responsibility (CSR) and Corporate Water Stewardship (CWS), which is the subject matter of this book. IWRM, CSR and CWS are not simply concepts any more. They have taken off from the drawing board and have become action plans and plans in action.

Water stress is not a natural phenomenon. The fulcrum on which the problem rests can be attributed to the two centuries of uncaring industrialisation. It is the responsibility of the corporate sector to redefine its priorities so that the world is not beset by water shortage and pollution. *Noblesse oblige*. That, in short, is the essence of IWRM, CSR and CSW; they have the power to bring divided worlds together and hopefully, they will.

Note

1. Guidelines are available at http://www.mdws.gov.in/sites/default/files/CSR_GuideLines_Water.pdf.

Bibliography

Ambuja Cement (2014) *CSR policy*. Available at http://www.ambujacement.com/Upload/PDF/CSR-Policy.pdf (accessed: 16 September 2016).

American Society for Quality (no date) *What is ISO 26000? Social responsibility guidance standard*. Available at http://asq.org/learn-about-quality/learn-about-standards/iso-26000/(accessed: 25 September 2016).

Ascoli, K. and Benzaken, T. (2009) *Public policy and the promotion of corporate social responsibility*. Available at https://www.bsr.org/reports/Public_Policy_Promotion_of_CSR.pdf (accessed: 24 September 2016).

Bichta, C. (2005) *Corporate social responsibility a role in government policy and regulation?* Available at http://www.bath.ac.uk/management/cri/pubpdf/Research_Reports/16_Bichta.pdf (accessed: 20 September 2016).

Coca-Cola Foundation (2016) *Grants paid in 2015—by organization*. Available at http://www.coca-colacompany.com/content/dam/journey/us/en/private/fileassets/pdf/our-company/2015-PIDC-Contributions-Report.pdf (accessed: 12 October 2016).

Commission of the European Communities (2002) *Corporate social responsibility: A business contribution to sustainable development*. Available at http://trade.ec.europa.eu/doclib/docs/2006/february/tradoc_127374.pdf (accessed: 24 October 2016).

Fenwick, T. (2007) 'Developing organizational practices of ecological sustainability', *Leadership & Organization Development Journal*, 28(7): 632–645.doi: 10.1108/01437730710823888.

Forte, A. (2013) 'Corporate Social Responsibility In The United States And Europe: How Important Is It? The Future Of Corporate Social Responsibility', *International Business & Economics Research Journal*, 12(7), pp. 815–824.

GAIL (no date) *Drinking water and sanitation*. Available at https://gailebank.gail.co.in/final_site/portable.html (accessed: 22 March 2017).

Holme, R. and Watts, P. (2000) *Corporate social responsibility: Making good business sense*. Geneva: World Business Council for Sustainable Development.

IISD (2013) *The European Union green paper on CSR*. Available at https://www.iisd.org/business/issues/eu_green_paper.aspx (accessed: 25 September 2016).

Insead Knowledge (2013) *When it comes to CSR, size matters*. Available at http://www.forbes.com/sites/insead/2013/08/14/when-it-comes-to-csr-size-matters/#4a9379f41b6f (accessed: 29 September 2016).

Investopedia (2007) 'Scalability'. Available at http://www.investopedia.com/terms/s/scalability.asp (accessed: 25 September 2016).

Lambooy, T. (2011) 'Corporate social responsibility: Sustainable water use'. *Journal of cleaner production*, 19: 852–866.

Mea, S. (2016) *The rationale behind ISO26000 – the standard on social responsibility*. Available at https://drmarkcamilleri.com/2016/08/15/the-rationale-behind-iso26000-the-standard-on-social-responsibility/(accessed: 25 September 2016).

Ministry of Drinking Water and Sanitation (no date) *CSR guideline for sanitation*. Available at http://www.mdws.gov.in/sites/default/files/Guide_Line_Sanitation_CSR.pdf (accessed: 12 March 2016).

OECD (2011) *OECD guidelines for multinational enterprises 2011 EDITION*. Available at http://www.oecd.org/corporate/mne/48004323.pdf (accessed: 29 September 2016).

Permasteelisa group (2011) *UN Global Compact*. Available at http://www.permasteelisagroup.com/sustainability/un-global-compact/ (accessed: 17 September 2016).

Postma, J. (2012) *Making business sustainable: corporate social responsibility in South Africa.* Available at http://essay.utwente.nl/61793/1/BSc_J_Postma.pdf (accessed: 25 September 2016).

Rand Water (no date) *Wetlands rehabilitation.* Available at http://www.randwater.co.za/CorporateResponsibility/Pages/WetlandsRehabilitation.aspx (accessed: 24 September 2016).

The Centre for Sustainable Practice in Arts (2009) *Cirque Du Soleil focus: water & CSR.* Availablr at http://www.sustainablepractice.org/2009/09/04/cirque-du-soleil-focus-water-csr/(accessed 27 March 2017).

UK Essays (2015) *CSR in Kenya: factors that drive and affect the business community.* Available at https://www.ukessays.com/essays/business/csr-in-kenya-factors-that-affect-business-community-business-essay.php (accessed: 25 September 2016).

UN Global Compact (2016a) *The ten principles.* Available at https://www.unglobalcompact.org/what-is-gc/mission/principles (accessed: 29 September 2016).

United Nations (2012) *Guiding principles on business and human rights implementing the United Nations 'protect, respect and remedy' framework.* Available at http://www.ohchr.org/Documents/Publications/GuidingPrinciplesBusinessHR_EN.pdf (accessed: 29 September 2016).

US Department of State (2011) *Corporate social responsibility.* Available at http://www.state.gov/e/eb/eppd/csr/(accessed: 19 September 2016).

Vijayara, A. (2011) *CSR initiatives focusing on water stewardship.* Available at http://www.justmeans.com/blogs/csr-initiatives-focusing-on-water-stewardship (accessed: 25 September 2016).

Visser, W. (2010) CSR 2.0 and the New DNA of Business, *Journal of business systems, governance and ethics*, 5 (3): 7. Also published on SSRN at: http://ssrn.com/abstract=1725159.

WHO. (2010). *The global annual assessment of sanitation and drinking-water (GLAAS).* Available at http://www.who.int/water_sanitation_health/monitoring/investments/glaas/en/(accessed.20 March 2017).

Williams, B.N. (2014) *Changing the face of corporate social responsibility.* Available at https://www.prsa.org/Intelligence/Tactics/Articles/view/10593/1091/Changing_the_Face_of_Corporate_Social_Responsibili#.V9s8tvl96Uk (accessed: 16 September 2016).

Glossary

The definitions/explanations incorporated in the glossary are sourced from USGS, US-EPA, EPA Ireland, Hoekstra (2011), Murali Krishna (2008), UN Water (2015), CEO Water Mandate (2013), Mayfield, (2010), World Water Council and WHO as referred in various chapters of this book.

100-year flood   This is not a flood cycle as the name suggests. It refers to a height of flood level that has 1% probability in given region in any year.

Aboriginal Water Right   An Aboriginal Water Right is a right to water for the indigenous people of America who resided in Canada prior to colonisation.

acid mine drainage   Sulphide ores exposed to air and water during mining produce sulphuric acid. Untreated waste from sulphide mines contaminates surface and groundwater. Common contamination sources are lead, zinc, copper ores.

activated carbon adsorption   Industrial waste containing refractory organics are difficult to remove by conventional biological treatment. These are removed by adsorption on activated carbon (or synthetic active-solid surface).

actual evapo-transpiration (ET)   Actual evapo-transpiration is the amount of water transpired and evaporated from the soil when the required amount of water is available from soil. ET = Potential evapo-transpiration (PET in mm) × Crop Area (ha) × Crop Coefficient (k_c) for each crop.

adaptive management   Adaptive management of water is a new paradigm that refers to a complete system of water management which includes all interconnected systems (Pahl-Wostl *et al.*, 2007)

aerobic biological treatment processes   Waste treatment process by aerobic water (i.e., in the presence of dissolved oxygen).

air photos   Photographs of land surface acquired through air-borne survey.

Industrial Water Resource Management: Challenges and Opportunities for Corporate Water Stewardship, First Edition. Pradip K. Sengupta.
© 2018 John Wiley & Sons Ltd. Published 2018 by John Wiley & Sons Ltd.

air stripping   Air stripping is removal of volatile pollutants in groundwater or effluent water by pumped air; commonly used for dissolved methane.

anaerobic biological treatment   Anaerobic treatment systems are based on a biological process operated and controlled under anaerobic conditions.

anemometer   Instrument for wind-speed measurement.

annual water resource   The average annual water availability of a country assessed in billions of cubic metres (BCM).

antecedent moisture condition (AMC)   AMC is an indicator of watershed wetness and soil moisture content prior to a storm.

aqueduct[1]   A pipe, conduit, or channel designed to transport water from a remote source, usually by gravity.

aqueduct[2]   Water-related data portal that accumulates and disseminates country-wise water information.

aquicludes   Impermeable geological formations that act as barriers to groundwater flow.

aquifer   Economically extractable undergroundwater stored in water saturated soil, alluvium and fractured rocks.

aquifer (confined)   *See* **confined aquifer**.

aquifer (unconfined)   *See* un**confined aquifer**.

aquifer parameter   Physical characteristics of aquifer on which storage, yield and flow of water depend.

aquifer performance test/aquifer test   Methods to determine aquifer parameters.

aquitard   An aquitard is a layer that acts like a confining bed but has a capacity to transmit water at a very slow rate in some conditions.

artesian condition   *See* **artesian water**.

artesian water   Groundwater under pressure in confined aquifers; when drilled for tube wells the water level rises above the level at which it is first encountered. It may or may not flow out at ground level. The pressure driving up water in the tube is called artesian pressure. *Also see* **flowing well**.

artificial recharge   Any process in which water is put back into groundwater storage from surface-water supplies such as irrigation or induced infiltration from streams or wells.

attached growth systems   Wastewater treatment processes in which the microorganisms and bacteria are attached to the media in a reactor tank. The waste flows over the media and reacts with the microorganisms.

base   A substance that has a pH of more than 7. A base has fewer free hydrogen ions (H^+) than hydroxyl ions (OH^-).

base flow   Sustained flow of a stream in the absence of direct runoff. Natural base flow is sustained largely by groundwater discharges.

baseline data   Environmental data on the basis of which impact of development or changes are assessed

baseline study   Baseline study is a component of information flow into the decision-support system based on which strategic models are made. Baseline studies are conducted for the current year to quantify changes in the system.

baseline water stress   Baseline water stress is the ratio of total annual water withdrawal (municipal, industrial and agricultural) and available annual blue water; it is expressed as a percentage, in which a higher percentage indicates more competition among users.

basin (in hydrology)   Also called a river basin. A tract of land containing a central river and all its tributaries from the source of the river to its mouth.

bedrock   A general term for solid rocks that lies beneath soil or alluvium.

benchmarking (water consumption)   Setting a measured standard for water consumption. Used for comparing water consumption patterns by similar industrial processes.

bioaccumulation   Processes by which pollutants progressively reach higher toxic levels in the food chain.

biological indicators   Direct measurements of population and health of fauna and flora in a body of water.

biological oxidation   Non-toxic organic matter in sewage can be fed to bacteria of suitable types in a *reactor*, so as to convert organic waste into carbon dioxide, water, methane, mercaptans, etc., which escape into the atmosphere. There are two types of biological oxidation, aerobic and anaerobic. In aerobic systems, free oxygen is required for the survival of bacteria. The aerobic process is conducted in aerated lagoons. In anaerobic processes, the bacteria can live without oxygen. This treatment is required for wastewater having a high BOD/COD ratio.

biological oxidation tank   Tanks through which air is blown to keep alive microorganisms used to treat waste.

biological/biochemical oxygen demand (BOD)   BOD is the amount of dissolved oxygen required by aerobic microbes used to treat organic wastes in water.

biomass   Organic waste (e.g., cattle droppings, crop-field wastes, municipal waste, timber mill waste), used as a source of fuel or energy. Biomass can be used directly by burning or processed into biofuels like ethanol and methane.

biophysical data   Data related to topography, soil, geology, vegetation and land use.

biosphere   Parts of the earth's surface and atmosphere occupied by living things.

blank pipe   Pipes of a tube well that are not slotted.

blue water   The water that can be withdrawn from nature by any device; blue water includes rainwater, surface water and groundwater, and is a visible component of the water cycle.

blue water footprint   The amount of surface or groundwater consumed to produce any goods or to render any services.

business water footprint   Total volume of freshwater used directly or indirectly to run and support a business.

cadastral map   Village maps showing individual land holdings.

capillary action   Liquid movement through narrow pores by forces of adhesion, cohesion and surface tension.

capillary fringe   The moist zone immediately above the water table.

capillary zone (capillary fringe)   A zone above aquifer that contains some water held in place by capillary suction.

catchment (river)   *See* **watershed**.

catchment (RWH)   For the purposes of RWH, a catchment is defined as an open place from which rainwater moves towards a single outlet point and can then be collected; the catchment must have a defined boundary so that there is no water loss due to an overflow across the boundary.

centrifugal pumping limit   Suction limit of a centrifugal pump (which is about 7 m above the watertable).

check dams   A small dam, sometimes temporary, constructed across small streams to reduce erosion by reducing downstream water flow.

chemical oxygen demand (COD)   Amount of oxygen consumed in oxidation of organic and inorganic chemicals.

clarification   Removal of suspended particles that cannot be removed by primary screening.

coefficient of permeability   *See* **hydraulic conductivity**.

collector well   Structures constructed to draw water from a river bed.

command area   The area covered under an irrigation facility.

common water-treatment plant   These plants are designed for treating wastewater for several industries at a single establishment to cut capital costs; used in small and medium industrial clusters.

conduits   Conduits are pipes that carry water from catchment to storage tanks or the recharging system.

cone of depression   Local conical depression in a water table formed during pumping.

confined aquifer   Aquifers bound above and below by impermeable confining beds. A confined aquifer is not recharged directly from local precipitation. Synonym: artesian.

confining beds   Confining beds are those rocks which act as a barrier between the atmosphere and the aquifer or between two aquifers. Confining beds do not allow water to pass through or move within the bed.

consent to operate   Government permission, often with conditions, to carry out industrial activity.

consumptive use   That part of water withdrawn that is evaporated, transpired by plants, incorporated into products or crops, consumed by humans or livestock, or otherwise removed from the immediate water environment. Also referred to as water consumed.

conveyance loss   Water lost from delivery pipelines and canals by leakage or evaporation.

corporate water assessment   Water resource assessment initiated by the industrial corporations.

corporate water disclosure document   Document prepared to apprise investors of the state of water in the industry.

corporate water stewardship (CWS) Management strategy for sustainable business and resolving water-related business risks, especially when corporate sector draws water from sources used by local people and livestock. CWS aims at socially equitable, environmentally safe and economically viable use of shared water sources.

cost-reflective tariff It is a method to generate sufficient rate of return by achieving equity and to recover the cost of supplying domestic water.

crop coefficient A property of plants on which the capacity of the plant to transpire water from soil depends. It is used for estimating actual evapotranspiration or ET.

current meter A suitably calibrated galvanometer used to measure stream flow velocity.

cytostatic Drugs that inhibit cell division (e.g., cancer cells).

cytotoxic Drugs that kill cells (e.g., cancer cells).

de-alkalisation (of water) Removal of hardness and alkalinity by weakly acidic cation resins.

decationisation Removal of all cations with strongly acidic cationexchange resins.

decision-support system Computer programmes for data analyses used by business organisations for decision making.

deep injection wells Injection wells for disposing of wastewater or water mixed with chemicals into porous geologic formations.

desalter water Water used in oil refineries for dissolving out salts from crude oil.

development of tube well Method of cleaning a newly constructed tube well. In this method, the bore is cleaned from drilling mud.

digital elevation models (DEM) Computer-generated three dimensional view of any terrain; e.g., maps showing lunar surface.

direct water pollution The act of releasing contaminated fluids directly into the water; e.g., releasing effluents into a river or sea.

discharge (hydrology) The volume of water pumped per unit time; usually expressed in cubic feet per second.

discharged water Water expelled from an industry.

discharging area The area where the water table meets the land surface.

disposal permits A type of licence awarded to an industry to dispose of solid waste or effluent in a particular place.

distribution mains Main pipelines that convey water from storage tanks to end-users.

drainage basin Land area where precipitation runs off into streams, rivers, lakes and reservoirs. It is a land feature that can be identified by tracing a line along the highest elevations between two areas on a map, often a ridge.

drawdown Lowering of the groundwater surface caused by pumping.

dynamic resource The amount of groundwater recharged annually from precipitation.

ecological flow   The quantity, timing, and quality of water flow into any freshwater ecosystem (e.g., rivers, lakes) or estuaries that is required to sustain life and human livelihood.

ecological identifier   Parameters by which ecological degradation can be identified.

ecosystem services   The benefits people obtain from ecosystems. There are four types of ecosystem services: (a) supporting, (b) provisioning, (c) regulating and (d) cultural services.

electrodialysis   Removal of salt ions by filtering through molecular sieves that use electrical potential difference as the propelling force.

empirical formulae   Formulae derived from physical observation rather than theory or pure logic.

engineered wetlands   Artificial wetlands used for wastewater treatment; insoluble pollutants are trapped, and soluble pollutants are absorbed by aquatic plants in the wetland.

environmental degradation   The deterioration of the environment through depletion of resources, such as air, water and soil; the destruction of ecosystems and the extinction of wildlife. It is defined as any change or disturbance to the environment perceived to be deleterious or undesirable.

equalisation tanks   Wastewater treatment processes work effectively in uniform conditions; when quality and quantity of wastewater vary with time, it is collected in equalization tanks for treatment.

eutrophication   Phenomenon caused by discharge of excessive nutrients (phosphate and nitrate) into bodies of water, which in turn causes algal bloom, choking higher aquatic life forms.

evapo-transpiration   Sum total of surface water evaporation and transpiration by vegetation.

exploitable water resources   Also called available water resources. The natural renewable water resources estimated for a 'water year' is not allowed to be exploited *completely* through human consumption like drinking, irrigation or industrial use.

field capacity   Amount of moisture retained in soil after excess water is drained by runoff and water percolation has decreased. Field capacity varies with soil structure and texture; it is reached usually 2–3 days after rain or irrigation.

first flush   The first rain with significant runoff washes dust and debris into the downpipe and ultimately to the storage tank. To avoid debris flow through the conduits, a first flush system is attached.

fixed spatial data   Spatial data to be generated only once.

flood plain   Relatively flat and normally dry land alongside river or lake that is inundated during floods.

flood stage   Elevated water level in which the inundated area is much lower than flood but sufficient to destabilise life close to riverbanks and lakes.

floodway   Flood-control channels or areas that must be kept free from anthropogenic obstructions to allow free passage of flood water.

flow-duration curve (FDC)   Cumulative frequency curve that shows the percentage of time during which the relevant discharge is equalled or exceeded in a given period; critical data for designing hydroelectric power plants.

flowing well   Groundwater under pressure in confined aquifers that rises without pumping in tube wells. If the water rises above the surface, it is known as a flowing well. *Also see* **artesian well**.

fly ash   Ash produced in small dark flecks by the burning of powdered coal or other materials and carried into the air.

fractional distillation   In this process, two liquids with different boiling points are separated.

geometry of the aquifer   This is defined by its depth, thickness, overlying and underlying geological formations and its areal extent, which are then shown on a map.

GIS   Geographical information system. It is a computer-aided system for processing spatial and non-spatial data for generating thematic maps.

gravity filter   Water-treatment process in which water flows through layers of sand by gravity.

gravity pipe   These are essentially closed conduits; water is forced by pressures higher than atmospheric pressure to flow against gravity.

green growth   Green growth is used in the policy statement of a government or a business to describe a path of economic growth that uses natural resources in a sustainable manner.

green water   Green water is that part of precipitation that goes back to the atmosphere through evaporation and transpiration.

green water footprint   Total rainwater evapo-transpiration (from fields and plantations) plus the water incorporated into harvested crops or timber.

grey water   Any polluted water discharged into the environment is called grey water (e.g., from industries, lavatories, washing machines etc).

grey water footprint   The volume of fresh water required to dilute the load of pollutants to meet with regulatory water quality standards.

groundwater recharge   Natural or artificial percolation of surface water into groundwater.

gutter   The lower part of the catchment where rainwater runoff accumulates.

hard water   Water with dissolved alkaline salts, mainly calcium and magnesium. Hard water requires more soap, detergent or shampoo to raise lather.

headwaters   The area of origin of rivers.

hydraulic conductivity   The capacity of a porous medium to transmit water. *Coefficient of permeability* is another term for hydraulic conductivity.

hydraulic gradient   Slope of groundwater table measured from a series of dug wells or tube wells.

hydraulic head   Hydraulic head or piezometric head is a specific measurement of liquid pressure above a geodetic datum. It is usually measured in units of length. Gradient is the pressure difference between two points on the piezometric surface or water table which governs the flow of groundwater

hydrogeology   Study of groundwater.

hydrologic cycle   The cyclic transfer of water vapour from the earth's surface via evapo-transpiration into the atmosphere, from the atmosphere via precipitation back to earth, and through runoff into streams, rivers, and lakes, and ultimately into the oceans.

hydrologic soil group   A system of classification of soil cover developed by the Soil Conservation Service (SCS) of the US Department of Agriculture (USDA) for use in rural areas. It is a versatile and widely used procedure for runoff estimation.

hydro-morphological impact   A set of impacts on the land surface due to the creation of water infrastructures that change direction, volume and hydrological conditions.

hydrosphere   Total water on earth's surface contained in oceans, rivers and lakes (some calculations include water in clouds).

industrial sludge   Concentrated, semi-liquid waste left after treatment of industrial water or wastewater.

injection wells   Wells constructed for injecting treated wastewater directly into the underground strata.

integrated water resources management   Integrated water resources management (IWRM) has been defined by the Global Water Partnership (GWP) as 'a process which promotes the coordinated development and management of water, land and related resources, in order to maximize the resultant economic and social welfare in an equitable manner without compromising the sustainability of vital ecosystems'.

inter-annual variability   The variation in water supply between years.

intermediate zone: The zone between the aerated soil and the water-saturated zone.

ion exchange   Ion exchange removes unwanted ions from water using resins. Alkaline resins are used for removing anions and acidic resins for cation removal.

Khosla's method   Khosla (1960) analysed the rainfall, runoff and temperature data for various catchments in India and the United States to arrive at an empirical relationship between runoff and rainfall.

leachate   *Leachate* is a generic term for water contaminated by leakage of fluids from landfill waste material.

leaky aquifer   A water-bearing formation with both upper and lower surfaces bound by aquitards, or one boundary is an aquitard and the other is an aquiclude.

levee   A natural or manmade earthen barrier along the edge of a stream, lake, or river to prevent annual flooding.

life cycle–based assessment   Assessment of water consumption of a product from procurement of raw material to consumption of the end product, or as it is said in business circles, 'from cradle to grave'.

lime treatment   Water passed through lime to kill bacteria by raising pH.

lithosphere   The rigid outer layer of earth.

lower catchment   Where sediment deposition is the major activity in a river system.

maximum contaminant level (MCL)   The allowable contaminant in drinking water without causing health risk.

millennium development goal,   World's time-bound and quantified targets for addressing poverty, hunger, disease and environmental degradation.

mineral dissolution   Recovery of metals by leaching mine dumps.

natural remediation capacity   The capacity of nature to control pollution.

natural renewable water resources   The sum total of a country's internal renewable resources. These include both groundwater and surface water, and the external resources it receives through trans-boundary inflow.

natural resource–based companies   Industries that use natural resources like mining, agriculture and forestry products as their raw materials.

nephelometric turbidity unit (NTU)   Unit of measure for the turbidity of water.

network of monitoring stations   An array of wells specially designed and dedicated for periodical monitoring of groundwater levels.

non-point source (NPS) pollution   Widespread pollution sources: e.g., acid rain, farming pesticides.

non-renewable water resources   Resources not renewable within lifetime of the user.

observation wells   Tube wells used to measure seasonal water level changes or changes caused by pumping.

operational water use   Operational water use includes the water used by an industry for running its production processes.

osmosis   Osmosis is a flow of a solvent through a semi-permeable membrane from lesser to higher concentration.

outfall   Place where a sewer, drain or stream discharges; the outlet or structure through which reclaimed water or treated effluent is finally discharged to a receiving body of water.

over-appropriation (of water)   Withdrawal of water beyond renewal capacity of nature. Groundwater is believed to be the most over-appropriated resource.

oxygen demand   Need for dissolved oxygen in water for biological and chemical processes; oxygen solubility in water is very small (40 mg/L at normal atmospheric pressure) but vital to aquatic life.

ozone treatment   Ozone in water breaks down into nascent oxygen, which is a powerful oxidising agent; kills anaerobic bacteria.

permeability   The ability of solid material to allow the passage of a liquid, such as water. Permeable materials, such as gravel and sand allow water to move quickly through them, whereas impermeable materials, such as clay, do not allow water to flow freely.

piezometric surface   *See* **potentiometric surface**.

point-source pollution   Water pollution from a single source; e.g., chromium-contaminated water from a tannery.

polychlorinated biphenyls (PCBS)   A group of synthetic, toxic industrial chemical compounds once used in making paint and electrical transformers, which are chemically stable and not biodegradable. PCBSs in industrial wastes contaminate surface and groundwater. Although virtually banned in 1979 with the passage of the Toxic Substances Control Act, they continue to appear in the flesh of fish and other animals.

porosity   The measure of water-bearing capacity of solids. Clay has much high porosity than sand but is a poor aquifer medium because it is impermeable.

potable standard   Water in which contaminants are within permissible limits.

potential evapo-transpiration (PET)   Potential evaporation or potential evapo-transpiration (PET) of a location is defined as the amount of evaporation that would occur if sufficient water were available in soil or in surface storage.

potentiometric surface   Potentiometric surface is an imaginary surface representing the static head of groundwater and defined by the level to which water will rise in a well under static conditions. The water table is the potentiometric surface for an unconfined aquifer.

precipitation (of water)   Rain, snow, hail, sleet, dew, and frost.

pressure conduits   Pipes in which water is delivered under pressure as in household plumbing.

pressure groups   Group of activists that influence administrative decision making to protect the interest of people: e.g., environmental pressure groups.

primary data   Data collected by the user.

primary screening   Primary screening eliminates floating material in water, like leaves, sticks, plastic, etc. The primary screen is a mesh of required size through which water is passed, leaving behind all floating materials.

primary sedimentation   In this process, water is stored in large tanks, called 'settling tanks', until most of the suspended particles (up to clay-sized grains) settle at the bottom of the tank.

primary wastewater treatment   The first stage of the wastewater-treatment process, in which mechanical methods, such as filters and scrapers, are used to remove pollutants. Solid material in sewage also settles out in this process.

problem-shed challenge   A problem-shed is defined as a geographic area that is large enough to encompass the issues but small enough to make implementation feasible.

qualitative analysis (of water) Identification of contaminants in water.

quantitative analysis (of water) Measuring the amount of contaminants in water.

rainfall infiltration method In this method, groundwater recharge is calculated as a percentage of annual precipitation.

rainfall intensity *See* **storm intensity**.

rainwater harvesting Rainwater harvesting is a technique of capturing rainwater for direct use or for groundwater recharge. A common and household method is harvesting rainwater from rooftops.

Ramsar sites Wetlands of international importance under the Ramsar Convention, which are protected for sustainable use.

rapid sand filter (RSF) Water cleared from suspended matter using sand beds through which water passes under gravity or pumped.

raw water Water in its natural state before entering a water-treatment plant.

recharge (of groundwater) Water added to an aquifer either naturally (e.g., by rain) or artificially.

recharging area A part of the land surface that constitutes the natural intake zone in a groundwater basin; water that enters the aquifer, then moves towards the lower hydraulic head.

recharge trench It is constructed for artificial groundwater recharge where there is a permeable strata of adequate thickness available at shallow depth (≤3 m).

reclaimed wastewater Effluents from wastewater (sewage) treatment plants diverted for beneficial use, e.g., in irrigation, factories, thermal power plants.

recycling of water Water that is used more than once before it is released into the natural water cycle.

regulatory risks Regulatory risks are experienced by industries when the regulatory authority modifies laws, licensing procedures and allocation policies without sufficiently advanced information to industries; it becomes difficult and cost intensive for industries to implement compliance at short notice.

renewable groundwater resources A part of the annual groundwater recharge that can be withdrawn for human consumption (domestic, industrial or irrigation use).

reputational risk Risk of loss resulting from damage to a firm's reputation.

reservoir A pond, lake, or basin, either natural or artificial, for storage and distribution of water.

residual drawdown The difference between the original static water level and the depth to water at a given instant during the recovery period after stoppage of pumping.

return flow Part of withdrawn water that is released for further use downstream or returned to the water cycle after use; e.g., from hydroelectric power plants, cooling towers, irrigation canals.

reverse osmosis (RO)   Reverse osmosis is a process where water is forced through a semi-permeable membrane (called RO membrane) that traps contaminants larger than water molecules.

right of way   Easement granted by the owner of land to lay water delivery pipelines.

Rio summit   Environmental summit held in Rio de Janeiro in 1992

riparian zone   Forested or wooded areas of land adjacent to a river, stream, pond, lake, marshland or estuary.

riparian rights   Special rights to withdraw water granted to residents living close to water bodies and granted only for domestic water use. This right is derived from traditional systems of water management.

river basin   *See* **watershed**.

river basin district   The area of land and sea, made up of one or more neighbouring river basins together with their associated groundwater and coastal waters, identified under Article 3(1) of Water Framework Directive of the European Union.

river linking   An Indian Government project for linking of rivers in different basins with canals for transferring water from one basin to another. The objective is to provide water to water-stressed regions by transferring water from water-abundant regions.

rock-catchment (RWH)   A rock-catchment harvesting system is practised in rocky uneven terrain. Runoff from natural rock surfaces is collected in large ponds.

runoff curve number (CN)   CN is an empirical parameter used in hydrology for predicting direct runoff or infiltration from a rainfall event.

safe category block   Localities in a water-scarce region where groundwater recharge rate is higher than in surrounding areas; generally happens due to percolation of irrigation water or localised rain; upgrading an area from 'water scarce' to 'safe category' results from regular monitoring.

saltwater intrusion   Intrusion of salt water or seawater into a freshwater aquifer; generally happens due to overdrawal from aquifers in coastal areas; can also be caused by coastal floods during a storm surge or tsunami.

saturated zone   The saturated zone is that part of the water-bearing formation in which all voids are filled with water. Fluid pressure is always greater than or equal to atmospheric pressure, and the hydraulic conductivity does not vary with pressure head.

scarcity index   A comparison of annual water withdrawal relative to total available water resources. A region is called 'water scarce' if annual withdrawal exceeds 20–40% of annual supply; and 'severely water scarce' if it exceeds 40%.

secondary data   Data that was collected by someone not under control of the user.

secondary wastewater treatment   Follows primary wastewater treatment. Removes suspended particles, colloids and dissolved organic matter by physical, chemical and biological processes. Reduces 80 to 95 percent of biochemical oxygen demand (BOD). Activated sludge and trickling filters are two of the most common means of secondary treatment. Disinfection is the final stage of secondary treatment.

self-supplied water   Homeowners getting their own water, e.g., from privately owned well.

semi-confined aquifer   *See* **leaky aquifer**.

separation of sludge   Removal of sludge from effluent treatment plants (ETP). ETP tanks need periodic sludge separation and dewatering of sludge.

service lines   Service lines carry water from a distribution main to buildings or industrial units.

sewage treatment plant   A facility designed to convert wastewater received from domestic sources into environmentally safe water.

shallow aquifer   Generally unconfined aquifers, or confined aquifers from which water can be drawn by hand pumps.

simulation of catchment behaviour   Analysing catchment behaviour through catchment modelling and estimating possible discharge of a stream under certain predetermined conditions.

sinkhole   A depression in the earth's surface caused by the dissolving of underlying limestone, salt, or gypsum. Drainage is provided through underground channels that may be enlarged by the collapse of a cavern roof.

skimming well   Skimming wells are constructed in aquifers where the water in the lower level is brackish or saline but the water in the upper part is relatively fresh.

sludge chamber   Chambers constructed for drying sludge, which is then handed over to recyclers.

softening water   Removal of calcium, magnesium and some metal cations present in hard water. Soft water extends life of water pipes and needs less soap for washing clothes. Water softening is usually done by ion-exchange resins.

soil moisture   The amount of water present at a depth of one metre of soil, which is expressed either in a percentage or in the unit of depth (in mm of water depth).

soil zone   The zone which extends up to a maximum depth of one metre or two and is criss-crossed by the roots of plants.

solute   Solids that dissolve in solvents.

solvents   Liquids that dissolve solid substances. Water dissolves more substances than any other, and is known as the 'universal solvent'.

source water   The term encompasses rivers, streams, natural springs, constructed canals, groundwater aquifers or any other freshwater source tapped by people.

spatial data   Any data that have fixed location and can be plotted on a map: e.g., land use patterns, villages, water wells, water divides, stream network configuration and their flood plain geometry, rocky outcrops, etc.

specific capacity (of water wells)   The specific capacity of a well is pumping rate (yield) divided by the drawdown. Used to determine the maximum yield possible from a well; varies with duration of discharge.

specific conductance   A measure of the ability of water to conduct an electrical current as measured using a 1-cm cell and expressed in units of electrical conductance, i.e., Siemens per centimeter at 25 degrees Celsius.

specific effluents   Effluents that need special treatment: e.g., heavy metals from electronics or electroplating industry.

specific retention   Hydrogeological parameters of an aquifer which determine how much water will be retained by the aquifer when pumped.

specific storage   Volume of water released per unit surface area of confined aquifer per unit decline in hydraulic head. Also known as storage coefficient.

specific yield   Specific yield is the fraction of drainable water yielded by gravity drainage when the water table declines. It is the ratio of the volume of water yielded by gravity to the volume of rock. Specific yield is equal to total porosity minus specific retention. It is dimensionless.

spray irrigation   Irrigation method in which water is shot from high-pressure sprayers onto crops. Because water is shot high into the air onto crops, some water is lost to evaporation.

stakeholder   Anyone who is both affected and/or interested by a problem or its solution. Stakeholders can be individuals or organisations and may include the public and communities

static reserve   Amount of water stored in an aquifer prior to commencement of monsoon recharge; variation in the static reserve is indicated by the static water level.

static water level   Water level of a tube well or observation well just before pumping is started.

storage coefficient   The storage coefficient, S, or storativity, is defined as the volume of water an aquifer releases from or takes into storage per unit surface area of aquifer per unit decline in hydraulic head. It is dimensionless.

storm intensity   Rainfall intensity expressed in mm per hour or inches per hour.

storm sewer   Sewer designed to carry surface runoff to prevent water logging. Storm sewers are completely separate from sanitary sewers that carry domestic and commercial wastewater.

stream order   Numerical scale to describe stream size; smallest tributaries are first-order streams, while the Amazon River is a twelfth-order waterway; first-, second- and third-order streams are called headwater streams.

subsidence   Depression of land surface caused by excessive pumping of groundwater; cracks and fissures can appear in land, posing serious threat to building foundations. Land subsidence is virtually an irreversible process.

surface runoff   Also called overland flow. Portion of precipitation (rain, snow melt etc.) that drains into rivers and streams.

surface water management   Management of a river basin or catchment with the objective to maximize availability of water and to minimize flooding problems.

suspended growth system Treatment of wastewater with microorganisms in which waste material flocculates and settles down.

suspended particles Minute solid particles (usually > 0.45 micron) that remain suspended in water due to movement of water molecules.

suspended-sediment concentration The ratio of the mass of dry sediment in a water-sediment mixture to the mass of the water-sediment mixture. Typically expressed in milligrams of dry sediment per liter of water-sediment mixture.

suspended-sediment discharge The quantity of suspended sediment passing a point in a stream over a specified period of time. When expressed in tons per day, it is computed by multiplying water discharge (in cubic feet per second) by the suspended-sediment concentration (in milligrams per liter) and by the factor 0.0027.

sustainable development goal Goals set by the United Nations in 2015 to achieve sustainable development.

sustainable yield The amount of water that can be extracted from a source without compromising the integrity of the concerned ecosystem.

system input volume (water audit) The total water input to a system (such as industry).

systemic knowledge asset Documents, monographs, reports, academic papers, etc., which are codified and stored for easy retrieval.

temporal data Location-specific variable data: e.g., rainfall, stream flow, groundwater level, user population, catastrophic changes brought about by natural disasters etc.

terrace River and stream terraces are remnants of the earlier valley level occurring well above the present floodplain; terrace formation is caused by river incision due to land uplift or lowering of sea level.

tertiary wastewater treatment Selected biological, physical and chemical separation processes to remove organic and inorganic substances that resist conventional treatment practices; the additional treatment of effluent beyond that of primary and secondary treatment methods to obtain a very high quality of effluent. There are several tertiary treatment methods: e.g., flocculation basins, clarifiers, filters, chlorine basins, ozone treatment, ultraviolet radiation etc.

thermal pollution A reduction in water quality caused by increasing its temperature, often due to disposal of waste heat from industrial or power-generation processes.

thermoelectric power water use Water used in the process of the generation of thermoelectric power. Power plants that burn coal and oil are examples of thermoelectric power facilities.

thermography A leak detection technique based on the principle that where there is a leak in the underground pipe the thermal signature of the surrounding soil will change. This thermal change is detected by a sensor.

topographical data Physical data in toposheets that are essential to terrain visualisation; they show both physiography (i.e., contours and drainage) and artefacts (i.e., roads, railways, habitations etc.).

total dissolved solids (TDS)   TDS is a parameter to indicate the total quantity of dissolved salts in water.

trans-boundary river basin   River basins that cut across state or national borders.

transmissivity   Transmissivity, T, also called transmissibility, is defined as the rate of flow of water through a vertical strip of aquifer one unit wide extending the full saturated thickness of the aquifer under a unit hydraulic gradient.

transpiration   Process by which water absorbed by plants, usually through the roots, evaporates into atmosphere from the plant surface, such as leaf pores. *See also* **evapo-transpiration**.

type curves   Curves generated from hypothetical data that represent ideal conditions.

ultrapure water   Water that has no measurably detectable impurity.

ultrasonic flow meter   A non-invasive flow meter that can generate data from outside on the rate of flow and the velocity of water flowing through the pipe.

ultraviolet disinfection   UV lamps of germicidal spectrum frequency are used for disinfection. The process kills 99% of germs in water.

unconfined aquifer   The dynamic upper surface of aquifers at atmospheric pressure (also called water table); water tables rise with rainwater infiltration.

unconfined groundwater   Water in an aquifer that has a water table; also, aquifer water found at or near atmospheric pressure.

unmet demand   A situation when the source water is not sufficient to meet a particular demand, say for irrigation.

unsaturated zone   The zone overlying water table in which pores contain both water and air.

upflow anaerobic sludge blanket (USAB) reactor   Water treatment that uses anaerobic reactions. Wastewater enters the reactor at the base, reacts with USAB and flows upward and out of the reactor tank.

upper catchment   A part of the river catchment at a higher altitude where erosion is the principal river activity.

Vadose zone   The underground zone between the land surface and the water table of the unconfined aquifer is called 'vadose zone'. This zone is undersaturated with water and plays a vital role in vertical movement of water.

value chain   The process or activities by which a company adds value to a product, including production, marketing, and the provision of after-sales service.

virtual water   Virtual water or embedded water is defined as the water consumed by plants to produce biomass or the water consumed for manufacturing a product or delivering a service

virtual water chain   The chain of production and consumption of water-intensive goods.

volumetric gauging   Measuring volume of water flow periodically.

voluntary corporate environmental behaviour   Actions by corporations that facilitate (1) public access to environmental information, (2) public participation in environmental planning and decision making, (3) access to justice in environmental matters and (4) corporate environmental responsibility.

wastewater   Water that has been used in homes, industries and businesses that is not for reuse unless it is treated.

wastewater-treatment return flow   Water returned to the environment by wastewater-treatment facilities.

water accounting   Initiatives taken by industries to quantify water resources, risks and opportunities.

water assessment tools   Part of Soil & Water Assessment Tools (SWAT). SWAT is a free and open source application using GIS for hydrological studies.

water audit   Accounting procedure to accurately determine the amount of unaccounted-for water (UAW) in a water distribution system; UAW is calculated from verifiable supply and consumption records.

water balance   Water balance is the balancing of water input and water output or consumption in an industrial or natural system. It also includes estimation of the status of water availability from different sources over a specific time period.

water budget   An accounting of all the water that flows into and out of a project area. This includes input water and water allocation for different units and also the discharged water.

water conflict   Conflict of interest over access to water.

water cycle   The continuous cycle of water going into the atmosphere by evaporation (mostly from oceans) and its return to surface as precipitation; also called the hydrological cycle.

water demand profile   A measure of average hourly water use throughout a 24-hour period.

water demand profile category   The specific purpose for which water is used (e.g., showers and outdoor irrigation).

water exported   The amount of water that is not a part of its own revenue water but is sold by an industry to the neighbouring water service providers

water footprint   Total water consumption by an individual, community, product, industry, state or a country.

water footprint of a consumer   Summation of water footprint in the production process of consumer goods and water used by the consumer.

water footprint of a nation   The measure of the total blue and green water consumed by the nation including the total national grey water footprint, plus the water footprint imported from other countries (by importing consumer products and other commodities) minus water footprint exported to other countries (by way of trade of commodities).

water footprint of a process step   Water footprint of a particular industrial process.

water footprint of a product   Each product has a discrete water footprint per unit of production.

water footprint of an individual   Summation of blue water consumed by an individual annually and the total virtual water content of everything used in daily life (e.g., food, energy, material and service) by the individual.

water footprint of industry   *See* **business water footprint**.

water governance   Water governance is the set of rules, practices and institutional arrangements that govern the decisions for the management of water resources and services in a country.

water holding capacity   Amount of water held between field capacity and wilting point; varies with soil porosity and thickness.

water intensity   A measure of either (1) the volume of water used per manufactured unit; or (2) the value of the product manufactured per 1000 litre of water consumed.

water level fluctuation method   The method of estimating groundwater recharge where seasonal fluctuation of the groundwater level is an input datum.

water profiling   Description of status of water resource of a basin or an industry that goes with the disclosure and sustainability report.

water quality   A term used to describe the chemical, physical and biological characteristics of water, usually with respect to its suitability for a particular purpose.

water quality standards   Standard parameters set for drinking water; e.g., by the World Health Organization.

water resource assessment (WRA)   Computing tool to evaluate water resources in relation to a reference frame, or to evaluate the dynamics of a given water resource in relation to human impacts.

water risk   Challenges to business arising from water conflict.

Water Risk Filter   It is computation tool for estimating water risk.

water security   Capacity of a population to safeguard acceptable-quality water in adequate quantities for sustaining livelihoods, human well-being, socio-economic development, protection against water pollution and related diseases, and for preserving ecosystems; sustainable development will not be possible without water security.

water stress   This occurs when demand for water exceeds the available amount or when poor quality restricts its use.

water table   The water saturated level below the vadose zone.

water use in supply chain   Water use in supply chain consists of the virtual water content of raw materials and the water consumed in the process of delivering (including transport) for processing.

water valuation   Referred as the amount of money an individual is willing to pay for water, regardless of whether it is for social or economic purposes.

water year   A time period of 12 months over which total precipitation is measured. Its beginning and end do not coincide with calendar year because the time period is based on the precipitation character of a country. Definition of water year varies with meteorological and geographical factors. It is used to compare annual precipitation.

watershed   Topographic feature identified by tracing a line along the highest elevations separating two river basins. It is also defined as the area of land where all of the water that falls in it and drains off of it goes to a common outlet. It is also termed as river basin and catchment.

watershed challenges   The challenges in identifying the boundary of IWRM activities which arise from an overlapping of natural, social and political boundaries.

wetland   A land area that is saturated with water, either permanently or seasonally.

wilting point   The minimum value of soil moisture content required by flora; plants begin to wilt when soil moisture is lower than the wilting point.

zero liquid discharge plant   The principle of zero discharge as the name implies, is recycling of all industrial wastewater within the industry; i.e., there is no wastewater exiting from the factory premises.

Annexure

Relation of units of hydraulic conductivity, transmissivity, recharge rates and flow rates.

Hydraulic conductivity (K)			
Metres per Day (m d⁻¹)	Centimetres per Second (cm s⁻¹)	Feet per Day (ft d⁻¹)	Gallons per Day per Square Foot (Gal d⁻¹ ft⁻²)
1	8.64×10^2	3.05×10^{-1}	4.1×10^{-2}
1.16×10^{-3}	1	3.53×10^{-4}	4.73×10^{-1}
3.28	2.83×10^3	1	1.34×10^{-1}
2.45×10	2.12×10^4	7.48	1

Transmissivity		
Square Metres per Day (m² d⁻¹)	Square Feet per Day (ft² d⁻¹)	Gallons per Day per Foot (Gal d⁻¹ ft⁻¹)
1	10.76	80.5
0.0929	1	7.48
0.0124	0.134	1

Industrial Water Resource Management: Challenges and Opportunities for Corporate Water Stewardship, First Edition. Pradip K. Sengupta.
© 2018 John Wiley & Sons Ltd. Published 2018 by John Wiley & Sons Ltd.

Recharge rate			
Unit Depth per Year	**Volume**		
	(m³ d⁻¹ km⁻²)	**(ft³ d⁻¹ mi⁻²)**	**(gal d⁻¹ mi⁻²)**
In millimetres	2.7	251	1874
In inches	70	6365	47748

Flow rate				
m³ s⁻¹	**m³ min⁻¹**	**ft³ s⁻¹**	**ft³ min⁻¹**	**gal min⁻¹**
1	60	35.3	2120	15800
0.0167	1	0.588	35.3	264
0.0283	1.70	1	60	449
0.000472	0.0283	0.0167	1	449
0.000063	0.00379	0.0023	0.134	1

Adapted from Heath, Ralph C., 1983, Basic ground-water hydrology: US Geological Survey Water-Supply Paper 2220.

Index

Industrial Water Resource Management: Challenges and Opportunities for Corporate Water Stewardship, First Edition. Pradip K. Sengupta.
© 2018 John Wiley & Sons Ltd. Published 2018 by John Wiley & Sons Ltd.